Enzyme Assays

The Practical Approach Series

Related **Practical Approach** Series Titles

- Proteolytic enzymes 2/e
- Protein purification applications 2
- Protein purification techniques 1
- Protein-ligand interactions: structure and spectroscopy
- Protein-ligand interactions: hydrodynamic and calorimetry
- Spectrophotometry & spectrofluorimetry
- Protein phosphorylation 2/e
- High resolution chromatography
- Protein expression
- Gel electrophoresis of proteins 3/e
- HPLC of macromolecules 2/e
- Affinity separations
- Subcellular fractionation
- Biological data analysis

Please see the **Practical Approach** series website at
http://www.oup.com/pas
for full contents lists of all Practical Approach titles.

No. 257

Enzyme Assays
Second Edition
A Practical Approach

Edited by

Robert Eisenthal

and

Michael J. Danson

Department of Biology and Biochemistry,
University of Bath, Bath, BA2 7AY U.K.

OXFORD
UNIVERSITY PRESS

Great Clarendon Street, Oxford OX2 6DP

Oxford University Press is a department of the University of Oxford.
It furthers the University's objective of excellence in research, scholarship, and education by publishing worldwide in

Oxford New York

Auckland Bangkok Bombay Buenos Aires
Cape Town Dar es Salaam Delhi Hong Kong Istanbul
Karachi Kolkata Kuala Lumpur Madrid Melbourne Mexico City
Mumbai Nairobi São Paulo Shanghai Singapore Taipei Tokyo Toronto

and an associated company in Berlin

Published in the United States
by Oxford University Press Inc., New York

First edition published 1992
Reprinted 1993 (with corrections), 1995
Second edition published 2002

A catalogue record for this title is available from
the British Library

Library of Congress Cataloging in Publication Data

Enzyme assays / edited by Robert Eisenthal and Michael J. Danson–2nd ed., rev.
(Practical approach ; 257)
"First edition published 1992, reprinted 1993 (with corrections), 1995."
1. Enzymes–Analysis. I. Eisenthal, Robert II. Danson, Michael J. III. Practical approach series ; 257.

QP601. E5153 2001 572′.7–dc21 2001036968

ISBN 0 19 963821 7 (Hbk)
ISBN 0 19 963820 9 (Pbk)

10 9 8 7 6 5 4 3 2 1

Typeset by Footnote Graphics
Printed in Great Britain
on acid-free paper by
The Bath Press, Avon

Preface

The explosion of genetic information over the past decade has led to the new challenge of identifying the functions of the thousands of proteins whose sequences have been revealed. For example, a significant proportion of the proteins identified as being coded on the human genome are predicted to be enzymes of known function. With the complexities of alternative splicing and post-translational modification, how many more as yet unrevealed catalytic activities exist is anybody's guess. The major task, of identifying the proteins, i.e. proteomics, complex and vast as it is, will be a simple job compared to the ultimate objective - this will be actually to define the *functional* roles of the components of the proteome in a way that relates to the entire system. In this task, the assay of enzyme activity will play an essential role.

Assay of enzyme catalytic activity is among the most frequently performed procedures in biochemistry, as it is involved with identification of enzymes, estimation of the amount of enzyme present, monitoring the purification of an enzyme, and determining its kinetic parameters. The determination of enzyme activity is a kinetic measurement - as such, there are many pitfalls for the unwary. Thus, the first chapter deals with the principles underlying enzyme assays. In doing so, deviations from apparently predicted behaviour are described so as to enable the investigator to distinguish between merely artefactual causes and those that arise from inherent properties of the enzyme under investigation.

The range of procedures used to measure the rate of an enzyme-catalysed reaction is limited only by the nature of the chemical change and the ingenuity of the investigator. Reflecting this, chapters 2–7 describe the most commonly-used techniques. The experimental approaches cover photometric, radiometric, electrochemical and HPLC methods. The theory underlying each method is outlined, together with a description of the instrumentation, sensitivity and sources of error. Although these methods are discussed in detail in many excellent texts and monographs, these do not in general address the unique problems arising from the use of these techniques in enzyme assays.

The identification of enzymes on electrophoretic gels is a technique that will assume increasing importance as proteomic and metabolomic investigations progress; the chapter covering this has been completely rewritten in this edition. The chapter on radiometric assay has also been rewritten to include coverage of the recently developed scintillation proximity assay (SPA), which obviates the need to separate radioactive substrate from product, and which is especially valuable in the assay of protein kinases.

Two new topics have been introduced. The chapter on high-throughput assays addresses the need to automate multiple assays efficiently, a problem that has assumed ever-greater importance in proteomics as well as drug screening programs. The other new topic is the determination of active site concentration. This is becoming essential as the availability of new enzyme activities increases due to emerging enzyme technologies, such as *in vitro* enzyme evolution, catalytic antibodies, imprinted polymers, and natural enzymes suspended in organic solvents and entrapped in porous plastics. We have also expanded the section on electrochemical methods to include the nitric oxide electrode, as the biological activity of NO is a topic that has grown exponentially over the decade since the appearance of the first edition of *Enzyme Assays*.

Publisher's limitations on length have meant that the inclusion of this new material has necessarily involved the omission of some topics that appeared in the first edition, such as polarographic methods, and the truncation of others. Thus, the sections on buffers and protein determination appear in an appendix on the chapter on extraction and fractionation, which now includes methods useful in extracting enzymes from commonly-used expression systems. Despite the universal availability of desktop computers and associated software for analysing enzyme kinetic data, our experience has shown that such software is often uncritically applied, leading at best to inappropriate analyses, and at worst to conclusions that mask interesting properties of an enzyme. Therefore, we have retained the chapter on statistical analysis of enzyme kinetic data, and this has been totally rewritten.

This book is not intended to provide a compilation of assay protocols for individual enzymes. As several thousand enzyme catalysed reactions are known, such a compilation would have been impossible in a book of this size. However, the techniques chapters contain experimental protocols that have been carefully chosen to represent the various types of enzyme-catalysed reactions amenable to assay by a particular technique - these can then be adapted to assay enzymes not specifically described. This generality notwithstanding, reference is made to over 250 individual enzymes or catalytic activities. These have been compiled in an enzyme index, which supplements the subject index, and which we hope will prove useful to all working in the molecular life sciences.

Bath R.E.
January 2002 M.J.D

To Janet and Janet

Contents

Protocol list

Absorption

Fluorescence

Radiometric Techniques

Application of HPLC to enzymatic analysis

Polarographic assays

Calibration

NO and cellular respiration studies

General pH-stat procedure and specific protocols for some individual enzymes

Abbreviations

ACV	δ(L-α-aminoadipyl) L-cystinyl-D-valine
ADP	adenosine diphosphate
AMP	adenosine monophosphate
AOL	agar overlay
APC	allophycocyanin
ATEE	acetyltyrosine ethyl ester
ATP	adenosine triphosphate
AU	absorbance units
BAEE	benzoylarginine ethyl ester
BBB	blood-brain barrier
Bis	*N,N'*-bis(2-hydroxyethyl)-2-aminoethanesulphonic acid
BIS	*N,N'*-methylenebisacrylamide
BSA	bovine serum albumin
bMBP	biotinylated myelin basic protein
cAMP	cyclic AMP
CCD	charge-coupled dipole
CDP	cytidine diphosphate
CI	covalently immobilized
CM	carboxymethyl
c.p.m.	counts per minute
DAD	diode array detector
DAP	diaminopimelic acid
dATP	deoxyadenosine triphosphate
DCI	3,4-dichloroisocoumarin
DEAE	diethylaminoethyl
dH_2O	distilled water
ddH_2O	double (or bi-) distilled water
DHF	dihydrofolate
DHQ	dihydroquinozolinium
DME	dropping mercury electrode
DMSO	dimethylsulphoxide
DNA	deoxyribonucleic acid
DOPA	3,4-dihydroxyphenylalanine
DTNB	5,5'dithiobis(2-nitrobenzoate)
DTT	dithiothreitol

EC	Enzyme Commission
EDTA	ethylenediamine tetra-acetic acid
EGTA	ethyleneglyco-bis(β-aminoethyl ether)*N*, *N*, *N*′-*N*′-tetraacetic acid
ELISA	enzyme-linked immunosorbent assay
FAD	flavin adenine dinucleotide
FCCP	carbonyl cyanide *p*-trifluoro-methoxyphenylhydrazone
FM	flow method
FMN	flavin mononucleotide
FRET	fluorescence resonance energy transfer
GT	glutamate oxaloacetate transaminase
HDL	high density lipoprotein
HEPES	4-(2-hydroxyethyl)-1-piperazine ethanesulphonic acid
HPLC	high performance liquid chromatography
HNBA	hydroxynitrobenzoic acid
HRP	horseradish peroxidase
HTRF	homogeneous time resolved fluorescence
HTS	high-throughput screening
ICE	interleukin converting enzyme
ID	internal diameter
kat	katal
LDH	lactate dehydrogenase
LDL	low density lipoprotein
LLD	lower limit of detection
M_r	relative molecular mass
MMP	matrix metalloproteinases
MOL	membrane overlay
MOPS	3-(*N*-morpholino)propanesulphonic acid
MPDP	1-methyl-4-phenyl-2,3-dihydropyridine
MTT	3-(4,5-dimethylthiazol-2-yl)2,5 diphenyltetrazolium bromide
NAD	nicotinamide adenine dinucleotide (oxidized)
NADH	nicotinamide adenine dinucleotide (reduced)
NADP	nicotinamide adenine dinucleotide phosphate (oxidized)
NADPH	nicotinamide adenine dinucleotide phosphate (reduced)
NAT	*N*-acetyltransferase
NCDC	2-nitro-4-carboxyphenyl-*N*,*N*′,-diphenylcarbamate
NPA	*p*-nitrophenylacetate
NPE	non-proximity effect
NSB	non-specific binding
OAB	*o*-aminobenzaldehyde
OAT	ornithine aminotransferase
OPA	*o*-phthaldehyde
PABA	*p*-aminobenzoate
PAGE	polyacrylamide-gel electrophoresis
PBS	phosphate buffered saline
PDE	phosphodiesterase
PEP	phosphoenolpyruvate
PFK	phosphofructokinase

P_i	inorganic orthophosphate
PK	pyruvate kinase
PMS	phenazine methosulphate
PMSF	phenylmethanesulfonylfluoride
PMT	photomultiplier tube
POL	paper overlay
PP_i	inorganic pyrophosphate
PS	polystyrene
PVP	polyvinylpyrrolidone
PVT	polyvinyltoluene
RI	refractive index
RAC	radioactive concentration
RNase	ribonuclease
RPC	reverse phase chromatography
RT	reverse transcriptase
SCE	standard calomel electrode
SDS	sodium dodecyl sulphate (sodium lauryl sulphate)
SEC	size exclusion chromatography
SEM	standard error of the mean
SM	sphingomyelin
SPA	scintillation proximity assay
SS	sum of squares
%T	(g acrylamide + g BIS)/100 ml (for PAGE)
TBA	*t*-butylammonium hydroxide
TCA	trichloracetic acid
TCC	2,3,5-triphenyltetrazolium chloride
TEMED	*N,N,N′,N′*-tetramethylethylene diamine
THF	tetrahydrofolate
TLC	thin layer chromatography
Tris	tris (hydroxymethyl) aminomethane
U	enzyme units (μmol/min)
UD	uridine diphosphate
UTL	ultrathin layer
UV	ultraviolet
VLDL	very low density lipoprotein
Yox	yttrium oxide

Chapter 1
Principles of enzyme assay and kinetic studies

Keith F. Tipton
Department of Biochemistry, Trinity College, Dublin 2, Ireland

1 Introduction

The activity of an enzyme may be measured by determining the rate of product formation or substrate used during the enzyme-catalysed reaction. For many enzymes there are several alternative assay procedures available and the choice between them may be made on the grounds of convenience, cost, the availability of appropriate equipment and reagents and the level of sensitivity required. It would not be possible in this account to give detailed descriptions of all the assay mixtures and procedures that have been devised for individual enzymes. The examples that will be presented here are intended to illustrate general features of specific types of assay. Convenient recipes for the assay of individual enzymes can be found in a variety of sources, such as *Methods in Enzymology* (1), *Methods in Enzymatic Analysis* (2) and *The Enzyme Handbook* (3), as well as in the original literature. On-line sources of references for the assay, and other properties, of most enzymes can be found in the BRENDA (http://www.brenda.uni-koeln.de) and EMP–Sekov (http://wit.mcs.anl.gov/EMP/indexing.html) databases.

This chapter will discuss the general principles of enzyme assay procedures and the problems that may arise in their application and interpretation. It may seem to be a catalogue of potential disasters, but it is essential to ensure that any assay procedure used gives a true measure of the activity of an enzyme. Far too many, otherwise carefully conducted, experimental studies have been rendered meaningless because of failure to ensure that the enzyme assay is giving valid results. All the potential problems may be avoided by careful experimental design and adequate controls. Furthermore, some apparently aberrant behaviour seen in enzyme assays can give valuable information on the properties of the enzyme being studied.

2 Behaviour of assays

2.1 Reaction progress curves

When the time-course of product formation, or substrate utilization, is determined a curve such as one of those shown in *Figure 1* is usually obtained. The time-course is initially linear but the rate of product formation starts to decline at longer times. There are several possible reasons for this departure from linearity and these will be discussed in turn.

2.1.1 Substrate depletion

The reaction may be slowing down because of substrate depletion. As the substrate concentration falls the enzyme will become less and less saturated and the velocity will fall, tending to zero as all the substrate is used. If the reaction is slowing down simply because of substrate

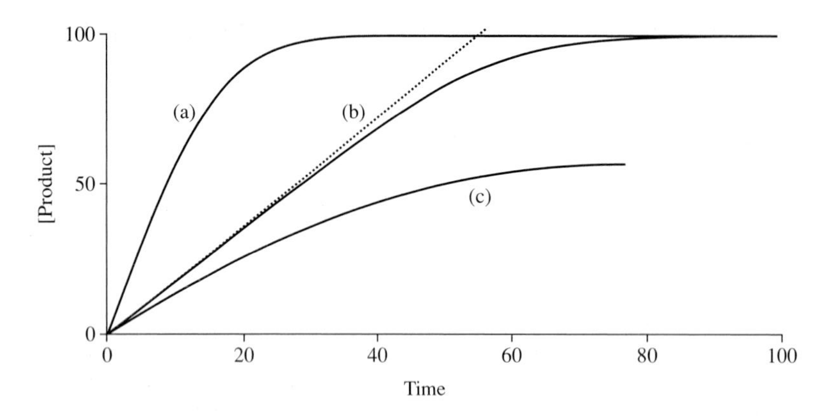

Figure 1 Typical progress curves of an enzyme-catalysed reaction. In curves (a) and (b) the reaction is slowing down because the substrate is being used up. In each case the initial substrate concentration is the same. In curve (a) the enzyme has high values of both K_m and V_{max}, such that the initial substrate concentration is only twice the K_m value and the rate of reaction is so rapid that the concentration of the enzyme–substrate complex (the degree of saturation of the enzyme) is continuously falling. In curve (b) the K_m and V_{max} values are both much lower, such that the initial substrate concentration is ten times the K_m value and the rate of reaction sufficiently slow to ensure that the concentration of the enzyme–substrate complex remains constant, and close to saturation, for an extended time. In curve (c) the reaction rate is slowing because the enzyme is unstable and loses activity at a constant rate with complete inactivation having occurred before all the substrate has been converted. Only in the case of curve (b) will it be possible to determine the initial reaction rate accurately by simply drawing a tangent to the initial, apparently linear, portion of the time-course (broken line).

exhaustion, the addition of more substrate should delay the fall-off. In some assays the substrate is continuously regenerated (Section 6.1.2 *iii*). Clearly, at any given enzyme concentration, the period of linearity would be expected to be longer at higher substrate concentrations. If initial substrate concentrations much below the K_m value are used it may be difficult to obtain a prolonged period of linearity unless highly sensitive assays are used to allow product formation to be detected under conditions where there is a negligible change in substrate concentration. It is a useful practice to calculate whether the total change observed corresponds to that expected from the amount of substrate initially present. Note that for a reaction involving more than one substrate this will correspond to the substrate that is present at the lowest stoichiometric concentration.

2.1.2 Equilibrium

A reversible reaction may be slowing down because it is approaching equilibrium, where the rate of the backward reaction (converting product to substrate) will increase until, at equilibrium, it is equal to the rate of the forward (substrate to product) reaction. A decline in rate due to this cause can be prevented by the presence of any system that removes the product. This might be achieved by the use of a second enzyme-catalysed reaction such as in a coupled enzyme assay (Section 6.1.2) or by the presence of a reagent which reacts with the products. For example, in the case of the oxidation of ethanol by alcohol dehydrogenase (EC 1.1.1.1):

$$CH_3CH_2OH + NAD^+ \rightleftharpoons CH_3CHO + NADH + H^+$$

the addition of semicarbazide to trap the acetaldehyde formed, as a semicarbazone, can reduce the curvature. This reaction produces hydrogen ions and rapidly approaches equilibrium at neutral pH values. Assay at higher pH values will prolong the linear phase. The addition of more substrate to a reaction that has ceased for this reason should also re-start it as it adjusts to a new equilibrium position.

2.1.3 Product inhibition

Products of enzyme-catalysed reactions are frequently reversible inhibitors of the reaction and a great deal of valuable information on the kinetic mechanism obeyed by an enzyme can be obtained from studying the nature of such inhibition (e.g. 4–7). As in the previous case, the use of a system that removes the product should prevent curvature due to this cause.

2.1.4 Instability

One of the components of the assay system may be unstable and be steadily losing activity or breaking down. This could be the enzyme itself or one of the substrates. The simplest way to check for this is to incubate the assay mixture for a series of times, under conditions identical to those used in the assay itself but without one of the components (enzyme(s) or substrates), before starting the reaction by the addition of the missing component. If the rates of the reaction are the same whichever component is missing during the pre-incubation period, a loss of linearity from this cause can probably be excluded. If they are not, this approach should indicate which of the components is unstable. A convenient method for determining whether an enzyme is stable during assay is described in Section 2.3.

It is important to ensure that the conditions of the pre-incubation are identical to those of the assay itself. For example, many compounds are light-sensitive and this can be a particular problem where relatively high intensities, such as are possible in fluorimetry, are used. Thus the pre-incubation should be carried out at the same level of illumination as in the assay. A further problem that can be encountered with optical assays is that the use of narrow slit widths can result in a localized destruction of only a small proportion of the material in the assay cuvette. For example, the fluorescence of tryptophan solutions may decline with time but removal of the cuvette and shaking it can result in an apparent return to the original level of fluorescence if the photo-destruction is limited to only a very small proportion of the total solution.

In some cases a component of the assay mixture may appear to be less stable under the pre-incubation conditions than it is in the complete assay mixture. This could result from the binding of substrate stabilizing the enzyme. In the case of light-sensitive compounds the absorbance of light by some other components of the assay may protect the photo-labile compound by decreasing the amount of light to which it is exposed. We have observed such behaviour in our studies on the oxidation of 1-methyl-4-phenyl-2,3- dihydropyridine (MPDP) by the enzyme monoamine oxidase (EC 1.4.3.4). MPDP absorbs at 340 nm and its oxidation may be followed by recording the decrease in absorbance at that wavelength. However, it is an extremely photolabile compound and is rapidly oxidized when illuminated at 340 nm. This can lead to a situation where the high rate of decline in absorbance at 340 nm observed in the absence of the enzyme actually decreases when crude preparations of the enzyme are added, because absorbance of the incident light by the enzyme preparation decreases that reaching the MPDP. When assays are carried out by alternative methods that do not involve irradiation of the substrate, the enzyme can indeed be shown to catalyse the oxidation of MPDP.

2.1.5 Time-dependent inhibition

An enzyme might be less stable when catalysing the reaction than it is under the pre-incubation conditions described in Section 2.1.4. Such an effect would result in a decline in the rate of the reaction with time, whereas the individual components of the assay mixture might appear quite stable during the pre-incubation experiments. In such cases the addition of more enzyme to the assay after the reaction had ceased would be expected to cause the reaction to restart. If the amount of enzyme added is the same as that originally used it would be expected that the resulting initial reaction rate would be the same as that obtained when the assay was originally started, unless there had been a significant depletion of substrate(s) or accumulation of inhibitory products during the reaction.

Several amino acid decarboxylases have been shown to give rise to progress curves such as that shown in *Figure 1* (curve c), because an occasional transamination reaction results in the conversion of the pyridoxal phosphate coenzyme to the pyridoxamine form during the progress of the assay. In this case the departure from linearity may be delayed by adding an excess of pyridoxal phosphate to the assay mixture (8).

Enzyme-activated irreversible inhibitors, which are also known as mechanism-based inhibitors, k_{cat} inhibitors or suicide inhibitors, are substrate analogues that are not intrinsically reactive but are converted by the action of a specific enzyme to a highly reactive species that combines irreversibly (or very tightly) with it (see 8–12 for reviews). Some inhibitors of this type react stoichiometrically with the enzyme to cause inhibition. However, since these compounds are substrate analogues, which must be involved in part of the enzyme-catalysed reaction in order to generate the reactive inhibitor, it is not surprising that others function as both substrate and inhibitor for the enzyme, according to the overall reaction:

$$\mathrm{E\ 1\ I} \rightleftharpoons \mathrm{E.I} \longrightarrow \mathrm{(E.I)^{*}} \begin{cases} \nearrow \mathrm{E{-}I} \\ \searrow \mathrm{E + Products} \end{cases} \qquad (1)$$

where I is the enzyme-activated inhibitor, E.I is the initial non-covalent complex (analogous to the enzyme–substrate complex) and (E.I)* represents an activated complex which can either react to give the irreversibly inhibited species (E–I) or break down to form products and the free enzyme (E). If the formation of products is followed, a curve such as that shown for the unstable enzyme case in *Figure 1* will result. Addition of more of the substrate/inhibitor would not restart the reaction, but addition of more enzyme would do so. Analysis of the behaviour of such systems can give the kinetic parameters describing the inhibitory process together with the *partition ratio*, which corresponds to the number of mol of product formed by one mol of enzyme before it is inhibited (for accounts of the kinetic analysis of such behaviour see 12–14). Several inhibitors of the enzyme monoamine oxidase have, for example, been shown to act in this way (15, 16). 2-Phenylethylamine, one of the amine substrates for that enzyme, has been shown to act as a time-dependent inhibitor at higher concentrations whereas lower concentrations of this substrate, where these time-dependent inhibitory effects are less important, the progress curves are non-linear because of substrate depletion (17). This type of behaviour emphasizes the necessity of checking the linearity of progress curves over a range of substrate concentrations, not just at the lowest substrate concentration that is to be used.

2.1.6 Assay method artefact

If the specific detection procedure used ceases to respond linearly to increasing product concentrations, this can lead to a decline in the measured rate of the reaction with time. In spectrophotometric or fluorimetric assays the absorbance of the product may reach such high levels that the apparatus no longer responds linearly to increasing concentrations (18,19). Many convenient enzyme assays involve the use of one or more auxiliary enzymes to allow the reaction to be followed (Section 6.1.2 *iii*). In such cases departure from linearity may result from failure of the auxiliary system to respond linearly to increasing rates of product formation. This could result from many of the causes described above or simply to it approaching its maximum velocity. Clearly, if such coupled-assay procedures are to be used, it is essential to perform careful control experiments to prove that the system is capable of providing a true measure of the activity of the enzyme being studied under all conditions that are to be used. This important aspect is discussed in more detail in Section 6.1.2 *iii*.

2.1.7 Change in assay conditions

If the assay conditions are not constant the rate of product formation might be expected to change. If, for example, the reaction under study involves the formation or consumption of hydrogen ions, the pH of the reaction mixture may change during the course of the reaction unless it is adequately buffered. If this resulted in a change of pH away from the optimum pH of the reaction this would lead to a decrease in the rate of the reaction. Clearly such a problem may be avoided by the use of adequate buffers, but it is important to check the pH of a reaction mixture before and at the end of a reaction time-course to ensure that such effects are not occurring. The practice of measuring the pH at the beginning and the end of a progress curve and assuming that the operating pH value is the mean between these two values is not valid because the pH may not change linearly during the assay. Furthermore, if the initial rate of the reaction is to be measured the operative pH should be that at the start of the reaction, not some arbitrary intermediate value occurring at a later stage.

2.2 Initial rate measurements

As can be seen from the above discussion, the decrease in the rate of product formation with time can be the result of one or more of a number of effects. At very short times, however, these effects should not be significant and thus if one measures the initial, linear, rate of the reaction by drawing a tangent to the early, linear, part of the progress curve (see *Figure 1*), these complexities should be avoided. Frequently the linear portion of an assay is sufficiently prolonged to allow the initial rate to be estimated accurately simply by drawing a tangent to, or taking the first-derivative of, the early part of the progress curve. Where loss of linearity occurs relatively rapidly because of depletion of substrate or approach to equilibrium (Sections 2.1.1 and 2.1.2) the period of linearity may be prolonged by decreasing the enzyme concentration, to slow down the rate of product formation, increasing the sensitivity of the assay method, if necessary. Methods for determining initial rates from such non-linear progress curves have been reviewed (20, 21) and some of these approaches will be discussed below.

It has often been assumed that restricting measurements of reaction rates to a period in which less than 10–20% of the total substrate consumption has occurred will provide a true measure of the initial rate. Consideration of the possible causes for non-linearity discussed in the previous section will show that such an approach may not be valid. Even if the only reason for departure from linearity were depletion of substrate, consideration of the Michaelis–Menten relationship will indicate that such an approach will only give a valid approximation if the initial substrate concentration is greatly in excess of the K_m value.

In cases where curvature makes it difficult to estimate the initial rate with accuracy, it may be possible to do so by fitting the observed time-dependence of product formation to a polynomial equation and deriving the initial slope at $t = 0$ (22). Graphs of [product]/time against either time or [product] will intersect the vertical axis at a point corresponding to the initial rate. Alternative, less sophisticated, approaches involve laying a glass rod or a small mirror approximately at right-angles to the early part of the progress curve. If the rod or mirror is moved until the reflection of the line is continuous with the line itself it will be exactly perpendicular to the progress curve. Thus, if a line is drawn along the surface of the rod or mirror, the initial-rate tangent should intersect with this line at 90°. Alternatively, the negative reciprocal of the line drawn to the surface of the rod, or mirror, will correspond to the initial rate. In either case it is important to check that the initial rate line passes through the origin (Product = 0 at $t = 0$).

Such approaches may be of value in several cases but in practice it may not be easy to estimate the zero time of the assay precisely. Starting an assay by adding one of the components and ensuring adequate mixing can lead to significant uncertainty about the exact time that the reaction was started. Furthermore, initial parts of a progress curve may be difficult to determine. For example, with spectrophotometric or fluorimetric assays of crude enzyme preparations there may be an appreciable period, during which particles are settling, before a rate can be accurately measured. Such problems can be further compounded in cases where there is either a burst or a lag before the true rate of the reaction is established (Section 2.4).

Because of these potential problems it is desirable, if at all possible, to adjust the conditions such that a linear response is maintained for a sufficient time to allow the direct measurement of initial rate. In cases where this cannot be achieved it is necessary to consider the possible causes of such non-linearity and to analyse the progress curves appropriately. For example, in the case of a compound acting as both a substrate and an enzyme-activated irreversible inhibitor (see Equation 1) a full analysis of the entire progress curve can be used, provided that the decline in velocity is solely due to such inhibition (10).

2.3 Integrated rate equations

If the decrease in the rate of an enzyme-catalysed reaction with time were solely due to the depletion of substrate, it would be possible to correct for this fall-off by use of the Michaelis–Menten relationship. Several attempts have been made to do this, but they will only be valid if substrate depletion is the sole cause of curvature in the time-course of product formation. It has sometimes been assumed that a more pronounced curvature of time-courses at lower substrate concentrations indicates that the decline in velocity is due to substrate depletion. However, such observations do not show whether substrate depletion is the only cause of the departure from linearity. Furthermore, a similar effect would be expected if the enzyme were unstable under the conditions of the assay, but was stabilized by its interaction with substrate. In such cases the stabilization would be greatest at higher substrate concentrations where the enzyme was more saturated. If depletion of substrate is the only cause of curvature in the time-course of the reaction it may be described by an integrated form of the Michaelis–Menten equation:

$$v = \frac{\mathrm{d}p}{\mathrm{d}t} = \frac{V_{\mathrm{max}}}{1 + (K_{\mathrm{m}}/s)} = \frac{V_{\mathrm{max}}}{1 + \{K_{\mathrm{m}}/(s_{\mathrm{o}} - p)\}} \tag{2}$$

where V_{max} and K_m are the maximum velocity and Michaelis constant, respectively, v is the initial velocity, p is the product concentration, s_o is the initial substrate concentration and s is the substrate concentration remaining at any time t.

Integration of this equation gives:

$$V_{\mathrm{max}}t = \{K_{\mathrm{m}} \ln[s_{\mathrm{o}}/(s_{\mathrm{o}} - p)]\} \tag{3}$$

and

$$\frac{2.303 \times \log[s_{\mathrm{o}}/(s_{\mathrm{o}} - p)]}{t} = \frac{V_{\mathrm{max}}}{K_{\mathrm{m}}} - \left(\frac{1}{K_{\mathrm{m}}} \times \frac{p}{t}\right) \tag{4}$$

Thus, if the amount of product formed is measured at a series of times and $(2.303/t)\log[s_o/(s_o - p)]$ is plotted against p/t a straight line will be obtained with a slope of $-1/K_m$ and an intercept on the base line of V_{max}. At low substrate concentrations, where $s_o << K_m$, this equation simplifies to:

$$\frac{2.303}{t} \times \log[s_o/(s_o - p)] = \frac{V_{max}}{K_m} \quad (5)$$

and thus a graph of 2.303 log $[s_o/(s_o - p)]$ against t will be a straight line of slope V_{max}/K_m which passes through the origin. Under these conditions it will not be possible to determine V_{max} and K_m separately.

The integrated rate equation has been extended to include cases where a reversible reaction approaches equilibrium and also to take account of inhibition by the products of the reaction (23 gives a detailed account). It is attractive in that it should allow full use to be made of all the data comprising the reaction progress curve, rather than just the small portion of it representing the initial rate of the reaction. Furthermore, it should allow detailed kinetic analysis to be undertaken with much less work than is required when initial rate measurements are used. Consideration of the form of the equation describing the reaction progress curve also indicates that the second-derivative of such a curve will show a minimum which corresponds, on the substrate concentration axis, to half the K_m value (24).

The problem of applying the integrated rate equation is that it is only valid if the departure from linearity is due only to substrate depletion or, in cases where more elaborate forms of the equation are applied (23), approach to equilibrium or product inhibition as well. However, as discussed in Section 2.1, the reasons for non-linear progress curves can be considerably more complex. Selwyn (25) has presented a valuable method for determining whether an enzyme is stable during assay. He pointed out that for any enzyme-catalysed reaction the rate of product formation will depend on the enzyme concentration (e) and some function (f) of the concentrations of substrate (s), product (p) and any inhibitor (i) or activator (a) present. Thus:

$$\frac{dp}{dt} = e \times f(s, a, i, p) \quad (6)$$

under conditions where a and i are constant and the concentration of substrate does not significantly change this equation can be integrated to give:

$$e \times t = f'(p) \quad (7)$$

where f′ is another function incorporating the terms the terms a, s and i. This indicates that the amount of product formed should depend only on the enzyme concentration and the time. Thus a graph of product concentration against $e \times t$ should describe the same curve whatever initial concentration of enzyme is used. However, if the enzyme is unstable during the course of the assay, Equation (7) will no longer hold since the active enzyme concentration will also be time-dependent. In such cases the graphs of p versus $e \times t$ will give different curves for each starting enzyme concentration. This latter behaviour would be expected if non-linearity of an assay were due to instability or time-dependent inhibition of the enzyme (Sections 2.1.4 and 2.2.5), whereas a single curve would be expected for cases outlined in Sections 2.1.1, 2.1.2 and 2.1.3.

2.4 Bursts and lags in progress curves

With some enzymes there may be either a burst of product formation or a lag before the linear phase of the reactions is obtained, as illustrated in *Figure 2*. This may be an artefact of the assay system being used but in other cases such behaviour can give interesting informa-

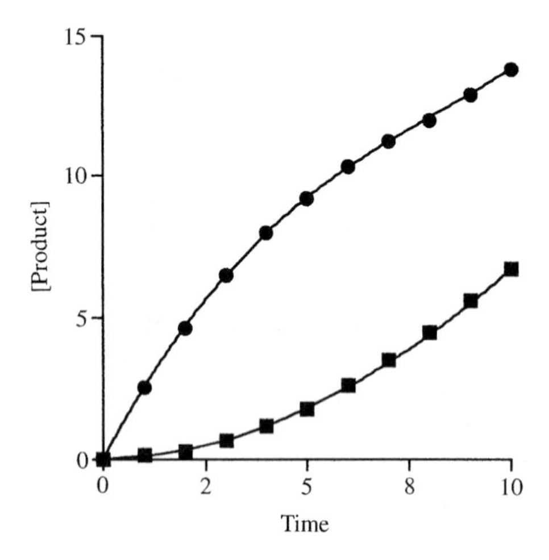

Figure 2 Time courses of enzyme-catalysed reactions showing burst or lag phases before the steady-state rate is obtained.

tion about the enzyme-catalysed reaction itself. The possible causes of such behaviour are listed below.

2.4.1 Inadequate temperature control

Frequently it is necessary to keep the enzyme solution and perhaps some other components of the assay mixture cold in order to ensure their stability. Addition of an ice-cold component to a reaction mixture that has been equilibrated to the assay temperature can lead to a drop in temperature and thus to a slower rate of reaction, which will increase as the temperature of the mixture rises to the equilibration value. It should also be remembered that reaction mixtures placed in temperature-controlled vessels do not immediately adjust to the new temperature and apparent lags may be seen if the mixture is not given an adequate time to adjust to the chosen temperature before the reaction is initiated. The converse behaviour can sometimes be seen if a reaction mixture is equilibrated in a water-bath which is also used for circulating a water jacket around the reaction vessel. In such cases there may be a significant drop in temperature between the water-bath and reaction vessel, leading to an apparent initial burst of activity before the temperature of the pre-equilibrated mixture falls to that in the reaction vessel.

If it is essential that a component of the assay be kept at a temperature different from that required for the assay, correction may be made for the effect on the temperature of the final assay mixture. Making the reasonable assumption that the heat capacities of all solutions making up the assay mixture are the same, the temperature of a mixture will be given by:

$$\mathbf{a}T_1 + \mathbf{b}T_2 = (\mathbf{a} + \mathbf{b})\,T_3$$

where T_1 and T_2 are the temperatures of the component solutions, T_3 is the resulting mixture temperature, and **a** and **b** are the volumes of the solution components. Using this relationship, one can calculate the initial temperatures of the assay components that will give, on mixing, the required assay temperature.

2.4.2 Settling of particles

When crude tissue preparations are assayed spectrophotometrically or fluorimetrically the measurements during the first few minutes of the assay may be erratic, owing to the settling of particles from the solution. This may sometimes be misinterpreted as a burst or lag phase in the reaction. In such cases remixing the contents of the assay cuvette after the reaction has become linear should result in a second phase of aberrant behaviour.

2.4.3 Slow detector response

A lag phase may result if the initial response of the detection system is too slow. This type of behaviour will be discussed in terms of coupled enzyme assays in Section 6.1.2.

2.4.4 Slow dissociation of a reversible inhibitor (or activator)

Although most reversible enzyme activators and inhibitors will dissociate extremely rapidly from the enzyme when the enzyme–inhibitor mixture is diluted, some, including those that show extremely high affinity for the enzyme, may bind to the enzyme and dissociate from it slowly (26, 27). In such cases dilution of the enzyme–inhibitor mixture into the assay may show a lag as the inhibitor slowly dissociates to its new equilibrium value. Conversely, if enzyme is added to a reaction mixture containing inhibitor the rate may slowly decrease until the binding equilibrium has been established.

2.4.5 Pre-steady-state transients

A burst or lag phase in the time-course of product formation can be due to the time taken for the concentrations of the intermediate enzyme–substrate and enzyme–product complexes to rise to their steady-state levels. Usually such transients are very rapid and only detectable by use of specialized equipment, such as a stopped-flow apparatus, which allows measurements of reaction to be made within milliseconds, or less, of mixing (28, 29). Occasionally, however, such processes can occur sufficiently slowly to be observed on the time-scale associated with normal enzyme assays. One of the best-known examples of such behaviour is the hydrolysis of *p*-nitrophenylacetate (NPA) by chymotrypsin (EC 3.4.21.1) (30). The enzyme (E–OH) reacts rapidly to form an acetyl-enzyme with the liberation of *p*-nitrophenol. The acetyl-enzyme is only slowly hydrolysed to regenerate the free enzyme with the release of acetate:

$$\text{E-OH} + \text{NPA} \rightleftharpoons \text{E-OH.NPA} \xrightarrow[\downarrow\ p\text{-nitrophenol}]{} \text{E-OCOCH}_3 \xrightarrow[\uparrow\ \text{H}_2\text{O}]{} \text{E-OH} + \text{CH}_3\text{COO}^- + \text{H}^+$$

Thus, if *p*-nitrophenol is monitored there is an initial rapid formation followed by a slower steady-state rate that is governed by the rate of hydrolysis of the acetyl-enzyme.

A rather more complicated transient phenomenon has been observed with the enzyme arylsulphatase A (EC 3.1.6.1). In this case the reaction between enzyme and substrate results in the formation of an inactive, covalently modified, form of the enzyme which is slowly hydrolysed to regenerate the active enzyme. Thus the initial rate slowly decays, in a first-order process, to give a final steady-state rate that depends on the partition between the active and inactive, covalently modified, enzyme forms (31).

2.4.6 Relief of substrate inhibition or activation

Many enzymes are inhibited by high concentrations of one or more of their substrates (5). If the initial substrate concentration added to an assay mixture is sufficient to cause some degree of inhibition, the rate of the reaction will tend to increase with time as substrate utilization decreases the inhibition. Alternatively, an initial burst phase in the progress curve can occur if the substrate also behaves as an activator at higher concentrations. Bursts or lags arising from such causes should be eliminated by reducing the initial substrate concentration to a level where inhibition, or activation, is not significant. High-substrate inhibition or activation should, of course, be readily detected by their characteristic effects on the dependence of initial velocity on substrate concentration (e.g. 5). A commonly used coupled assay for phosphofructokinase (EC 2.7.1.11) involves the use of phosphoenolpyruvate, pyruvate kinase, NAD and lactate dehydrogenase (Section 6.1.2). However, the enzyme from some sources is allosterically inhibited by phosphoenolpyruvate, which can lead to a lag in the progress curve (32, 33).

A similar lag phase in the reaction progress curve can occur if the substrate solution is contaminated by a small amount of another substrate for the enzyme which has a higher affinity for it but is broken down rather slowly (6, p. 72).

2.4.7 Activation by product

A progress curve that curves upward may be observed if one of the products of the reaction is an activator. This type of behaviour can, for example, occur in the assay of phosphofructokinase, which is activated by the product fructose-1,6-bisphosphate (32). In this case, however, there is a further complication because the substrate ATP is an allosteric inhibitor of the enzyme which can lead to lag phases, as discussed above (Section 2.4.6).

2.4.8 Substrate interconversions

If a compound exists in more than one form, only one of which is an effective substrate for the enzyme, a slow interconversion between these forms can lead to burst or lag phases in progress curves. For example, a lag in the progress curve of the reaction catalysed by fructokinase (EC 2.7.1.3) may be observed when freshly prepared solutions of fructose are used as substrate. This is because in such solutions the sugar is essentially all in the pyranose form, which is not a substrate, and only mutarotates rather slowly to give the active furanose form. If fructose solutions are allowed sufficient time for the mutarotation equilibrium between the two forms to be established the lag phase in the progress curve is no longer apparent (34). Such effects can also give rise to burst phases if a substrate exists in a slow equilibrium between active and inactive forms where an initial rapid phase, corresponding to the utilization of the active form of the substrate, would be followed by a slower phase determined by the rate of isomerization from inactive to active forms. Clearly, in cases of substrates that can exist in different isomeric forms, hydration states or polymeric forms that interconvert relatively slowly, such effects should be taken into account if bursts or lags are observed.

2.4.9 Hysteretic effects

Frieden (35) used the term hysteresis to refer to burst or lag phases in progress curves resulting from slow isomerization of the enzyme. He argued that such behaviour may have important regulatory significance. Detailed treatments of the behaviour of such systems have been presented (36–38).

Hysteretic effects can yield a number of differently shaped reaction progress curves. However, it is important to exclude the other possible causes of bursts or lags before concluding that the effect is due to hysteresis. If it is possible to monitor changes in the conformation of the enzyme in solution, correlation of the time-courses of such effects with those seen in the progress curves may provide evidence for hysteretic behaviour. If the effect is due to a slow conformational change induced by one of the substrates, the behaviour may depend on the way in which the reaction is started. Thus, hysteretic effects might be observed if the reaction is started by the addition of enzyme to a mixture containing all the other substrates, whereas, if the enzyme were preincubated with the substrate responsible for inducing the conformational change before starting the reaction with another substrate no such effect might be seen.

Hysteresis may occur if an enzyme exists in a slow association–dissociation equilibrium in which the two polymerization states differ in their activities. In this case the magnitude of the burst or lag may depend on the enzyme concentration since this will affect the degree of association. Furthermore, the hysteresis may be dependent on how the reaction is initiated. If the reaction is started by addition of a sample of enzyme from a concentrated stock solution the effects might be different from those observed when the enzyme is diluted into an incomplete assay mixture and allowed to equilibrate before starting the reaction with another component. Such behaviour has, for example, been shown to account for the hysteresis observed with hexokinase (EC 2.7.1.1) (39) and glutaminase (EC 3.5.1.2) (40). The mitochondrial form of aldehyde dehydrogenase (EC 1.2.1.3) can show extremely long lag phases before the reaction becomes detectable (41). With preparations from some species it appears that an enzyme association–dissociation phenomenon may contribute to the lag (42) whereas the polymerization state does not appear to be a factor with the enzyme from some other species (43).

2.4.10 Summary

With assays that show burst or lag phases, it is important to determine the cause in order to know which phase of the reaction corresponds to the true 'initial rate' of the reaction. The term initial-rate is normally used to refer to the steady-state rate of the reaction that is established after any pre-steady-state events have occurred. In the cases described in Sections 2.4.1, 2.4.2 and 2.4.3 the initial rate corresponds to the linear phase of the reaction that is established *after* any apparent burst or lag. The same would apply for the case in Section 2.4.4 but analysis of the behaviour could give valuable information on the rates of ligand association and dissociation. Where a lag or burst results from pre-steady-state transients (Section 2.4.5), the initial rate (steady-state) is that obtained after the transient phase, although more complete analysis of the curve can give valuable information about the values of individual rate constants (28, 29). In contrast, the true initial rate is that obtained at the start of the reaction for the case in Section 2.4.6, where the rate at the substrate concentration initially present is required. Similarly, in the case in Section 2.4.7 the initial rate corresponds to that at the start of the reaction since, by definition, no significant product formation should occur during this phase. Where slow substrate interconversions occur (Section 2.4.8) the problem becomes one of determining the true substrate concentration at which the initial rate has been measured.

Genuine hysteretic effects (Section 2.4.9) are much more difficult to analyse since the various phases of the reaction progress curve may be controlled by different conformational or aggregation studies of the enzyme. If, for example, there are two forms of the enzyme that have different activities, it might be possible to obtain rate data for both species from the different phases of the progress curve, perhaps by the use of computer-aided curve fitting procedures (44). In practice, however, the results of such an analysis might be difficult to interpret since true initial rate conditions may not apply at the later stages of the progress curve. Furthermore, a more detailed knowledge of the mechanisms underlying the observed transients would be necessary before any such analysis could yield meaningful results.

2.5 Blank rates

2.5.1 Possible causes

It is not uncommon to observe an apparent rate of reaction in the absence of one of the components of the complete assay mixture. It is important to understand the causes of such blank rates in order to make appropriate corrections, since for any accurate studies it is essential to ensure that the determined rates are due only to the specific enzyme-catalysed reaction under investigation. It is possible that a blank rate will only occur with certain components of an incomplete assay mixture and thus it is necessary to test for such rates using different combinations of the system, for example by omitting the enzyme and each of the substrates in turn. Some of the more common causes of blank rates are listed below:

i. Settling of particles

Spectrophotometric and fluorimetric assays of enzyme activities in crude tissue preparations, such as homogenates or subcellular organelle preparations, will be affected by the settling of particles causing changes in absorbance and light-scattering. After a sufficient time for the particles to settle these changes should cease, but they will start again on mixing the assay system again, as will occur when the full reaction is initiated by the addition of the missing component. It may be possible to use detergents to reduce this problem by rendering the particles soluble, but it will, of course, be necessary to check whether the detergent used has any effect on the activity of the enzyme under study.

ii. Precipitation

Gradual precipitation of material in the assay mixture can lead to similar problems in optical assays as those caused by settling of particles. In some cases such effects may be confused with genuine reaction rates. It is thus important to be aware of possible artefacts of this type and to inspect the assay cuvette at the end of the 'reaction' for signs of turbidity or visible precipitate formation. Changes in absorbance at wavelengths distant from those where any reaction-dependent changes should occur can be used to monitor turbidity changes directly. Such effects may result from the enzyme or another component of the mixture not being fully soluble under the assay conditions or from interactions between different components leading to precipitation. Magnesium or calcium ions are added to many assay mixtures because they are essential for the activity of a number of enzymes. However, if such mixtures contain strong phosphate buffer precipitation will occur when the solubility product of calcium or magnesium phosphate is exceeded. A more confusing situation can occur if one of the products of the reaction is not very soluble and precipitates during the later stages of the reaction, giving rise to accelerating progress curves. Provided that all other components of the reaction are soluble, a fall in absorbance following centrifugation of the reaction mixture may indicate precipitation to be affecting the results. Although the blank rates arising from precipitation directly affect optical assays, such behaviour could also invalidate the results obtained with other assay procedures.

iii. Contamination of one of the components of the assay mixture

The presence of one of the substrates in the enzyme solution can give a blank rate with an incomplete reaction mixture. Crude tissue preparations may contain endogenous substrates and this will lead to a reaction in the absence of added substrate. If the degree of contamination is quite small, the blank rate from this source would be expected to be non-linear and to cease when the endogenous substrate is exhausted. If the enzyme is stable under the assay conditions it may be possible to wait until the blank rate dies away before starting the assay. Alternatively, if the contaminating substrate is a small molecule, it should be possible to remove it by dialysis or gel filtration. Problems from this source would be expected to decrease on purification of the enzyme. However, some commercially available enzyme preparations contain substrate, which has been added for stability; it thus may be necessary to remove such material, for example by dialysis or gel filtration, before assay.

The possibility of contamination of reagents with substrate cannot be excluded. For enzymes which use CO_2 or bicarbonate as a substrate great care must be taken to remove all such material from each component of the mixture. Volatile substrates such as ammonia or aldehydes can be particularly difficult sources of contamination in laboratories where such compounds are in frequent use. Cross-contamination can also occur unless care is taken to ensure that a different dispenser is always used for each component of the assay and that reaction vessels are thoroughly cleaned.

iv. Adsorption to assay vessels

Many proteins adhere to glass and, in cases where a vessel has already been used for one assay, this can result in the presence of sufficient adsorbed enzyme to give a rate in a subsequent assay in the absence of added enzyme. Adsorption can be so strong that rinsing with distilled water is insufficient to remove the bound enzyme and more vigorous procedures such as acid washing are required. The use of silicone-treated glass or plastic vessels may minimize this problem but we have found that not all plastics are inert in this respect. If they are suitable for the assay, disposable plastic cuvettes are recommended.

Contamination of the assay vessels with one of the substrates can also lead to blank values. For example, in radiochemical assays it is necessary to ensure that apparently clean reaction vessels or scintillation vials do not contain any significant amounts of adsorbed radioactive material.

v. Non-enzymic reactions

Solutions of NAD(P)H are unstable at pH values below neutrality, leading to a spontaneous fall in absorbance at 340 nm. Similarly, many *p*-nitrophenyl esters that are used as esterase substrates are relatively unstable in aqueous solution and steadily hydrolyse to liberate *p*-nitrophenol. In these cases the blank rates due to the non-enzymic reactions should be subtracted from the rates given in the presence of enzyme. In spectrophotometric assays correction can most conveniently be done by using a double-beam (or ratio-recording) spectrophotometer that automatically records the difference between the absorbance of the sample and that of the blank. A steady drift in the response of the recording apparatus can also give rise to an apparent blank rate and it is important to check the stability from time to time in the absence of reactants. The reaction of exogenous factors can also lead to blank rates; for example, in poorly buffered solutions the absorption of CO_2 will lead to a drift to lower pH values, which would be reflected as blank rates when enzymes are assayed by determining changes in pH or by use of a pH-stat. Reaction between different components of an assay mixture can also give rise to blank rates. For example, aldehydes can react non-enzymically with NAD^+ to give a product that has a similar absorbance to NADH. This reaction can cause significant problems in determining the activity of aldehyde dehydrogenase at alkaline pH values (45) but it is not significant at neutral or acid pH values.

vi. Contaminating enzymes

The presence of another enzyme in the preparation which catalyses an interfering reaction can give rise to a blank rate. If the substrate for the contaminating enzyme is also a contaminant of the preparation it may be possible to remove it by dialysis or gel filtration. However this is not always possible; for example, the assay of dehydrogenases in crude tissue preparations may be difficult because of the presence of NADH–cytochrome-*c* reductase (EC 1.6.99.3) and in this case cytochrome-*c* is not readily removed by dialysis. In such cases it may be necessary to use an inhibitor of the contaminating enzyme, for example rotenone, to inhibit the mitochondrial form of that enzyme, taking care to ensure that it has no effect on the enzyme under study, or to purify the enzyme in order to remove the contaminating material. It may not be satisfactory simply to use an alternative assay procedure that does not detect the activity of the contaminating enzyme because the latter reaction may result in significant depletion of the substrate.

In some cases the contaminating enzyme may require no substrates other than those present for the assay of the enzyme under study. For example, an assay for the enzyme pyruvate carboxylase (EC 6.4.1.1) involves the use of malate dehydrogenase to couple the oxaloacetate produced to the oxidation of NADH, which may be followed spectrophotometrically (*Figure 3*). If the enzyme preparation is contaminated with lactate dehydrogenase this will also catalyse the oxidation of NADH in converting pyruvate to lactate. Clearly in this case it is not possible to exclude pyruvate from the assay mixture, because it is a substrate for the enzyme being assayed. It would, however, be possible to use an alternative assay, such as the incorporation of radioactively labelled bicarbonate into oxaloacetate, because the interference from lactate dehydrogenase might not be expected to be important in the absence of added NADH. The coupled assay can only be used satisfactorily if the pyruvate carboxylase preparation is purified to a state where it is free from contaminating lactate dehydrogenase.

2.5.2 Correction for blank rates

As will be clear from the above discussion, it is important to understand the cause of a blank rate before one may make the appropriate corrections for it. In many cases it is possible to obtain the true rate of the enzyme-catalysed reaction simply by subtracting the blank rate given in a suitable incomplete mixture from that obtained with the full assay. This approach assumes that the blank rate is an artefact that is unconnected with the activity of the enzyme under study, that it continues linearly for the total period of the assay and that it will be unchanged in the full assay. In cases where these assumptions are valid, failure to subtract the blank rate will yield apparent anomalies in kinetic behaviour. If the blank rate occurs in the absence of the enzyme, failure to subtract it will give a plot of initial velocity against enzyme concentration that does not pass through the origin but shows a finite activity at zero enzyme concentration (*Figure 4*). Failure to subtract a blank rate that occurs in the absence of one of the substrates can give behaviour that does not conform to the Michaelis–Menten equation (Section 5.2).

If an apparent blank rate is due to the settling of particles (Section 2.5.1 *i*), subtraction of the initial blank rate from the initial rate obtained after starting the reaction may be adequate, but such rates are normally irregular and it would be better to await the decline of the blank rate and the stabilization of the assay before measuring the rate. It would be inappropriate to subtract the blank rate if it were due to contamination of the enzyme (Section 2.5.1 *iii*) or assay vessel (Section 2.5.1 *iv*) with one of the substrates and the complete assay mixture contained saturating concentrations of that substrate. In these cases the blank rate is due to the enzyme itself and its subtraction would therefore result in an underestimation of the activity.

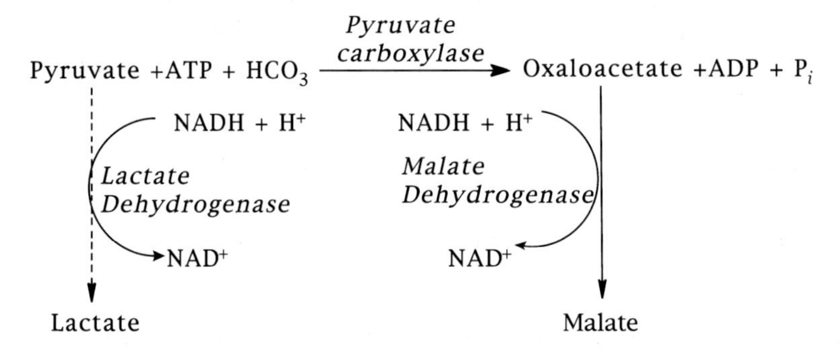

Figure 3 A coupled assay for pyruvate carboxylase. The broken line shows the interfering reaction that will take place in the presence of contaminating lactate dehydrogenase.

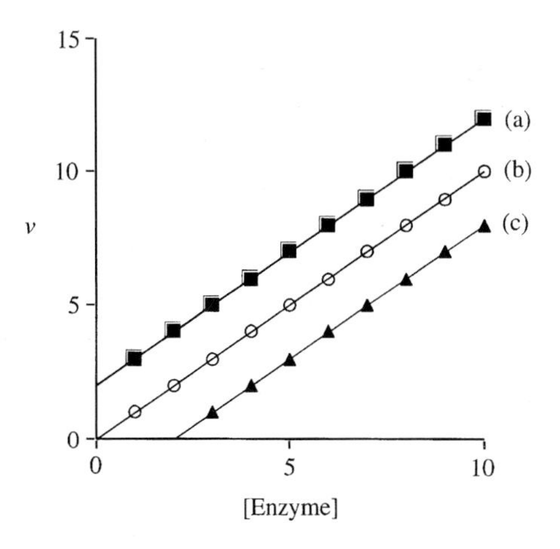

Figure 4 The dependence of initial velocity on the enzyme concentration. Line (b) shows the expected dependence; lines (a) and (c) show possible results from incorrect treatment of blank rates. In line (a) a blank rate occurring in the absence of the enzyme has not been subtracted and in line (c) a blank rate that occurs in the absence of one of the substrates, but is suppressed in the full assay, has been subtracted.

A more complicated system, where it is inappropriate to subtract an apparent blank rate, can occur if an enzyme can catalyse the decomposition of one of its substrates alone but that reaction is suppressed by the presence of the second substrate. This can, for example, occur in the assay of pyruvate carboxylase (see *Figure 3*). The enzyme has a relatively weak ATPase activity and will catalyse the hydrolysis of ATP in the absence of the other substrates. However, there is competition between this and the full reaction and it is effectively suppressed in the complete assay mixture (46). Thus, if the enzyme were assayed by measuring the formation of ADP or inorganic phosphate, there would be a blank rate, which should not be subtracted from the rate seen in the full mixture. Subtraction of the blank rate would be expected to give a dependence of initial velocity on enzyme concentration which did not pass through the origin, giving an activity of zero at a finite enzyme concentration (*Figure 4*).

2.5.3 Masking of an assay

In some cases the activity of a contaminating enzyme may interfere with the assay of an enzyme. The blank rates that can occur in the assay of enzymes utilizing NADH in the presence of contaminating NADH-cytochrome-*c* reductase was discussed in Section 2.5.1 *vi*. Such contamination would, of course, affect attempts to assay dehydrogenases in the direction of NADH formation by catalysing the reoxidation of the NADH formed. This could lead to an underestimation of the true reaction rate or even a complete masking of the reaction. In such cases it would be necessary to work in the presence of an inhibitor of the contaminating enzyme. As discussed in Section 2.5.1 *vi*, use of an alternative assay that does not rely on measurement of NADH formation may not be a satisfactory alternative in such cases.

3 The effects of enzyme concentration

3.1 Direct proportionality

As enzymes are catalysts, the initial velocity of the reaction would be expected to be proportional to the concentration of the enzyme. This is indeed the case for most enzyme-catalysed reactions, where a graph of initial velocity against total enzyme concentration will be a straight line passing through the origin (zero activity at zero enzyme concentration - see *Figure 4*). There are, however, some cases where this simple relationship does not appear to hold and it is thus important to check for linearity in all studies. In some cases departure from linearity may be artefactual resulting from, for example, changes of the pH or ionic strength of the assay mixture as increasing amounts of the enzyme solution are added, and it is important to check that such effects are not occurring. In other cases the behaviour can be more interesting. A graph of initial velocity against enzyme concentration can show either upward or downward curvature, as illustrated in *Figures 5* and *6*. Some common causes that can result in such behaviour are considered below.

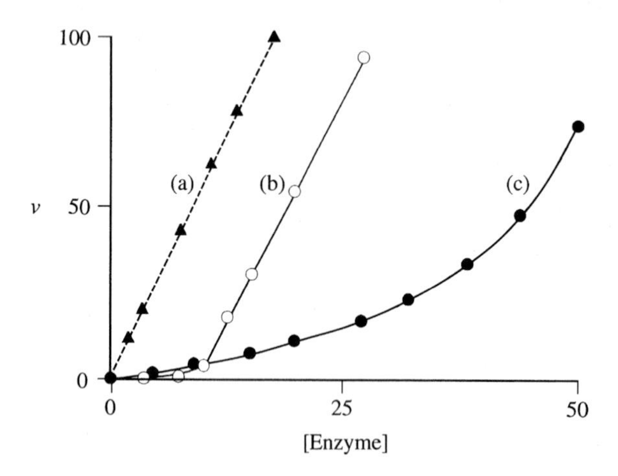

Figure 5 Upwardly curving dependence of initial velocity on enzyme concentration. Curve (a) shows the normally expected relationship; curve (b) represents the case where there is an irreversible inhibitor contaminating the assay mixture and curve (c) shows the possible behaviour if there were a reversible activator present in the enzyme preparation.

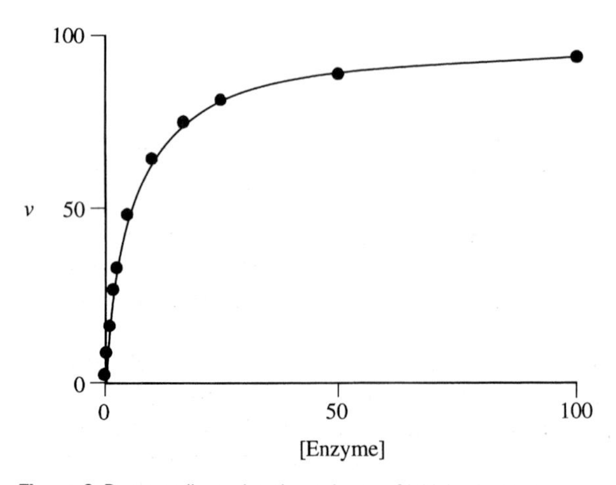

Figure 6 Downwardly curving dependence of initial velocity on enzyme concentration.

3.2 Upward curvature

There are two common causes for this type of behaviour.

3.2.1 The presence of a small amount of an irreversible inhibitor of the enzyme in the assay mixture

In this case small amounts of enzyme added will be completely inhibited and activity will only be detected after sufficient enzyme has been added to react with all the inhibitor present. This will give rise to a curve of the type shown in *Figure 5*, curve (b). A number of enzymes are irreversibly inhibited by heavy metal ions and contamination of buffer or substrate solutions by these is a common cause of such behaviour. Since such irreversible inhibition is time-dependent, the order of addition of components to the assay mixture may affect the observed results. If the enzyme is preincubated in an incomplete assay mixture before starting the reaction by addition of a substrate, a curve such as (b) of *Figure 5* might result. However, if the reaction were started by the addition of enzyme, a non-linear time course of the reaction would be expected (Section 2.1) and if it were possible to estimate the initial velocity before the irreversible inhibition became significant, a linear dependence on enzyme concentration would be expected to result.

3.2.2 The presence of dissociable activator in the enzyme solution

This can be represented by the equilibrium between activator (A) and enzyme (E):

$$\mathrm{E} + \mathrm{A} \underset{}{\overset{K_a}{\rightleftharpoons}} \mathrm{EA} \tag{8}$$

where the dissociation constant K_a will be given by:

$$K_a = \frac{[\mathrm{E}][\mathrm{A}]}{[\mathrm{EA}]} \text{ and thus: } [\mathrm{EA}] = \frac{[\mathrm{E}][\mathrm{A}]}{K_a} \tag{9}$$

In these equations *a* and *ea* are the concentrations of activator and EA complex, respectively and *e* is the concentration of free enzyme. The total enzyme concentration e_T will be:

$$[\mathrm{E}]_T = [\mathrm{E}] + [\mathrm{EA}] \tag{10}$$

Since the activator is present in the enzyme solution it will be present in a constant proportion to the enzyme concentration and thus its concentration can be expressed as $x \times e_T$. Thus, substituting into Equation (9) gives:

$$[\mathrm{EA}] = \frac{[\mathrm{E_T}]^2}{[\mathrm{E_T}] + (K_a/x)} \tag{11}$$

and thus the concentration of the EA complex will not increase linearly with the enzyme concentration, giving rise to a curve such as that shown in *Figure 5*, curve (c). The precise shape of the curve obtained will depend on the kinetic mechanism of activation and whether the free enzyme has any significant activity. If an excess of the activator were added to the assay mixture, this should displace the dissociation equilibrium so that essentially all the enzyme would exist as the EA complex and lead to a linear velocity–enzyme concentration relationship. There are several examples of this type of behaviour (6, p. 48).

A particular case occurs when an enzyme exists in an associating equilibrium, with the aggregated form being the more active. Thus, the endogenous activator may be regarded as being the enzyme itself. Ox heart phosphofructokinase has been shown to behave in this way (47) giving a curve similar to (c) of *Figure 5*. In this case the allosteric activator AMP promotes association and a linear dependence of initial velocity on enzyme concentration is observed in the presence of high concentrations of AMP. In contrast, the allosteric inhibitor citrate promotes disaggregation and results in a more pronounced curvature.

3.3 Downward curvature

There are three common cases that can give rise to curves such as that shown in *Figure 6*, where the reaction rate reaches an apparent maximum at higher enzyme concentrations.

3.3.1 The detection method may become rate-limiting at higher enzyme concentrations

For example, if a coupled assay (Section 6.1.2) is used, the activity of the coupling enzyme could become limiting, such that the addition of further amounts of the enzyme under study would not result in further increases in the measured velocity. Similar effects may be obtained with other assay methods if there is a failure of the detection system to respond linearly to increasing reaction rates (Section 2.1.6). An over-damped recorder or over-long time constant may lead to a maximum rate of response. In optical assays high enzyme concentrations can increase the initial absorbance of the solution to such a high level that the instrument no longer responds linearly (Chapter 2).

3.3.2 Failure to measure the true initial rate of the reaction

This can lead to an apparent maximum value in the velocity–enzyme concentration curve. If a discontinuous assay, which measures the amount of product formed after a fixed time, is used it may be that at higher enzyme concentrations the reaction has gone to completion or reached equilibrium within the assay time used. In that case the addition of more enzyme would not result in any increase in the measured product formation. This further emphasizes the necessity of ensuring that the assay procedure measures the initial rate of the reaction under all conditions used.

3.3.3 The presence of a dissociable inhibitor in the enzyme solution

This is the converse of the dissociable activator case discussed in Section 3.2.2. Increasing the enzyme concentration will lead to a proportional increase in that of the inhibitor and hence the amount of enzyme that is in the inactive enzyme–inhibitor complex will increase. If the complex is completely inactive, a graph of initial velocity against enzyme concentration will tend to an apparent maximum, as shown in *Figure 6*, whereas if it has a finite activity the curve will tend to a constant slope that is less than the initial slope. There are several well-documented examples of such behaviour (e.g. 6, p. 51). If the inhibitor is a small molecule it should be possible to remove it from the enzyme solution, for example by dialysis or gel filtration, but this would not be possible if the concentration-dependent inhibition resulted from reversible polymerization of the enzyme to give an inactive, or less active, form.

4 Expression of enzyme activity

In order to express the activity of an enzyme in absolute terms it is necessary to ensure that the assay procedure used is measuring the true initial velocity and that this is proportional to the enzyme concentration. Under these conditions the ratio (velocity/enzyme concentration) will be a constant that can be used to express the activity of an enzyme quantitatively. This can be valuable for comparing data obtained with the same enzyme from different laboratories, assessing the effects of physiological or pharmacological challenges on cells or tissues, monitoring the extent of purification of enzymes and comparing the activities of different enzymes, or of the same enzyme from different sources or with different substrates.

4.1 Units and specific activity

The activity of an enzyme may be expressed in any convenient units, such as absorbance change per unit time per mg enzyme protein, but it is preferable to have a more standardized unit in order to facilitate comparisons. The most commonly used quantity is the *Unit*, sometimes referred to as the International Unit or Enzyme Unit. One Unit of enzyme activity is defined as that catalysing the conversion of 1 μmol substrate (or the formation of 1 μmol product) in 1 min. The specific activity of an enzyme preparation is the number of Units per mg protein. Since some workers use the term unit to refer to more arbitrary measurements of enzyme activity, it is essential that it is defined in any publication.

If the relative molecular mass of an enzyme is known it is possible to express the activity as the *molecular activity*, defined as the number of Units per μmol of enzyme; in other words the number of mol of product formed, or substrate used, per mol enzyme per min. This may not correspond to the number of mol substrate converted per enzyme active-site per minute since an enzyme molecule may contain more than one active site. If the number of active sites per mol is known the activity may be expressed as the *catalytic centre activity*, which corresponds to mol substrate used, or product formed, per min per catalytic centre (active site). The term turnover number has also been used quite frequently but there appears to be no clear agreement in the literature as to whether this refers to the molecular or the catalytic centre activity.

4.2 The katal

Although the Unit of enzyme activity, and the quantities derived from it, have proven to be most useful, the Nomenclature Commission of the International Union of Biochemistry has recommended the use of the katal (abbreviated to kat) as an alternative. This differs from the units described above in that the second, rather than the minute, is used as the unit of time in conformity with the International System of units (SI Units).

One katal corresponds to the conversion of 1 mol of substrate per second. Thus it is an inconveniently large quantity compared to the Unit. The relationships between katals and Units are

$$1 \text{ kat} = 60 \text{ mol min}^{-1} = 6 \times 10^7 \text{ Units}$$
$$1 \text{ Unit} = 1 \ \mu\text{mol min}^{-1} = 16.67 \text{ nkat}$$

In terms of molecular or catalytic centre activities the katal is, however, not such an inconveniently large quantity and it is consistent with the general expression of rate constants in s^{-1}.

4.3 Stoichiometry

When expressing the activity of an enzyme it is important to bear in mind the stoichiometry of the reaction. Some enzyme-catalysed reactions involve two molecules of the same substrate. For example, adenylate kinase (EC 2.7.4.3) catalyses the reaction:

$$2\text{ADP} \rightleftharpoons \text{AMP} + \text{ATP}$$

and carbamoylphosphate synthetase (ammonia) (EC 6.4.3.16) catalyses:

$$\text{HCO}_3^- + 2\text{ATP} + \text{NH}_4^+ \longrightarrow \text{Carbamoylphosphate} + 2\text{ADP} + \text{P}_i$$

In the former case the activity will be twice as large if it is expressed in terms of ADP utilization than if expressed in terms of the formation of either of the products. In the latter case the value expressed in terms of disappearance of ATP or formation of ADP would be twice that obtained if any of the other substrates or products were measured. Thus it is important to specify the substrate or product measured and the stoichiometry when expressing the specific activity of an enzyme.

4.4 Conditions for activity measurements

Although the quantity velocity/enzyme concentration is a useful constant for comparative purposes, it will only be constant under defined conditions of pH, temperature and substrate concentration. A temperature of 30 °C has become widely used as the standard for comparative purposes, but in some cases it may be desirable to use a more physiological temperature. There is no clear recommendation as to pH and substrate concentration except that these should be stated and, where practical, should be optimal. However, it would be more appropriate to use physiological pH values, which may differ from the optimum pH, if the results are to be related to the behaviour of the enzyme *in vivo*. Since the activities of some enzymes are profoundly affected by the buffer used and the ionic strength of the assay mixture, the full composition of the assay should be specified. In cases where there is a nonlinear dependence of initial velocity upon enzyme concentration, and the artefacts referred to in Section 3 are excluded, it may be possible to work in the presence of an excess of the appropriate activator to obtain linearity (Section 3.2.2). However, when enzyme association–dissociation is involved it may not be possible to find an appropriate effector to displace the equilibrium towards the active polymerization state. An alternative approach would be to investigate conditions of pH, buffer composition or ionic strength under which the strength of the inter-subunit interactions are such that the polymerization/ depolymerization becomes unimportant over the concentration range used. In the case of phosphofructokinase mentioned in Section 3.2.2, for example, linearity can be achieved either by inclusion of an excess of the allosteric activator AMP or by working at a higher pH value.

5 The effects of substrate concentration

5.1 The Michaelis–Menten relationship

The Michaelis–Menten equation predicts a hyperbolic relationship between initial velocity and substrate concentration, and the kinetic behaviour of many enzymes is described by this relationship. It would not be possible to provide a detailed treatment of enzyme kinetics within this account and this section will concentrate on the practical problems that can arise when apparently complex behaviour is observed. A discussion on the analysis of kinetic data is included in Chapter 10 of this book and the reader is referred to several comprehensive accounts for detailed treatments of the steady-state kinetics of enzyme-catalysed reactions (4–7, 48–51). Although the double- reciprocal plot is recognized to be a poor procedure for determining enzyme kinetic parameters (Chapter 12), it is useful for illustrative purposes and will be used for such in this account. The direct-linear plot, which is a superior procedure for the calculation of data, is less clear for their presentation and therefore will not be used here.

Many of the pitfalls to be avoided in studies of the variation of the initial velocity over a range of substrate concentrations are similar to those discussed in terms of studies of the effects of variation of enzyme concentration (Section 3). Thus it is important to ensure that other factors, such as pH and ionic strength, which may affect the activity of the enzyme, remain constant. Although it is generally a simple matter to ensure that changing the substrate concentration does not affect the pH of the reaction mixture, it may be less easy to control the ionic strength if the substrate is a charged, or multi-charged, species. It may be possible to work at such high ionic strengths that the changes due to substrate addition are insignificant. Where such an approach is not possible it will be necessary to perform separate control experiments on the effects of ionic strength on the activity of an enzyme. Changes in the dielectric of the reaction mixture should also be controlled in cases where one of the substrates is non-polar or if it is added in solution in an organic solvent. In the latter case, it will also be necessary to check that the solvent does not itself affect the activity of the enzyme.

5.2 Failure to obey the Michaelis–Menten equation

Departure from the simple hyperbolic behaviour predicted by the Michaelis-Menten equation can result from a number of different causes. In each case it is necessary to ensure that the behaviour seen is a genuine phenomenon rather than an artefact. In this section some of the common cases of such apparently complex behaviour will be discussed in turn.

5.2.1 High-substrate inhibition

It is not uncommon for enzymes to be inhibited by high concentrations of one, or more, of their substrates, leading to kinetic plots such as those shown in *Figure 7*. Such behaviour can be useful in helping to deduce the kinetic mechanism involved (5–7,52) but it can restrict the

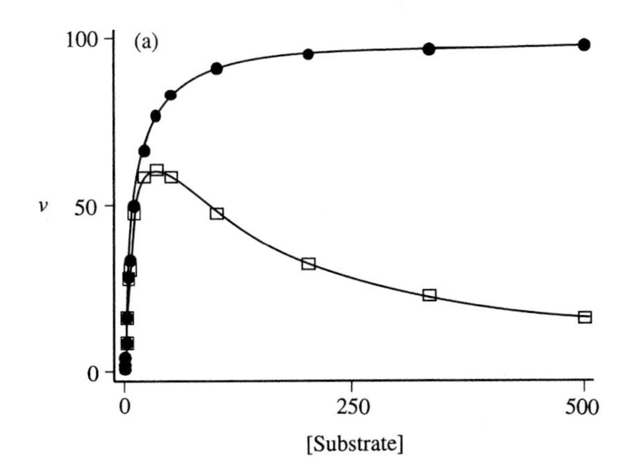

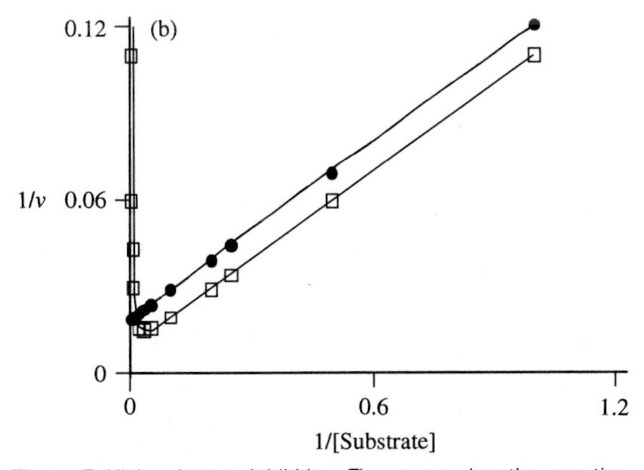

Figure 7 High-substrate inhibition. The curves obey the equation:

$$v = V_{max}/[1 + (K_m/S) + (S/K_i)]$$

where for the closed symbols $K_i = \infty$ and for the open symbols is set to 100. In both cases V_{max} and K_m are 100 and 10, respectively. Data are shown as Michaelis–Menten plots (a) and as double-reciprocal plots (b). The curves in (b) have been displaced from one another for clarity.

range of substrate concentrations that can be used for determining K_m and V_{max} values. The treatment of high-substrate inhibition data has been discussed in detail (6, p. 126). If such inhibition is observed, it is necessary to carry out appropriate controls to ensure that it is a property of the enzyme and its substrate rather than an artefact arising from failure to control the pH, ionic strength or dielectric of the assay medium correctly. It is also necessary to show that the inhibition is due to the substrate itself rather than to an inhibitory contaminant since, as discussed in Section 6.4.3, contamination of the substrate with a compound that is either a non-competitive (mixed) or uncompetitive inhibitor of the enzyme will lead to behaviour resembling high-substrate inhibition. A particular situation to guard against concerns substrates that chelate metal ions. If the enzyme requires free metal ions for activity, an excess of a chelating substrate, such as ATP or citrate, may reduce their concentrations to levels where the enzyme is unable to function. This may be remedied by ensuring that the metal ions are always present in excess. If that is not possible, it may be necessary to calculate the concentrations of free and complexed species (53) to ensure that the free metal ion concentration remains sufficient for activity. The complexities that can arise when the metal–substrate complex is the true substrate for the enzyme will be discussed in more detail in Section 5.2.2 *vi*.

5.2.2 Sigmoid kinetics

A sigmoid dependence of initial velocity upon substrate concentration (*Figure 8*) may indicate that the enzyme obeys cooperative kinetics. However, there are several other possible causes of such behaviour and it is necessary to carry out careful control experiments before ascribing such behaviour to this cause. Some effects that may result in such kinetic behaviour will be considered below:

i. True cooperativity

Cooperativity, as strictly defined, is a phenomenon reflecting the equilibrium binding of substrates, or other ligand (54) where the binding of one molecule of a substrate to an enzyme can either facilitate (positive cooperativity) or hinder (negative cooperativity) the binding of subsequent molecules of the same substrate. Positive cooperativity will give rise to kinetic behaviour such as that shown in *Figure 8* (lines corresponding to $h = 2$). In order to ensure that such behaviour is due to positive cooperativity, it would be necessary to perform substrate-binding studies under equilibrium conditions, since the steady-state initial velocities may not bear any simple relationship to the equilibrium saturation curve for substrate binding.

Most cooperative enzymes also exhibit allosteric behaviour. That is, their activities are affected by the binding of molecules (allosteric effectors) to sites distinct from the active site. Allosteric effects are, however, distinct from cooperativity and may occur in enzymes that show no cooperativity. Full discussions of the methods available for analysing cooperative behaviour and distinguishing between the possible models that may account for such effects are available elsewhere (e.g. 6, p. 399; 49, p. 151; 55, 56).

It is common to present data for cooperative enzyme in terms of the simple model advanced by Hill (57) in an attempt to explain the sigmoid saturation curve for oxygen binding to haemoglobin:

$$Ys = \frac{[S]^h}{K + [S]^h} \tag{12}$$

where [S] is the substrate concentration, Y_s represents the fractional saturation of the enzyme with substrate and h (also referred to as, among other things, n, n_H or n_h) is the Hill constant, which does not necessarily correspond to the number of substrate-binding sites present in

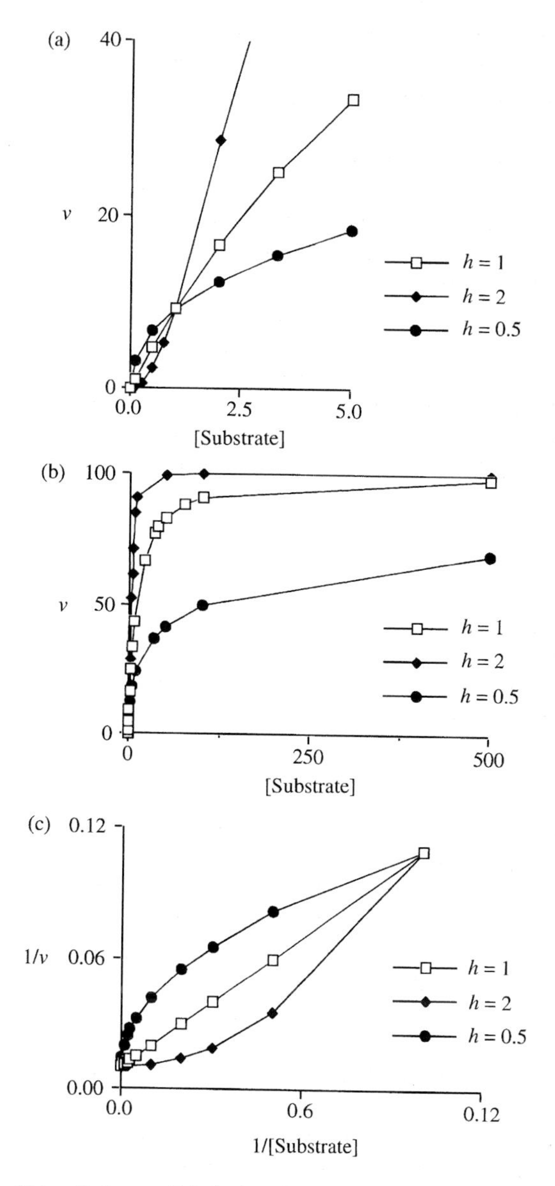

Figure 8 Cooperativity in the effects of substrate concentration on initial velocity. The curves were fitted to the Hill equation (Equation 13) with V_{max} and K being equal to 100 and 10, respectively. The behaviour at $h = 1$ (no cooperativity), $h = 2$ (positive cooperativity) and $h = 0.5$ (negative cooperativity) is shown. Panels (a) and (b) show the same data over different ranges of substrate concentration; panel (c) shows the data in double-reciprocal form. The curves have been displaced from one another for clarity.

the molecule (6 p.399; 55, 56). In the case of a cooperative enzyme where initial velocities are measured the corresponding Hill equation can be written as:

$$v = \frac{V_{max}[S]^h}{K + [S]^h} \tag{13}$$

which can be transformed to a linear relationship as:

$$\log\left(\frac{v}{V_{max} - v}\right) = h\log[S] - \log K \tag{14}$$

Thus a graph of log $\{v/(V_{max} - v)\}$ against log[S] should give a straight line of slope = h and intercept on the y-axis = − log K. Such plots are known as Hill plots. Although the Hill equation has been shown to be based on an inadequate model, because it envisages the simultaneous binding of all substrate molecules to the enzyme (e.g. 6, p. 399; 55, 56) the plot is still widely used to express cooperativity. A Hill constant of greater than unity indicates positive cooperativity whereas one of less than unity is given in cases of negative cooperativity. If there is no cooperativity the value of h will be unity and Equation 13 will reduce to the simple Michaelis–Menten equation.

Despite its widespread use, the Hill equation is an invalid model for cooperative systems and it has been shown that for any system that involves the sequential binding of substrates, the plot will be linear only over a restricted range of substrate concentrations with the slopes tending to unity at very high and very low substrate concentrations (see 6, p.399; 55, 56 for further discussion). Thus, departure from the linearity predicted by Equation 14 may be expected and the Hill constant is calculated from the region of maximum slope.

ii. Alternative pathways in a steady-state system

Any enzyme reaction mechanism in which there are alternative pathways by which the substrates can interact to give the complex that breaks down to give products will, under steady-state conditions, give rise to a complex initial rate equation. The simplest example concerns an enzyme with two substrates, Ax and B, which are converted to the products A and Bx, respectively. A reaction mechanism in which the two substrates can bind to the enzyme in a random order:

$$\begin{array}{ccccc} & \text{EAx} & & & \\ \nearrow\!\!\swarrow & & \searrow\!\!\nwarrow & & \\ \text{E} & & & \text{EAxB} \dashrightarrow \text{E + A + Bx} & \\ \searrow\!\!\nwarrow & & \nearrow\!\!\swarrow & & \\ & \text{EB} & & & \end{array} \tag{15}$$

will, under steady-state conditions, give an initial-rate equation of the form

$$v = \frac{p[\text{Ax}][\text{B}] + q[\text{Ax}]^2[\text{B}] + r[\text{Ax}][\text{B}]^2}{s + t[\text{Ax}] + u[\text{B}] + v[\text{Ax}][\text{B}] + w[\text{Ax}]^2 + x[\text{B}]^2 + y[\text{Ax}]^2[\text{B}] + z[\text{Ax}][\text{B}]^2} \tag{16}$$

where the constants p–z are combinations of rate constants. At a fixed concentration of one of the substrates, for example B, this fearsome equation may be simplified to:

$$v = \frac{b_1[\text{Ax}] + b_2[\text{Ax}]^2}{a_0 + a_1[\text{Ax}] + a_2[\text{Ax}]^2} \tag{17}$$

An equation of this form, containing squared terms in the substrate concentration, can give rise to a variety of curves describing the variation of initial velocity with substrate concentration, as shown in *Figure* 9, depending on the values of the individual rate constants (58). The behaviour of this system has been considered in detail by Ferdinand (59) who pointed out that sigmoidal behaviour would be expected if the rates through the two alternative pathways (through EAx and through EB) leading to the EAxB ternary complex were sufficiently different. Under such conditions a sigmoid dependence of initial velocity upon substrate concentration will result if the concentration of one of the substrates is varied at a fixed, non-saturating concentration of the other, whereas if the concentration of the other substrate is varied at a fixed, non-saturating concentration of the first, a curve such as that shown in *Figure 9* line (■) will result. Several other mechanisms in which there are alternative (two, or more) ways in which substrates can bind to an enzyme have been shown to give rise to complex steady-state rate equations (60–63).

The random-order mechanism will yield simple Michaelis–Menten behaviour if the rate of breakdown of the EAxB ternary complex to give products is slow relative to its rate of breakdown to the binary (EAx and EB) complexes so that the system remains in thermodynamic equilibrium (see 4–7, 48–51).

iii. The reaction involves two, or more, molecules of the same substrate

For a reaction of the type:

$$2Ax \longrightarrow Ax_2 + A$$

the initial rate equation for a two-substrate reaction in which the substrates bind randomly under equilibrium conditions, or in an ordered mechanism under steady-state conditions, the rate equation becomes:

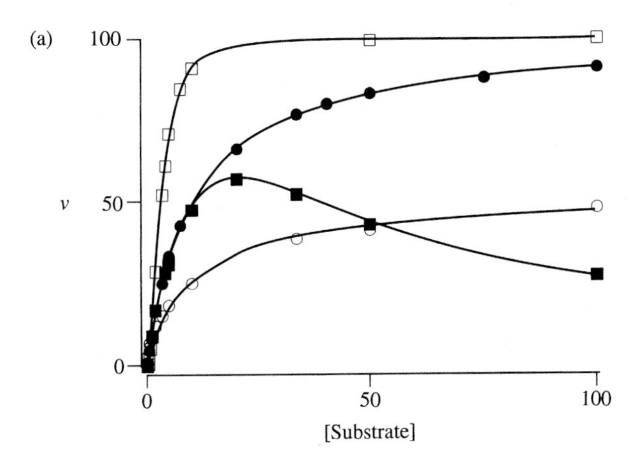

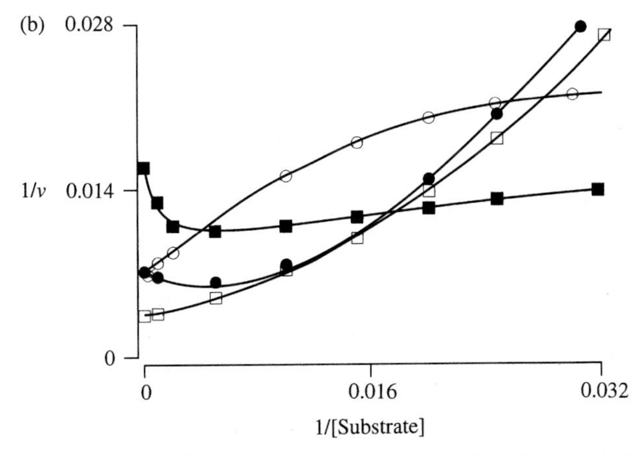

Figure 9 Possible initial velocity–substrate concentration relationships that can be obtained with systems obeying Equation 16. The data are shown as Michaelis–Menten (a) and double-reciprocal plots (b). The curves have been displaced from one another for clarity.

$$v = \frac{V_{max}}{1 + \frac{K_{m1}^{Ax}}{[Ax]} + \frac{K_{m2}^{Ax}}{[Ax]} + \frac{K_s^{Ax} K_{m2}^{Ax}}{[Ax]^2}} \quad (18)$$

where the K_m values for the first and second molecules of Ax to bind to the enzyme are designated K_{m1} and K_{m2}, respectively. Equation 18 predicts a sigmoidal dependence of initial velocity on substrate concentration. The degree of sigmoidicity will depend on the values of the individual constants. If the two substrate binding steps are separated by an irreversible step, such as in a double-displacement (ping-pong) mechanism, the value of the constant K_s^{Ax} will be zero and thus the equation will simplify to (6, p.79):

$$v = \frac{V_{max}}{1 + \frac{K_{m1}^{Ax}}{[Ax]} + \frac{K_{m2}^{Ax}}{[Ax]}} \quad (19)$$

and hyperbolic kinetics will result. The enzyme hydroxymethylbilane synthase (EC 4.2.1.24) involves the interaction of four identical substrate molecules to yield the product porphobilinogen. Since this enzyme exhibits simple Michaelis–Menten kinetic behaviour, there must be steps that are essentially irreversible between the binding of each of the substrate molecules to the enzyme (64). An alternative way in which an essentially irreversible step may occur is if the binding of the two identical substrate molecules is separated by the binding of a different substrate. In that case very high concentrations of the latter substrate would render its binding essentially irreversible, resulting in an irreversible step between the binding of the two identical molecules, and hyperbolic kinetics would be obtained (65).

Determination of the stoichiometry of the reaction catalysed should establish whether sigmoid initial velocity curves are likely to result from this cause. For example, such curves are given by the enzyme carbamoylphosphate synthetase (ammonia) when ATP is the variable substrate, because the reaction catalysed involves two molecules of this substrate (65 and Section 4.3).

iv. Enzyme isomerization

Several possible mechanisms in which an enzyme exists in two, or more, forms which interconvert relatively slowly can give rise to kinetic equations of the form of Equation 17. Different systems of this type have been presented in several publications (66–69). If the rates of the isomerization steps are sufficiently slow, hysteric effects (Section 2.4.9) may also be observed. Mechanisms of this type do not require the presence of multiple binding sites for the same substrate and thus may not show cooperative substrate binding.

v. Failure to determine initial rate

If an enzyme is assayed by determining the extent of reaction after an arbitrarily fixed time, it may be that the reaction has proceeded to completion, or equilibrium within that time at the lower substrate concentrations. This would result in an underestimation of the true initial velocity at the lower substrate concentrations and hence a curve which might appear sigmoid. A similar effect may occur if an enzyme is unstable in the reaction mixture but is stabilized by the binding of substrate. In this case the effectiveness of stabilization will increase with substrate concentration as the enzyme tends to become saturated with substrate. This will lead to an increasing underestimation of the true initial velocity as the substrate concentration decreases below that required to saturate the enzyme. Such an effect, leading to apparently sigmoid curves of velocity against substrate concentration, has, for example, been reported for the enzyme threonine deaminase (EC 1.1.1.103) (70).

vi. Failure to take account of substrate–activator complexes

The true substrate for a number of enzymes is a complex between substrate and an activator, usually a divalent metal ion. One of the most widely studied cases is the complex between ATP and magnesium ions, which is the true substrate for many enzymes catalysing reactions involving ATP. In such cases it is necessary to calculate the concentration of the complexed form, as the true substrate, for kinetic studies. In general the concentration of the metal-substrate complex, [MS], will be given by the equation:

$$[\mathrm{MS}] = \frac{1}{2}\{([\mathrm{S_T}] + [\mathrm{M_T}] - K_\mathrm{d}) - \sqrt{([\mathrm{S_T}] + [\mathrm{M_T}] + K_\mathrm{d})^2 - 4[\mathrm{M_T}][\mathrm{S_T}]}\} \quad (20)$$

where $[\mathrm{S_T}]$ and $[\mathrm{M_T}]$ are the total concentrations of substrate and metal ion, respectively, and K_d is the dissociation constant of the (MS) complex.

This equation does not predict a linear dependence of the concentration of the MS complex upon the concentrations of M plus S if these are mixed together and varied in a fixed ratio. Several kinases have been erroneously reported to show cooperative behaviour because sigmoid kinetics were observed when the ATP and magnesium were mixed together in a fixed ratio and the 'substrate concentration' was varied by the addition of varying amounts of this mixture. In fact such behaviour is predicted by the relationship shown in the above equation and has nothing to do with cooperativity (65, 71–73). The simplest way to overcome this problem is to work at such high concentrations of the metal ion, compared to the K_d value, that the substrate remains essentially fully in the complexed form at all concentrations used. However, this may not be possible, for example if the free metal ion inhibits the enzyme. In such cases it will be necessary to calculate the complex concentration at each concentrations used. This is no easy task because the interactions between metals and ligands will be affected by the pH, temperature and ionic strength (74). A useful procedure and computer program to calculate the binding of metal ions to ATP has been presented by Storer and Cornish-Bowden (53).

It should be remembered that the substrate may not be the only species in solution that binds metal ions. Many buffer species, such as phosphate or citrate buffer, are metal chelators and allowance should be made for this when calculating the amount of complexed species in solution. A number of buffers which do not bind magnesium ions, or bind them only very weakly, are available (75).

The special problems of studying the kinetics of enzyme reactions in which a metal–substrate complex may be involved are beyond the scope of the present account and the reader is referred to several fuller descriptions of the procedures involved (73, 76–78). Although the above analysis has concentrated on metal–substrate complexes, the possibility that other activator–substrate complexes might be involved in the activity of an enzyme should not be excluded. For example, the true substrate for formaldehyde dehydrogenase (EC 1.2.1.1) is the reversible adduct formed between formaldehyde and glutathione (79). The analysis of the behaviour of such systems would be similar to that discussed above.

5.2.3 Apparent negative cooperativity

Negative cooperativity, in which the binding of one molecule of substrate to an enzyme decreases its affinity for binding subsequent molecules of the same substrate, leads to a Hill constant of less than unity and a double-reciprocal plot that curves downward (*Figure 8*). The curve of velocity against substrate concentration may, at first sight, appear to be normal, but, as shown in *Figure 8*, this is illusory. As discussed in Section 5.2.2, true negative cooperativity is a substrate-binding phenomenon which should be reflected in similar behaviour if equilibrium binding is studied. There are, however, a number of other possible causes of such effects which may be difficult to distinguish from negative cooperativity.

i. Alternative pathways in a steady-state system

As discussed in Section 5.2.2 *ii*, mechanisms of this type can yield initial velocity–substrate concentration curves that resemble those seen with negative cooperativity (see *Figures 8* and 9). However, because this effect arises from steady-state, rather than substrate-binding, complexities the substrate binding curves determined at equilibrium should be rectangularly hyperbolic.

ii. The presence of more than one enzyme catalysing the same reaction

If the same reaction is catalysed by two enzymes which have different K_m values, the rate of the overall reaction will be given by:

$$v = \frac{V^a}{1 + K_m^a/[S]} + \frac{V^b}{1 + K_m^b/[S]} \qquad (21)$$

where $K_m{}^a$ and V^a are the Michaelis constant and maximum velocity of one enzyme and $K_m{}^b$ and V^b are the corresponding constants for the other. This equation will result in a behaviour similar to that of negative cooperativity. Thus it is always necessary to check for the presence of more than one enzyme if such behaviour is observed. This is particularly important because if the two enzymes had different binding affinities for their substrates, as well as different K_m values, similar behaviour would be shown in studies of the equilibrium binding of substrate.

Equation 21 predicts a smooth curve when the data are plotted in double-reciprocal form (*Figure 8*) without sharp breaks. It is not possible to obtain accurate estimates of the two K_m and V_{max} values by extrapolation of the apparently linear portions of the double-reciprocal plots at very high and very low substrate concentrations . Such an approach does not yield accurate values (80) unless the difference between the K_m values is greater than 1000-fold. It is possible, however, to determine the individual values using an iterative procedure that fits the data points to the sum of two rectangular hyperbolas (e.g. 81, 82). It is always good practice to construct a curve from the calculated values using Equation 21 and compare it with the data points to ensure that a good fit has been obtained.

iii. Failure to subtract a blank rate

If a blank rate is not subtracted from the rates observed with the full reaction mixture (see Section 2.5) a curve resembling negative cooperativity (*Figure 8*) can result because the rate will appear to be finite when the substrate concentration is zero. A similar distortion can occur in equilibrium studies if there is nonspecific binding of the substrate that is not corrected for. If there is a blank rate that is proportional to the concentration of the substrate, failure to subtract it will give rise to a downwardly curved double-reciprocal plot that passes through the origin ($1/v = 0$ when $1/[S] = 0$) since the velocity will tend to become infinite as the substrate concentration is raised towards infinity (83).

iv. High-substrate activation

A mechanism in which the binding of substrate to an enzyme–product complex facilitates the release of the product, and thus accelerates the reaction, has been proposed to account for the downwardly curving double-reciprocal plots seen with aldehyde dehydrogenase (84). In this case, since the second substrate molecule binds to an enzyme–product complex, normal saturation curves would be expected from equilibrium binding studies with substrate or substrate analogues in the absence of products.

v. Enzyme isomerizations

Systems involving isomerization of the enzyme, such as those discussed in Section 5.2.2 *iv*, can give rise to curves resembling negative as well as positive cooperativity.

vi. Tight binding

If an enzyme has a very high affinity for its substrate it may be necessary to use substrate concentrations that are similar to those of the enzyme in binding, or initial-rate kinetic, studies. Under these conditions the free substrate concentration will be significantly altered by enzyme–substrate complex formation and the dissociation constant (K_s) for the reaction:

$$E + S \rightleftharpoons ES$$

will be:

$$K_s = \frac{([E] - [ES])([S] - [ES])}{[ES]} \tag{22}$$

This relationship leads to an equation of the same form an Equation 20 and does not predict a simple binding curve. Methods for analysing the behaviour of such systems have been presented elsewhere (85, 86).

5.2.4 Even more bizarre curves

Complexities of the steady-state reaction mechanism (Section 5.2.2 *ii*) can result in multiple inflection points or waves in the curves of initial velocity against substrate concentration (62). True cooperativity, in which an enzyme shows a mixture of positive and negative cooperativity for the successive binding of molecules of the same substrate, can also give multiple inflection points and plateaus in the saturation, or velocity–substrate concentration curve (87). Such behaviour could also arise from a mixture of two enzymes, one exhibiting positive cooperativity and the other no cooperativity. This might come about if the properties of a cooperative enzyme became modified during the purification in such a way that a proportion of the molecules had lost their ability to interact cooperatively with the substrate. Complex negative cooperativity for multiple binding sites can result in curves of velocity against substrate concentration which may appear to be composed of linear sections with apparently sharp breaks between them (88, but see also 89).

6 Experimental approaches

6.1 Type of assay

Although many enzyme-catalysed reactions result in changes in the properties of the reactants that are relatively easy to measure directly and continuously, others do not and in such cases it is necessary to use an indirect assay method that involves some further treatment of the reaction mixture. In some cases it may be possible to use such indirect assays to monitor the progress of the reaction continuously, but in others it is necessary to stop the reaction before further treatment of the assay mixture to allow the extent of reaction to be determined. Continuous assays have the advantage of allowing progress curves for the reaction to be followed directly and should thus make it a relatively simple matter to determine initial rates and see any deviations from the initial linear phase of the reaction or any of the anomalous types of behaviour discussed earlier. Because discontinuous assays will give the extent of a reaction after a chosen fixed time, it may be tempting to select a reaction time and assume that initial rate conditions will hold for that time. I hope that the discussion in the previous sections has made it clear that such assumptions can lead to gross errors and that it is necessary to show that product formation proceeds linearly for the time used in such assays, under all conditions employed.

6.1.1 Direct continuous assays

Any difference between the properties of the substrates and products that can be directly measured may be used to provide the basis for direct assays. Changes in absorbance, fluorescence, pH, optical rotation, conductivity, enthalpy, viscosity or volume of the reaction mixture have all been used to assay the activities of individual enzymes. Provided that the sensitivity is sufficiently high and the procedure does not impose undesirable limitations on the assay conditions that can be used, direct continuous assays are always to be preferred, because they allow observation of the progress curve, which simplifies the estimation of initial rates and allows detection of any anomalous behaviour. Some commonly-used procedures are mentioned below (see also 1–3).

Spectrophotometric assays are probably the most widely-used procedures (2, 90). The high standards of accuracy and reliability of many commercially-available spectrophotometers make such assays particularly convenient when the reaction catalysed involves an absorbance change. In cases where the reaction results in a change in fluorescence of one of the substrates, fluorimetric assays can be used. This usually results in a considerable gain in sensitivity and can be particularly valuable when only small amounts of the enzyme are available or if it has a very low K_m value for its substrates. The applications of spectrophotometric and fluorimetric assays, and some of the precautions that should be taken in their use, are discussed in more detail in Chapter 2.

Reactions that involve the release or uptake of hydrogen ions can be assayed directly in unbuffered, or weakly buffered, solutions by following the change in pH with a glass electrode. The use of such assays should be restricted to a pH range over which the change does not appreciably affect the activity of the enzyme. Changes in hydrogen ion concentrations may also be followed spectrophotometrically by use of an indicator which changes its absorbance with protonation state (see Chapter 2). An alternative method which avoids significant change of pH during the assay is to use a pH-stat, which titrates the reaction mixture with either acid or alkali to keep the pH constant whilst recording the rate of addition (91, and Chapter 7). Ion-sensitive electrodes or gas electrodes can be used for monitoring changes in the concentrations of some other reactants, such as ammonia or CO_2 (see 1). Reactions involving the uptake or output of oxygen can be followed polarographically by means of an oxygen electrode (92 and Chapter 6).

6.1.2 Indirect assays

These involve some further treatment of the reaction mixture either to produce a measurable product or to increase the sensitivity or convenience of the assay procedure. In some cases it is possible to use indirect assays continuously to monitor the progress of the reaction whereas in others the method can only be used discontinuously.

i. Discontinuous indirect assays

These assays, which may also be termed sampling assays, involve stopping the reaction after a fixed time and treating the reaction mixture to separate a product for analysis or to produce a change in the properties of one of the substrates or products, which can then be measured. Examples of the former type of assay include radiochemical assays, which are discussed in detail in Chapter 3. The application of liquid chromatographic systems for rapid separation and quantitation of reactants has allowed many assays to be devised that are based on this technique. These are discussed in Chapter 4.

Measurement of luminescence can form the basis of highly sensitive assay procedures (97). The formation or disappearance of ATP can be determined by measuring the light emission in the presence of fire-fly luciferase, which catalyses the reaction:

$$\text{ATP} + \text{luciferin} + O_2 \longrightarrow \text{oxyluciferon} + PP_i + CO_2 + \text{light}$$

Similarly, NAD(P)H can be determined using the bacterial luciferase system, which catalyses the reactions:

$$NADPH + H^+ + FMN \longrightarrow NADP^+ + FMNH_2$$
$$FMNH_2 + RCHO + O_2 \longrightarrow H_2O + RCOOH + light$$

where RCHO is a long-chain (8–14 carbons) aliphatic aldehyde. This reaction has also been used to determine the oxidation of aliphatic amines to the corresponding aldehydes by the enzyme monoamine oxidase (93). Mutants of the bacteria that require the presence of fatty acids have been used to determine lipase and phospholipase activities (94), and a cyclic AMP requiring mutant has also been used to determine that compound (95).

The earthworm luciferase system requires H_2O_2 and may be used as a basis for determining the formation of that compound (96). There are also several chemiluminescence reactions that can be used for the determination of H_2O_2 formation. These include the light-emitting reaction with luminol (3-aminophthalazine-1,4-dione), which is catalysed by metal-ion complexes at alkaline pH values or by peroxidase (EC 1.11.1.7).

Manipulations to render a product detectable include, for example, adjustment of the pH to alkaline values to allow the *p*-nitrophenol, produced by the action of acid phosphatase on *p*-nitrophenylphosphate, to be detected, and the use of colour reactions to determine inorganic phosphate (e.g., 98). The use of discontinuous assays to increase the sensitivity of detection can be illustrated by the use of strong alkali to convert NAD^+ or $NADP^+$ into highly fluorescent derivatives. These products fluoresce at 460 nm when excited at 360 nm with an intensity that is about 10-fold higher than that given by NAD(P)H (90, 99). Although the reduced coenzymes do not react in this way, it is necessary to remove them to avoid interference from their fluorescence. This may be achieved by prior treatment with 0.2 M HCl, which destroys the reduced coenzymes without affecting the oxidized forms. To apply this method for the determination of NAD(P)H, the oxidized coenzymes can be destroyed by treatment with dilute alkali and the reduced forms can then be reoxidized with H_2O_2 before the strong alkali treatment.

With all discontinuous assays it is important to ensure that the procedure used to terminate the reaction does so instantaneously. Methods involving rapid mixing with acid or alkali to alter the pH to a value where the enzyme is inactive are usually effective but methods involving transfer of the reaction vessel to an ice or boiling-water bath may be less satisfactory if the volume of the assay mixture is relatively large. It is essential to check that the method used does, in fact, stop the reaction instantaneously. This can be often be done by comparing the results given by samples in which the reaction is stopped at zero time with those given by samples from which either the enzyme or one of the substrates has been omitted or the enzyme has been inactivated in some way, such as by heat treatment or incubation with an irreversible inhibitor, before it is added to the assay mixture.

ii. Continuous indirect assays

Assays of this type involve carrying out the manipulations necessary to detect product formation, or product remaining within the assay mixture, in such a way as to allow the change to be followed continuously as it occurs. Such assays should allow progress curves to be determined in a single assay and thus they may be less prone to errors arising from the sample manipulations necessary in discontinuous assays.

Reagents that react with one of the products of the reaction to form a detectable compound can be included in the assay mixture. If such an assay is to give valid results the detection reaction must occur so rapidly that the enzyme-catalysed reaction is always rate-limiting, so that the rate determined will correspond to the activity of the enzyme under study. It is also necessary that the reagent used has no effect on the activity of the enzyme and that it does not react with any of the other components of the system. An example of this type of reaction is the detection of carnitine acyltransferases (EC 2.3.1.7, EC 2.3.1.21 and EC 2.3.1.137) (100). These enzymes will catalyse the transfer of acyl groups from coenzyme A to carnitine. This process results in the liberation of the free sulphydryl group of CoASH, which reacts extremely rapidly with the reagent 5, 5'-dithiobis-2-nitrobenzoate (Nbs_2) releasing a yellow-coloured compound whose formation can be followed at 412 nm. A combination of this approach with the use of a synthetic substrate can be used for the assay of acetylcholine esterase (EC 3.1.1.7). In this assay Nbs_2 is included in the assay mixture and acetylcholine is replaced by acetylthiocholine, which is hydrolysed by the enzyme to yield acetate plus thiocholine (101).

iii. Coupled assays

The most commonly used assays of this type involve the use of one, or more, additional enzymes to catalyse a reaction of one of the products to yield a compound that can be detected directly. Assays of this type are known as coupled assays and the auxiliary enzymes added are frequently referred to as coupling enzymes. A number of such assays have been listed by Rudolph *et al.* (102). Coupled assays often involve the reduction of $NAD(P)^+$, or the oxidation of the corresponding reduced coenzymes, because these processes can be readily determined spectrophotometrically or fluorimetrically. However, there are many other possibilities. For example, some of the luminescence systems discussed in the previous section may be adapted for continuous use under appropriate conditions (103, 104).

A simple coupled assay for the determination of hexokinase activity by coupling the formation of glucose-6-phosphate to the reduction of $NADP^+$ in the presence of glucose-6-phosphate dehydrogenase (EC 1.1.1.49) is shown in *Figure 10* and a coupled assay for pyruvate carboxylase was discussed in Section 2.5.1 *vi*. It is not necessary for coupled assays to be restricted to a single coupling enzyme. *Figure 11* shows two alternative assays for phosphofructokinase. That involving the reaction of the ADP produced with phosphoenolpyruvate in the presence of pyruvate kinase (EC 2.7.1.40) has the advantage that the ATP used in the reaction is continuously regenerated, which should prevent any fall-off in the reaction velocity due to depletion of this substrate (Section 2.1.1). However, it has the limitation that phosphofructokinase from some sources is allosterically inhibited by phosphoenolpyruvate (78).

If assays of this type are to yield valid results it is essential that the coupling enzyme(s) used never becomes rate-limiting so that the measured rate is always determined by the activity of the enzyme under study. The velocity of the reaction catalysed by a coupling enzyme will depend upon the substrate concentration available to it. Since this is produced by the activity of the enzyme under study, there will be very little of it available during the early part of the reaction and, thus, the coupling enzyme will be functioning at only a small fraction of its maximum velocity. As the reaction proceeds the concentration of the intermediate

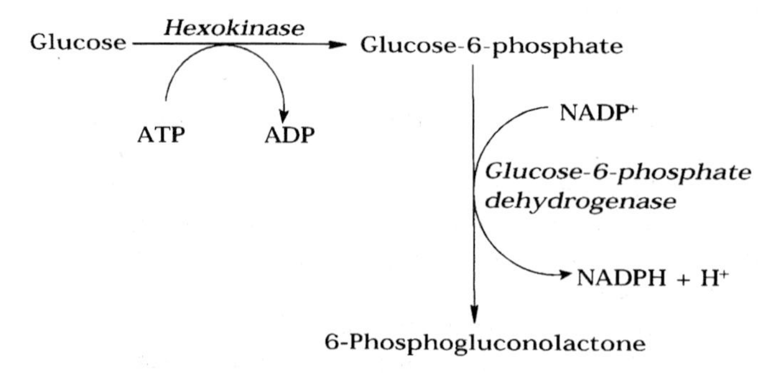

Figure 10 A coupled assay for hexokinase.

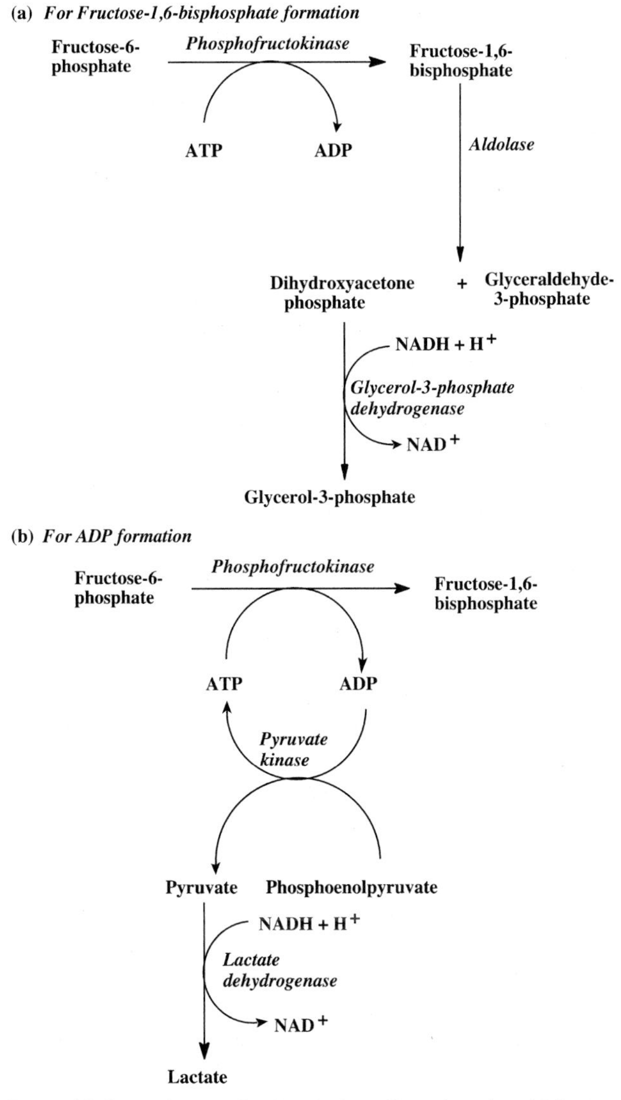

Figure 11 Coupled assays for phosphofructokinase based on (a) fructose-1,6-bisphosphate production and (b) ADP formation.

substrate will increase which will, in turn, allow the coupling enzyme to work faster. Thus the rate of the coupling reaction will increase with time until it equals the rate of the reaction catalysed by the first enzyme. At this stage the concentration of the intermediate substrate will remain constant because of a balance between the rate of its formation by the enzyme under study, and the rate of its removal by the coupling enzyme. This behaviour of coupled enzyme assays is illustrated in *Figure 12*. It results in a lag in the rate of formation of the product of the coupled reaction. It is, of course, necessary to minimize this lag period, which is often referred to as the coupling time, because of the possibility that the reaction catalysed by the first enzyme will have started to slow down (Section 2.1) before the coupling

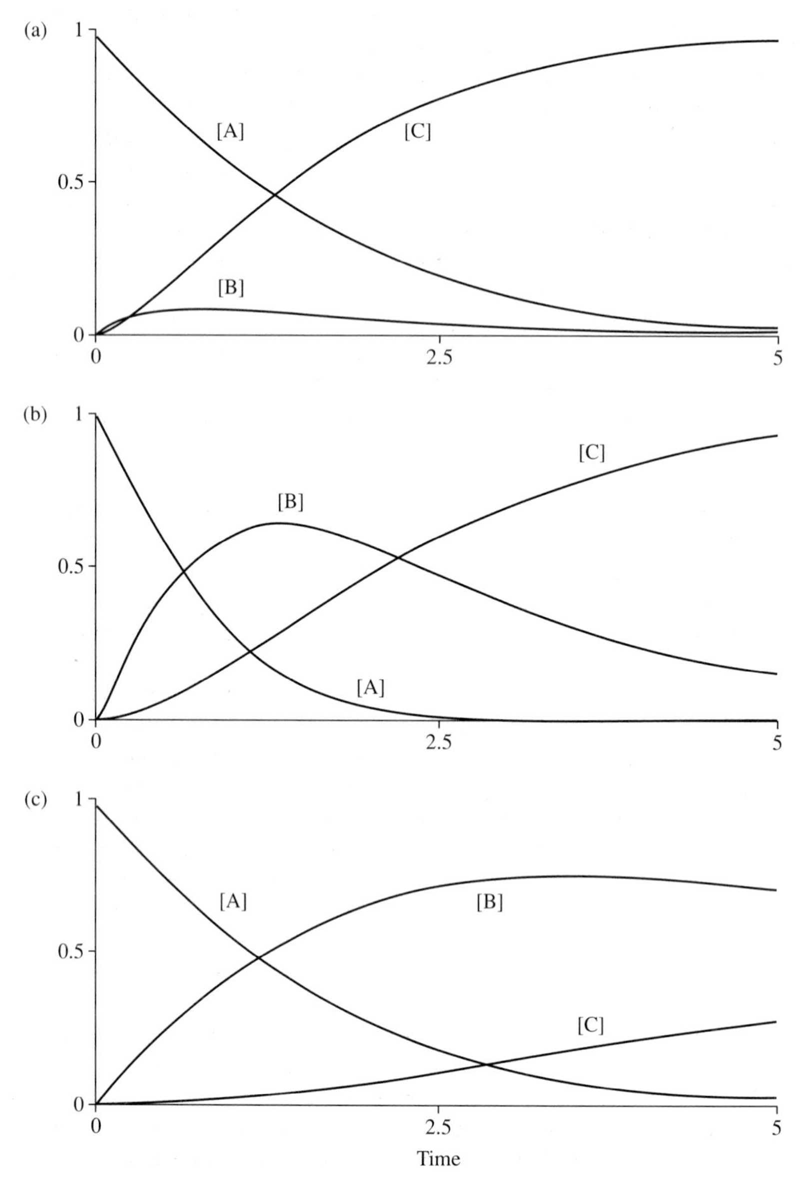

Figure 12 Time-course of a coupled enzyme assay involving a single coupling enzyme. The lag in the appearance of the product formed by the coupling enzyme corresponds to the period in which the concentration of the intermediate product, which is the product of the reaction catalysed by the first enzyme and the substrate for the coupling enzyme, rises to a constant (steady-state) concentration. Three simulated examples of the variation of substrate (A), intermediate product (B) and final product (C) for the simple model:

$$\mathbf{A} \xrightarrow{e_1} \mathbf{B} \xrightarrow{e_2} \mathbf{C}$$

are shown. In (a) The maximum velocity of the enzyme being assayed (e_1) and that of the coupling enzyme (e_2) present in the assay were 1 and 50 μmol min^{-1}, respectively. In (b) the V_{max} values for e_1 and e_2 were both 5 μmol min^{-1} and in (c) they were 5 and 0.5, respectively. The K_m values for e_1 and e_2 were 1 and 10 μM, respectively, and the initial substrate concentration was 1 μM, in all three cases.

enzyme has reached its steady-state velocity. In that case the coupled assay would never give an accurate measure of the activity of the enzyme under study.

The efficiency with which a coupling enzyme can function will depend on its K_m value for the substrate being formed. The lower its K_m value the more efficiently it will be able to work at low substrate concentrations. The lag period can also be reduced by increasing the amount of the coupling enzyme present so that it can catalyse the reaction more rapidly at low substrate concentrations. The higher the K_m value of the coupling enzyme the greater the amount of it will be required to produce the same lag period. Thus it is necessary to characterize the performance of a coupled assay to ensure that it gives an accurate measure of the activity of the enzyme under study. Usually this can be done experimentally by checking that the measured velocity is not increased by increasing the amount of the coupling enzyme present and is proportional to the amount of the first enzyme present at all substrate concentrations, and under all conditions, that are to be used. Generally this is achieved by having a very large excess of the coupling enzyme(s) present. It is possible to calculate the amount of a coupling enzyme that must be added to give any given coupling time (102) and such calculations may be useful in saving the expense of adding too much reagent. It must, however, be remembered that it will be necessary to recheck that a coupled assay is performing correctly each time the assay conditions are altered since these may affect the behaviour of the coupling enzyme(s). The purity of the coupling enzyme(s) used should also be checked. Since these are used in relatively high concentrations, even a small degree of contamination with other enzymes, or substrates, that might affect the reaction under study could become important.

A rather unusual type of coupled assay is the cycling assay for alcohol dehydrogenase activity (105). In this reaction, shown in *Figure 13*, the NADH formed by the oxidation of ethanol is used to reduce lactaldehyde to propanediol in a reaction catalysed by the same enzyme. Here, NAD(H) functions catalytically and one mol of propanediol is produced per mol of ethanol oxidized. The rates of such cycling assays can be much faster than conventional assays because the rate-limiting step in the reaction catalysed by this enzyme is the dissociation of the product NADH, which does not need to occur in the cycling system. Because the coenzyme functions catalytically it is only necessary for it to be present at low concentrations and this minimizes competition from other enzymes that use NAD(H) (Section 2.5.3). The cycling assay illustrated in *Figure 13* is performed discontinuously by measuring the amount of propanediol formed chemically after stopping the reaction. However, if the lactaldehyde is replaced by an aldehyde that undergoes an absorbance change on reduction to the corresponding alcohol, a continuous assay is possible (106). Cycling assays of this type have also been used as the basis of sensitive methods for determining metabolite concentrations (90).

Another type of coupled assay is the, so-called, 'forward-coupled' assay. In this system the enzyme reaction that is directly monitored catalyses a reaction that provides the substrate for

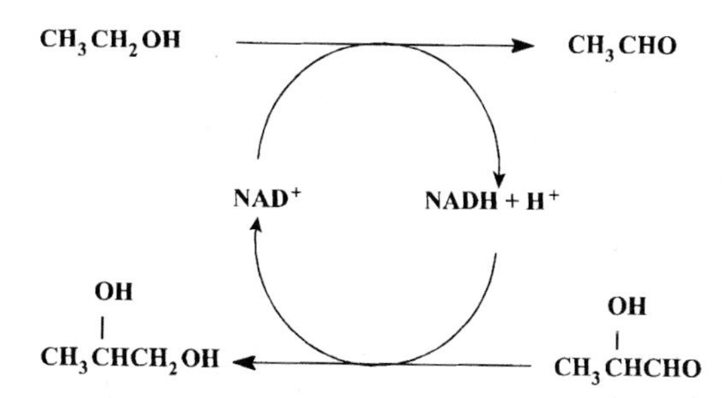

Figure 13 A cycling assay for alcohol dehydrogenase.

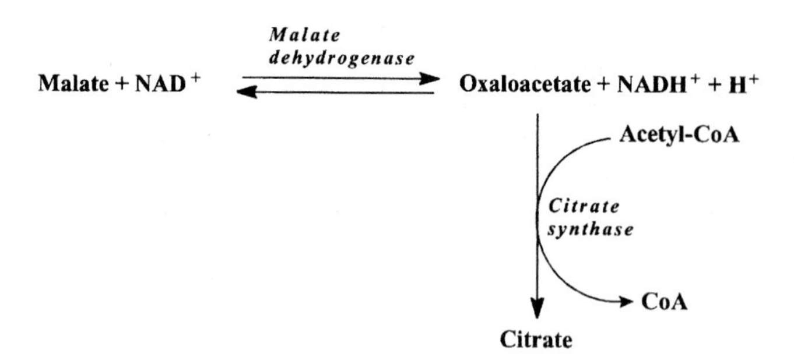

Figure 14. The forward-coupled assay for citrate synthase.

the enzyme whose activity is being determined. An example of this is the assay of citrate synthase (EC 4.1.3.7) using malate dehydrogenase (EC 1.1.1.37) as the coupling enzyme (107). This is illustrated in *Figure 14*. If the reaction catalysed by the 'forward-coupling' enzyme, malate dehydrogenase, is at equilibrium, removal of oxaloacetate by citrate synthase should result in an increase in NADH concentration as the equilibrium is re-established and this may be used to follow the reaction.

The behaviour of such systems can be complex and the appearance of NADH may not be simply related to rate of the reaction by the enzyme under study (108). This can be illustrated by considering the simple case where the forward coupling enzyme has only a single substrate (A) and product (B) and the latter serves as a substrate for the enzyme under study. This can be represented by the simple scheme:

$$A \overset{K}{\rightleftharpoons} B \longrightarrow C \tag{23}$$

where the disappearance of A is used as a measure of the activity of the enzyme that converts B to C. In this system the formation of C will result in a decrease in the concentration of B which will, in turn, be compensated by a decrease in the concentration of A as the equilibrium of the first reaction is maintained. The concentration of C formed will thus be given by the sum of the decreased in the concentrations of A and B.

$$-(\Delta[A] + \Delta[B]) = \Delta[C] \tag{24}$$

Since the equilibrium of the forward coupling reaction will be given initially by $K = [B]/[A]$, after the formation of some of the product C this will become:

$$K = \frac{([B] - \Delta[B])}{([A] - \Delta[A])} \tag{25}$$

Thus the relationship between the formation of C and the decrease in the concentration of A will depend on the value of the equilibrium constant, K. If $K = 1$, $\Delta[A] = \Delta[B]$ and therefore $\Delta[A] = \Delta[C]/2$. If the equilibrium position is far in the direction of B the decrease in the concentration of A will only be a small fraction of the amount of C formed. For example if $K = 100$, $\Delta[B]/\Delta[A]$ will be equal to this value and the change in the concentration of A will only be 1% of the increase in [C]. Only when the equilibrium position is far over in the direction of A will the decrease in its concentration approach the increase in the concentration of C. At an equilibrium constant of 0.01, for example, the relationship will give $\Delta[C] = 99\%\ \Delta[A]$.

This analysis may be extended to enzymes involving the interaction between two substrates (108), such as the use of malate dehydrogenase as the forward-coupling enzyme for the assay of citrate synthase (*Figure 14*). In this case the relationship between NADH formation and citrate production will be more complex, since the equilibrium concentration of oxaloacetate initially present will depend on the concentrations of the other reactants. The full equations for this system show that, at any fixed pH value, the formation of NADH will approach that of citrate only when the initial concentration of the former compound is sufficiently high for the initial malate dehydrogenase equilibrium to be far over towards malate (108).

Assay systems of this type can be useful if the substrate for the enzyme under study is unstable but can be formed in the assay mixture by the action of another enzyme. Such a system is shown in *Figure 15* where N^t-methylimidazole acetaldehyde is generated as a substrate for aldehyde dehydrogenase by the action of amine oxidase (EC 1.4.3.6) on N^t-methylhistamine (109). If the formation of NADH is monitored under conditions where there is an appreciable lag-phase in the progress curve (see *Figure 12*), it is possible to determine the K_m and maximum velocity of aldehyde dehydrogenase for this aldehyde by analysis of the approach to steady-state conditions [109]. Under appropriate conditions, the analysis of the progress curves for coupled assay of this type may allow the kinetic parameters of both enzymes involved to be determined [110]. It is important to remember that, as with other procedures for the analysis of reaction progress curves, the precautions and limitations discussed in Section 2.1 must be taken into account if valid results are to be obtained.

6.1.3 Automated assay procedures

There are a number of procedures that automate the assay of enzymes in order to allow large numbers of samples to be assayed rapidly and efficiently. Many of these involve the determination of the product formation after a fixed time from the start of the reaction. It will therefore be necessary to ensure that the values obtained represent a true reflection of the initial rate of the reaction. The use of flow-systems involving multi-channel pumps to mix reactants and determine the product formation after a fixed time have been discussed in detail [111] and the basic principles have not changed significantly since then. Immobilized enzymes may also be used to determine product formation by automated procedures (e.g. 112). Flow systems may also be used where the detection system has a slow response. A classical example of this is the work of Roughton and Rossi-Bernardi [113], who were able to make measurements of the extent of a reaction at times considerably shorter than the response-time of a CO_2 electrode by using a flow system in which the 'age' of the reaction mixture was constant as it flowed past the detector.

Many modern spectrophotometers can be equipped with multiple cuvette holders which allow several reactions to be followed by determining the absorbance in each sequentially

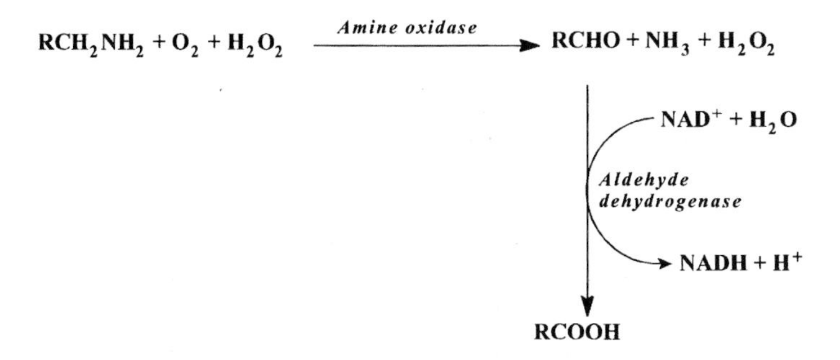

Figure 15 The use of an amine oxidase to generate substrate for the reaction catalysed by aldehyde dehydrogenase.

and repetitively. Typically, four or six samples, with appropriate blanks, may be studied in this way. Micro-titre plate readers with appropriate wavelength selection and temperature control offer the possibility of assaying many more samples and, by taking rapid multiple readings of the absorbance in each sample well, reaction progress curves may be determined for them all. The centrifugal fast analyser offers an alternative method for determining reaction time-courses and has been applied to the assay of several enzymes (e.g. 114).

A rather different application of automated assay procedures is to use a pumping system to generate a linearly increasing gradient of substrate concentration which is then mixed with enzyme and the extent of reaction is measured after a fixed time [115, 116]. Such a procedure allows the curve for the dependence of velocity upon substrate concentration to be generated in a single experiment. Since this procedure determines the extent of reaction after a fixed time it is, of course, necessary to ensure that true initial rate conditions apply at all substrate concentrations. However, it is relatively easy to check this by stopping the flow of enzyme and substrates and observing the time-course of the reaction as the mixture 'ages' in the detector (115).

6.2 Choice of assay method

There is often a variety of different assay methods available for an enzyme. Provided that adequate controls are carried out to ensure that they do in fact determine the initial velocity of the reaction and the measurements are free from the potential artefacts discussed earlier, the choice may simply depend on convenience and the availability of the appropriate materials and apparatus. However, there are some general considerations that may be helpful in choosing between alternative methods.

In the assay of enzymes with high K_m values towards their substrates, it may be necessary to work at high substrate concentrations to ensure a sufficient period of linearity of the reaction progress curve. In such cases it may be more accurate to follow the reaction by determining the extent of product formation rather than the disappearance of substrate, since the latter would involve measurement of decreases from very high initial values.

Many assay types impose some limitations on the assay conditions. This may be due to substrate instability under some conditions, such as the breakdown of NAD(P)H at acid pH values, which makes it difficult to estimate the rates of the enzyme-catalysed reaction accurately, or the requirement for alkaline conditions to observe the liberation of *p*-nitrophenol spectrophotometrically. In coupled assays one of the products is continuously removed and thus such methods will not be suitable for studies on the inhibition of the enzyme activity by that product. Other assays that involve determination of the formation of a specific product may become less practicable if large amounts of that product are added for inhibition studies. In coupled assays it must always be remembered that any change in the assay conditions may affect the behaviour of the coupling enzyme(s) as well as those of the enzyme under study.

The above considerations mean that it is often necessary to use more than one type of assay in a complete study of the behaviour of an enzyme. During the purification of an enzyme it is convenient to have a procedure that can rapidly give an estimate of the activity. This may make direct assay procedures more useful than discontinuous ones, which involve time-consuming separation or detection procedures. However, the possibility that contaminating activities in impure tissue preparations may prevent the accurate application of some procedures (Sections 2.5.1 *vi* and 2.5.3) can also affect the choice under these conditions.

Many discontinuous assay procedures appear to be particularly attractive because, once a reliable method has been obtained, it is often possible to perform a large number of incubations at the same time. Such procedures are often used in micro-titre plate readers as the basis of high-throughput, rapid-screening assays (see Chapter 12). However, it is essential to remember the necessity of ensuring that the procedure is giving a true measure of the initial rate of the reaction, by determining the time-course of the reaction by stopping it at different times, either after removal of samples from a single assay mixture or by treating individual incubation samples, for determination of the extent of the reaction. It will be necessary to repeat such experiments each time the assay conditions are changed. *Figure 16* shows the time-courses of an assay that slows down because of substrate depletion, determined in the presence of different concentrations of either a competitive or an uncompetitive inhibitor. Clearly a short reaction time is necessary to get a true indication of the potency of the inhibitor; after a extended assay time of 60 min the inhibitor at twice its K_I concentration has little effect on the extent of the reaction in either case.

6.3 The effects of pH

Many studies of the behaviour of enzymes involve investigations of the effects of variations of pH on their activities. Correctly designed and analysed studies of this type can yield a great deal of valuable information about the mechanisms involved in the catalytic process. Space does not permit a detailed consideration of the effects of pH on enzyme activity and the reader is referred to more complete treatments (117–119). Other studies are simply aimed at determining the optimum pH of the reaction. It is important to show whether the effects of

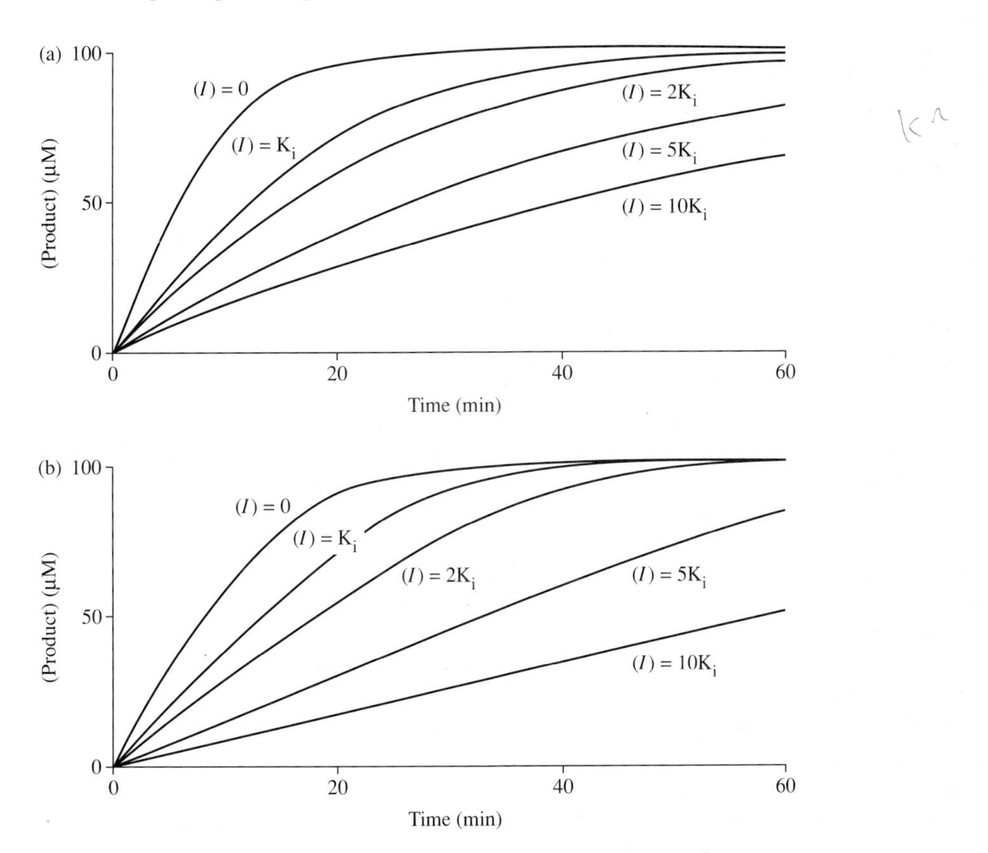

Figure 16 Time-courses for an enzyme-catalysed reaction in the presence of different amounts of a competitive inhibitor (a) and an uncompetitive inhibitor (b). The initial substrate concentration was 100 μM in each case, and the V_{max} of the enzyme in the assay was 10 μmol min^{-1}. The K_m and K_i values were both 50 μM.

pH are reversible or whether they result from irreversible changes in the activity of the enzyme. This can usually be achieved by adjusting the pH to different values and incubating under the assay conditions in the absence of one of the substrates for the period of an assay before readjusting to a value where the enzyme is known to be stable and determining the activity. Many enzymes are unstable at the more extreme pH values and some may precipitate. Clearly any data obtained in regions where the enzyme is rapidly losing activity may be difficult to interpret. A more detailed consideration of buffers is given in Chapter 9.

6.4 Practical considerations

It is not possible to cover all the factors that may affect the performance and validity of enzyme assays and this section will be restricted to a general consideration of some of the more important aspects.

6.4.1 Enzyme nature and purity

The state of purity of an enzyme preparation affects the ease with which it may be studied. Some optical assays may be difficult to perform accurately with impure preparations which contain large amounts of absorbing, and possibly particulate, material. Furthermore, the presence of contaminating activities (see Sections 2.5.1 *vi* and 2.5.3) or substrates may make it difficult to perform certain assays, and in studies on the inhibition by products it is also necessary to ensure the absence of contaminating enzymes that might react with the added product. Such problems often make it necessary to purify an enzyme, at least partly, in order to obtain reliable data on its activity.

Membrane-bound enzymes pose a particular problem since removal of the enzyme from its membrane environment can lead to changes in properties ranging from complete loss of activity (120) to changes in the kinetic mechanism obeyed (121). Furthermore, the surface charge on the membranes can have profound effects on the accessibility of charged substrates to the active sites of enzymes associated with them (122).

If a published assay procedure cannot be made to work, it is possible that the enzyme under study has not been identified correctly. There is considerable scope for confusion since there are many cases where more than one enzyme will catalyse a similar reaction. For example, the amine oxidases EC 1.4.3.4 and 1.4.3.6 catalyse similar reactions but have very different substrate specificities, and the two carbamoyl phosphate synthases (EC 6.3.5.5 and 6.4.3.16) differ in the source of ammonia that they can use. There are several examples in the literature where such pairs of enzymes have been confused. The use of the EC nomenclature (123) can help to avoid such confusion, but it should also be remembered that there may be considerable species differences and differences in the specificities of isoenzymes. Thus, for example, the standard assay of alcohol dehydrogenase with ethanol as substrate may not detect some of its isoenzymes, since they have low activities towards that alcohol (124).

6.4.2 Enzyme stability

In any series of studies with an enzyme preparation it is essential to assay its activity at regular intervals under standard conditions to check that it remains constant. If an enzyme is unstable and steadily loses activity with time it may be possible to correct all the values obtained to those that existed at the start of the studies if a series of standard assays have been carried out at different times to allow a calibration curve of activity against time be constructed. It is a much better procedure, however, to try to find conditions under which the stock enzyme solutions may be stored without appreciable loss of activity over the time involved (see 125 for a discussion of enzyme stability). Blank rates (Section 2.5) may develop during enzyme storage and it is important to check for their appearance.

6.4.3 The substrates

It is essential that pure substrates are used. The presence of contaminants will lead to incorrect estimates of the concentrations of solutions prepared by weight and it will also lead to errors in the determination of kinetic constants if any of the impurities is inhibitory. If a substrate is contaminated by a competitive inhibitor, the inhibition equation can be written as:

$$v = \frac{V_{max}}{1 + \frac{K_m (1 + [I]/K_i)}{[S]}} \tag{26}$$

where K_i is the inhibitor constant (6). If the inhibitor is present in the substrate solution its concentration will be some fixed proportion (x) of that of the substrate. Thus we can write

$$[I] = x[S]$$

Substituting $x[S]$ for [I] in Equation (26) gives:

$$v = \frac{V_{max}}{1 + \frac{K_m(1 + x[S]/K_i}{[S]}} = \frac{V_{max}}{1 + \frac{K_m}{[S]} + \frac{xK_m}{K_i}} \tag{27}$$

This will result in a proportional decrease in both K_m and V_{max}, as shown in *Figure 17* (126,127). Several stereospecific enzymes are competitively inhibited by other stereoisomers of the substrate and thus the use of racaemic mixtures, rather than the correct enantiomer, may yield incorrect estimates of both V_{max} and K_m values. If an *R*, *S*-substrate is used with an enzyme that is only active towards one of the enantiomers, it cannot be assumed that the effective substrate concentration can be taken as one-half of that prepared by weight. This analysis has been extended to cases where the inhibitory contaminant is an uncompetitive

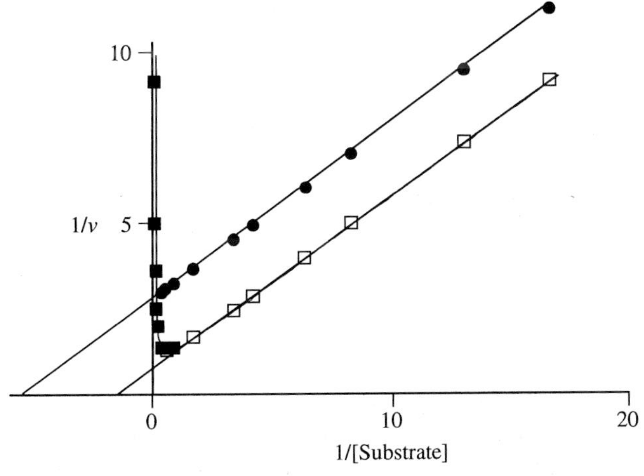

Figure 17 The possible effects of inhibitory contaminants in the substrate solution. The straight line (□) shows the behaviour in the absence of the contaminant, line (●) would result if the contaminant were a competitive inhibitor. If the contaminant were an uncompetitive inhibitor the behaviour would be the same as in the absence of inhibitor at low substrate concentrations but would deviate as indicated by (■) at higher concentrations. The behaviour of a mixed or non-competitive inhibitor would have the features of both these types of contaminating inhibitor curves.

or a noncompetitive (mixed) inhibitor of the enzyme (128). In the uncompetitive case the relationship becomes:

$$v = \frac{V_{max}}{1 + \frac{K_m}{[S]} + \frac{x[S]}{K_i}} \tag{28}$$

This equation predicts that the K_m and V_{max} values will not be affected by the contaminant, provided that the true substrate concentration is known, but that there will be apparent inhibition at high substrate concentrations, as shown in *Figure 17*. Since noncompetitive and mixed inhibition can be regarded as a combination of competitive and uncompetitive effects, the contaminant would result in a decrease in both K_m and V_{max} values as well as the appearance of inhibition at high substrate concentrations (see Equations 27 and 28 and *Figure 17*).

The purity of substrate solutions should be checked chromatographically or enzymically (2, 90). If the latter procedure is used a discrepancy between the enzymically-determined substrate concentration and that calculated on a weight basis may indicate whether purification is necessary. The stability of the substrate solutions is also important. Many biochemicals are relatively unstable and their breakdown would lead to changes in the true substrate concentrations. Thus it is important to use freshly prepared substrate solutions and to assay their concentrations at intervals to check for any changes. Since the breakdown of substrates in solution may give rise to the formation of inhibitors, it may not be satisfactory simply to correct for changes in substrate concentration with time by using data on the rate of its decomposition. The storage conditions can affect the stabilities of substrate solutions and data on this, and the most satisfactory storage conditions for a number of commonly used substrates have been listed (2, 90) and are also available in the literature from a number of suppliers of biochemicals.

Although it may be relatively simple to purify substrates that are available in large amounts, the small quantities of radioactively labelled substrate used in many radiochemical assays, together with their high cost, often leads to the possibility that they may contain impurities that are being ignored. One way of testing for the presence of kinetically significant contaminants in radiochemicals is to compare the kinetic behaviour obtained from assays where the radioactive substrate is varied as a fixed proportion of the unlabeled substrate (constant specific radioactivity) with that observed when the amount of radioactivity is kept constant whilst the total substrate concentration is varied (constant radioactivity) (129).

6.4.4 Solvents and buffers

It is of course essential to ensure that all solvents used in enzyme assays are free from contaminating material that could affect the activity of the enzyme. Heavy metals are inhibitors of many enzymes and great care is necessary to exclude them. For example, the persistent claim that EDTA was a specific activator of fructose-1,6-bisphosphatase was eventually shown to be merely an effect of this compound chelating heavy metal contaminants (130). Some substrates are not readily soluble in water and are added to the enzyme assay in solution in an organic solvent. It is, of course, necessary in such cases to carry out adequate control experiments to check that the solvent itself has no effect on the activity of the enzyme under study or on the assay method being used. There is also a possibility that changes in solution dielectric may affect enzyme activity. In any enzyme assay it is important to check that the addition of the other components does not affect the pH of the assay. It is often convenient to adjust the pH of enzyme and substrate solutions to the same value as that of the assay buffer, but this may not always be possible if any of these are not stable for long periods at that value. As discussed in Chapter 9, the choice of buffer solution may be of great importance.

6.4.5 Assay mixtures

For assay mixtures that contain a number of different components it may be possible to make up a bulk mixture containing all of them except the enzyme to be assayed. The use of such a 'cocktail' can be particularly useful if a number of assays are to be performed under identical conditions, such as during enzyme purification. If such a cocktail gives rise to a blank rate, it will of course be necessary to find out the component(s) responsible by using incomplete mixtures. More seriously, the use of such cocktails does not allow one to check for blank rates involving the enzyme in an incomplete reaction mixture (Section 2.5). Thus it is necessary to carry out adequate controls to test for this before relying on such an assay cocktail.

Although the use of assay cocktails can save time, it must be remembered that the different components may differ in their stabilities. For example, one of the substrates of an enzyme involved in a coupled assay may decay more rapidly than the other components, thus limiting the useful lifetime of the entire mixture. The decay of one of the components of an assay cocktail may lead to the formation of inhibitory products so it may not be possible to 'revitalize' the mixture by adding more of the component that has degraded. Thus, if a cocktail is to be used, it is preferable to premix only those components that are stable under the storage conditions, in order to avoid the expense of having to discard it before it has all been used. In the case of coupled assays, if no activity of the enzyme under study can be detected it will be possible to check that the coupling system is working adequately by seeing whether there is a response when the product that is being determined is added to the system.

6.4.6 Mixing

If there are no significant blank rates in any of the incomplete reaction mixtures, the choice of whether to start the assay by the addition of enzyme or one of the substrates may not be important. If the enzyme is unstable in the assay mixture it would be preferable to start the reaction by adding it last. In other cases, however, it might be preferable to preincubate the enzyme in the incomplete assay mixture and start by the addition of one of the substrates. This, for example, might be the case if there were hysteretic effects resulting from the dilution of the enzyme (Section 2.4.9). It is often necessary to store the enzyme preparation and substrates in ice to ensure their stability. Thus, it is important to keep the final volume of material used to start the reaction as small as possible to avoid too great a fall in the temperature of the assay mixture when it is added (Section 2.4.1). If the effects of varying the concentrations of enzyme are to be studied it may be possible to keep the volumes of these added so low in relation to that of the total assay mixture that the differences between the volumes added are negligible. However, if this is not possible, it will be necessary to adjust the volume of the assay mixture to a constant final value in each case, using the same buffer as that in which the varied enzyme or substrate is contained.

It is essential to ensure that the components are properly mixed in the assay mixture. However, some enzymes are denatured by too vigorous shaking. With an assay vessel that is not mechanically stirred or shaken it is usually possible to ensure adequate mixing by covering the top with parafilm and inverting the tube or cuvette a few times. This can be achieved relatively quickly but more rapid, and just as effective, mixing can be achieved by using a stirrer made from a small piece of plastic that has been bent over and flattened at one end. A few vertical strokes of the stirrer should be adequate and in spectrophotometric or fluorimetric assays it is often possible to do this without removing the cuvette from the apparatus. If the volume of the material used to start the reaction is relatively small it may be possible to place it on the flattened end of the stirrer so that addition and stirring can be performed in a single operation. It is a simple matter to make such stirrers in the laboratory although they are also available commercially. If it is necessary to start a reaction by adding a component to a blank and reaction cuvette simultaneously, it is possible to mix them both at once using a stirrer made from a U-shaped plastic rod flattened at both ends. Particular care in mixing is necessary if the enzyme is in a dense medium, such as glycerol or sucrose solution, which are sometimes used to enhance stability (125).

6.5 Conclusions

As stated earlier, it is not possible to recommend the ideal type of assay method to use since this will depend on the nature of the enzyme being studied, its degree of purity and the purpose of the assays. Indeed it may prove necessary to use more than one assay method in a complete study of an enzyme, if for example product inhibition is to be studied and a coupled assay involving that product has been routinely used, or if the assay of choice cannot be used during the early stages of the purification.

It has not been possible to cover all aspects of enzyme assay and kinetics in the space available for this review, but references given in the appropriate sections of the text should fill any gaps. The subject of enzyme inhibition has been treated in detail in several works (4–6, 48) and an account of some of the pitfalls and problems in such studies has been presented (131). This account has concentrated on the difficulties that can be encountered with enzyme assays because the results from invalid determinations are of no use to anyone. By taking care to validate the assay procedures used, under all conditions of the studies, it is possible to ensure that meaningful results are obtained.

References

1. Colowick, S. P., Kaplan, N.O. , *et al.* (1955ff). *Methods in enzymology.* Vol. 1 and continuing. Academic Press Inc., New York.
2. Bergmeyer, H-U (1983ff). *Methods of enzymatic analysis*, Vols 1–10 (3rd edn). Verlag Chemie, Weinheim.
3. Barman T. E. (1969 & 1974). *Enzyme handbook.* Springer-Verlag, Heidelberg.
4. Segel, I. H. (1975). *Enzyme kinetics.* Wiley-Interscience Inc., New York.
5. Tipton, K. F. (1996). In *Enzymology LabFax.* (ed. P. C. Engel), p. 115. Bios Scientific Publishers, Oxford & Academic Press, San Diego.
6. Dixon, M. and Webb, E. C. (1979). *Enzymes* (3rd edn). Longman Ltd, London.
7. Cleland, W. W. (1970). In *The enzymes* (ed. P. D. Boyer), p. 1. Academic Press Inc, New York.
8. Seiler, N., Jung, M. J., and Koch-Weser, J. (1978). *Enzyme activated irreversible inhibitors.* Elsevier-North Holland B. V., Amsterdam.
9. Palfreyman, M. G. (1987). In *Essays in biochemistry* (ed. R. D. Marshall and K. F. Tipton), Vol. 23, p. 28. Academic Press Ltd, London.
10. Tipton, K. F. (1989). In *Design of enzyme inhibitors as drugs* (ed. M. Sandler and H. J. Smith), p. 70. Oxford University Press, London.
11. Silverman, R. B. and Hoffman, S. J. (1984). *Med. Res. Rev.*, **44**, 415.
12. Tipton, K. F. (1980). In *Enzyme inhibitors as drugs* (ed. M. Sandler), p. 1. Macmillan Ltd, London.
13. Waley, S. G. (1985). *Biochem. J.*, **277**, 843.
14. Tatsunami, S., Yago, M., and Hosoe, M. (1981). *Biochim. Biophys. Acta*, **662**, 226.
15. Tipton, K. F., Fowler, C. J., McCrodden, J. M., and Strolin Benedetti, M. (1983). *Biochem. J.*, **209**, 235.
16. Tipton, K. F., McCrodden, J. M., and Youdim, M. B. H. (1986). *Biochem. J.*, **240**, 379.
17. Kinemuchi, H., Arai, Y., Oreland, L., Tipton, K. F., and Fowler, C. J. (1982). *Biochem. Pharmacol.*, **31**, 959.

18. Donovan, J. W. (1973). In *Methods in enzymology* (ed. C. W. Hirs and S. N. Timasheff), p. 497. Academic Press Inc, New York.
19. Undenfriend, S. (1962). *Fluorescence assay in biology and medicine.* Academic Press Inc., New York.
20. Wharton, C. W. (1983). *Biochem. Soc. Trans.,* **11**, 817.
21. Waley, S. G. (1981). *Biochem. J.*, **193**, 2009.
22. Nicholls, R. G., Jerfy, M., and Roy, A. B. (1974). *Anal. Biochem.*, **61**, 93.
23. Orsi, B. A. and Tipton, K. F. (1979). In *Methods in enzymology* (ed. D. L. Purich), p. 159. Academic Press Inc., New York.
24. Wharton, C. W. and Szawelski, R. J. (1982). *Biochem. Soc. Trans.*, **10**, 233.
25. Selwyn, M. J. (1965). *Biochim. Biophys Acta*, **105**, 193.
26. Tipton, K. F. and Fowler, C. J. (1984). In *Monoamine oxidase and disease* (ed. K. F. Tipton, P. Distert, and M. Strolin Benedetti), p. 27. Academic Press Ltd., London.
27. Cha, S. (1976). *Biochem. Pharmacol.*. **25**, 1561 and 2695.
28. John, R. A. (1985). In *Techniques in the life sciences* (ed. K. F. Tipton). Vol. BI/II, Supplement BS 118, p. 1. Elsevier Ltd, Ireland.
29. Hiromi, K. (1980). In *Methods of biochemical analysis* (ed. D. Glick), Vol. 26, p. 137. Wiley, New York.
30. Hartley, B. S. and Kilby, B. A. (1954). *Biochem. J.*, **56**, 288.
31. O'Fagain, C., Bond, U., Orsi, B. A., and Mantle, T. J. (1982). *Biochem. J.*, **201**, 345.
32. Söling, H. J., Bernhard, G., Kuhn, A., and Luck, H. J. (1977). *Arch. Biochem. Biophys.*, **182**, 563.
33. Cronin C. N. and Tipton, K. F. (1985). *Biochem. J.*, **227**, 113.
34. Rauschel, F. M. and Cleland, W. W. (1977). *Biochemistry*, **16**, 2169.
35. Frieden, C. (1970). *J. Biol. Chem.*, **245**, 5788.
36. Kurganov, B. I., Dorozhko, A. I., Kagan, Z. S., and Yakovlev, V. A. (1976). *J. Theor. Biol.*, **60**, 247, 271 ,287 and **61**, 531.
37. Neet, K. E. and Ainslie, G. R. (1980). In *Methods in enzymology* (ed. D. L. Purich), Vol. 64, p. 192. Academic Press Inc., New York.
38. Frieden, C. (1979). *Ann. Rev. Biochem.*, **48**, 471.
39. Williams, D. C. and Jones, J. G. (1976). *Biochem. J.*, **155**, 661.
40. Nimmo, G. A. and Tipton, K. F. (1981). *Biochem. Pharmacol.*, **30**, 1635.
41. Allanson, S. and Dickinson, F. M. (1984). *Biochem. J.*, **223**, 163.
42. Dickinson, F. M. and Allanson, S. (1985). In *Enzymology of carbonyl metabolism* (ed. T. G. Flynn and H. Weiner), Vol. 2, p. 71. A. R. Liss Inc., New York.
43. Pietruszko, R., Ferenca-Biro, K., and McKerrell, A. D. (1985). In: *Enzymology of carbonyl metabolism* (ed. T. G. Flynn and H. Weiner, H), Vol. 2, p. 29. A. R. Liss Inc., New York.
44. Bates, D. J. and Frieden, C. (1973). *J. Biol. Chem.*, **248**, 7878 and 7885.
45. Duncan, R. J.S. and Tipton, K. F. (1971). *Eur. J. Biochem.*, **22**, 257.
46. Estabrook-Smith, S. B., Hudson. P. J., Gross, N. H., Keech, D. B., and Wallace, J. C. (1976). *Arch. Biochem. Biophys.*, **171**, 709.
47. Hulme, E. C. and Tipton, K. F. (1971). *FEBS Lett.*, **12**, 197.
48. Cornish-Bowden A. (1995). *Fundamentals of enzyme kinetics.* Portland Press, London.
49. Wong, J. T-F. (1975). *Kinetics and enzyme mechanism.* Academic Press Inc., New York.
50. Fromm, H. J. (1975). *Initial rate enzyme kinetics* Springer-Verlag, Heidelberg.
51. Roberts, D. V. (1977). *Enzyme kinetics.* Cambridge University Press, Cambridge.
52. Dalziel, K. (1957). *Acta Chem. Scand.*, **11**, 1706.
53. Storer, A. C. and Cornish-Bowden, A. (1976). *Biochem. J.*, **159**, 1.
54. Monod, J., Wyan, J. and Changeux, J. P. (1966). *J. Mol. Biol.*, **12**, 88.
55. Whitehead, E. P. (1970). *Prog. Biophys. Mol. Biol.*, **21**, 323.
56. Tipton, K. F. (1979). In *Companion to biochemistry* (ed. A. T. Bull, J. R. Lagnado, J. O. Thomas and K. F. Tipton), p. 327. Longman Ltd., London.
57. Hill, A. V. (1910). *J. Physiol.* (London), **40**, 4.
58. Hearon, J. Z., Bernhard, S. A., Friess, S. L., Botts, D. J., and Morales, M. F. (1959). In *The enzymes*, 2nd edn (ed. P. D. Boyer, H. A. Lardy, and K. Myrback), p. 49. Academic Press Inc., New York.
59. Ferdinand, W. (1966). *Biochem. J.*, **98**, 273.
60. Sweeny, J. R. and Fisher, J. R. (1968). *Biochemistry,* **7**, 561.

61. Whitehead, E. P. (1976). *Biochem. J.*, **159**, 449.
62. Bardsley, W. G. and Childs, R. E. (1975). *Biochem. J.*, **149**, 313.
63. Wong, J. T-F., Gurr, P. A., Bronskill, P. M., and Hanes C. S. (1972). In *Analysis and simulation of biochemical systems* (ed. H. C. Hemker and B. Hess), p. 327. North Nolland B. V., Amsterdam,
64. Williams, D. C., Morgan, G. S., McDonald, E., and Battersby, A. R. (1981). *Biochem. J.*, **193**, 301.
65. Elliott, K. R. F. and Tipton, K. F. (1974). *Biochem. J.*, **141**, 807.
66. Rabin, B. R. (1967). *Biochem. J.*, **102**, 22C.
67. Frieden C. (1967). *J. Biol. Chem.*, **242**, 4045.
68. Ainslie G. R., Shill, J. P., and Neet, K. E. (1972). *J. Biol. Chem.*, **247**, 7088.
69. Ricard, J., Meunier, J. C., and Buc, J. (1974). *Eur. J. Biochem.*, **49**, 195.
70. Harding, W. M. (1969). *Arch. Biochem. Biophys.*, **129**, 57.
71. Blair, J. McD. (1969). *FEBS Lett.*, **2**, 245.
72. Purich, D. L. and Fromm, H. J. (1972). *Biochem. J.*, **130**, 63.
73. Answorth, S. (1977). *Steady-state enzyme kinetics*, p. 62. MacMillan, London,
74. O'Sullivan, W. J. and Smithers, G. W. (1979). In *Methods in enzymology* (ed. D. L. Purich), Vol. 63, p. 294. Academic Press Inc., New York.
75. Good, N. E. and Izawa. S. (1972). In *Methods in enzymology* (ed. A. San Pietro), Vol. 24, p. 53. Academic Press, New York.
76. Tipton, K. F. (1985). In *Techniques in the life sciences* (ed. K. F. Tipton), Vol B1/II Supplement BS 113, p. 1. Elsevier Ltd, Ireland.
77. Morrison, J. F. (1979). In *Methods in enzymology* (ed. D. I. Purich), Vol. 63, p. 257. Academic Press Inc., New York.
78. Cronin, C. N. and Tipton, K. F. (1987). *Biochem. J.*, **247,** 41.
79. Uotila, L. and Mannervik, B. (1979). *Biochem. J.*, **177**, 869.
80. Dixon, H. B. F. and Tipton, K. F. (1973). *Biochem. J.*, **133**, 837.
81. Burns, D. J. W. and Tucker, S. A. (1977). *Eur. J. Biochem.*, **81**, 45.
82. Spears, G., Sneyd, J. T., and Loten, E. G. (1971). *Biochem. J.*, **125**, 1149.
83. Denizeau, F., Wyse, J., and Sourkes, T. L. (1976). *J. Theor. Biol.*, **63**, 99.
84. Henehan, G. T.M. and Tipton, K. F. (1991). *Biochem. Pharmacol.*, **42**, 979
85. Williams, J. W. and Morrison, J. F. (1979). In *Methods in enzymology* (ed. D. L. Purich), Vol. 63, p. 437. Academic Press Inc., New York.
86. Henderson, P. J. F. (1973). *Biochem. J.*, **135**, 101.
87. Teipel, J. and Koshland, D. E. (1989). *Biochemistry*, **8**, 4656.
88. Dalziel, K. and Engel, P. C. (1968). *FEBS Lett.*, **1**, 349.
89. Cornish-Bowden, A. (1988). *Biochem. J.*, **250**, 309.
90. Lowry, O. H. and Passonneau, J. V. (1972). *A flexible system of enzymatic analysis.* Academic Press Inc., New York.
91. Jackobson, C. F., Leonis, J., Linderstrom-Lang, K., and Ottesen, M. (1957). *Meth. Biochem. Anal.*, **4**, 171.
92. Lessler, M. A. and Vrierley, G. P. (1969). *Meth. Biochem. Anal.*, **17**, 1.
93. Youdim, M. B. H. and Tenne, M. (1987). In *Methods in enzymology* (ed. S. Kaufman), Vol. 142, p. 617. Academic Press Inc., New York.
94. Ulitzer, S. and Hastings. J. W. (1978). In *Methods in enzymology* (ed. M. A. De Luca), Vol. 57, p. 189. Academic Press Inc., New York.
95. Hastings, J. W. (1978). In *Methods in enzymology* (ed. M. A. De Luca), Vol. 57, p. 125. Academic Press Inc., New York.
96. Mulkerrin, M. G. and Wampler, J. E. (1978). In *Methods in enzymology* (ed. M. A. De Luca), Vol. 57, p. 375. Academic Press Inc., New York.
97. Campbell, A. K. (1989). In *Essays in biochemistry* (ed. R. D. Marshall, and K. F. Tipton), p. 41. Academic Press Ltd, London.
98. Lebel, D., Poirier, G. G., and Beaudoin, A.R (1978). *Anal. Biochem.*, **85**, 86.
99. Kaplan, N. D. Colowick, S. P., and Barnes C. C. (1951). *J. Biol. Chem.*, **191**, 461.
100. Fritz, I. B., Schultz, S. K., and Srere, P. A. (1963). *J. Biol. Chem.*, **238**, 2509.
101. Ellman, G. L., Courtney, K. D., Andres, V., and Featherstone, R. M. (1961). *Biochem. Pharmacol.*, **7**, 88.

102. Rudolph, F. B., Baugher, B. W., and Beissner, R. S. (1979). In *Methods in enzymology* (ed. D. L. Purich), Vol. 63, p. 22. Academic Press Inc., New York.
103. Wulff, K. (1983). In *Methods of enzymatic analysis* (ed. H-U. Bergmeyer), Vol. 1, p. 340. Verlag Chemie, Weinheim.
104. De Luca, M. A. (ed.) (1978). *Methods in enzymology*, Vol. 57. Academic Press Inc., New York.
105. Raskin, N. H. and Sokoloff, L. (1970). *J. Neurochem.*, **17**, 1677.
106. Dunn, M. F. and Bernhard, S. A. (1971). *Biochemistry*, **10**, 4569.
107. Tubbs, P. K. and Garland, P. B. (1964). *Biochem. J.*, **93**, 550.
108. Pearson, D. J. (1965). *Biochem. J.*, **95**, 23C.
109. Gitomer, W. L. and Tipton, K. F. (1983). *Biochem. J.*, **211**, 277.
110. Duggleby, R. G. (1983). *Biochim .Biophys. Acta*,. **744**, 249.
111. Roodyn, D. B. (1970). *Automated enzyme assay*. Elsevier BV, Amsterdam.
112. Wienhausen, G. and De Luca, M. (1986). In *Methods in enzymology* (ed. M. A. De Luca and W. D. McElroy), Vol. 133, p. 198. Academic Press Inc., New York.
113. Roughton, F. J.W. and Rossi-Bernardi, L. (1964). *Proc. Roy. Soc. B,* **164**, 381.
114. Boghosian, R. A. and McGuinness, E. T. (1981). *Int. J. Biochem.*, **13,** 909.
115. Illingworth, J. A. and Tipton, K. F. (1969). *Biochem. J*,. **115,** 511.
116. Gurr, P. A., Wong, J. T-F., and Hanes, C. S. (1973). *Anal. Biochem.*, **51**, 584.
117. Tipton, K. F. and Dixon, H. B. F. (1979). In *Methods in enzymology* (ed. D. L.Purich), Vol. 63, p. 183. Academic Press Inc., New York.
118. Laidler, K. J. and Bunting, P. S. (1973). *The chemical kinetics of enzyme action*. Oxford University Press, Oxford.
119. Cleland, W. W. (1982). In *Methods in enzymology* (ed. D. L. Purich), Vol. 87, p. 390. Academic Press Inc., New York.
120. Bock, H-G. O. and Fleischer, S. (1974). In *Methods in enzymology* (ed. S. Fleischer and L. Packer), Vol. 32, p. 374. Academic Press Inc., New York.
121. Houslay, M. D. and Tipton, K. F. (1975). *Biochem. J.*, **145**, 311.
122. Wojtczak, L. and Nalecz, M. J. (1979). *Eur. J. Biochem.*, **94**, 99.
123. Boyce, S and Tipton, K. F. (2001) *Encyclopedia of life sciences*. Macmillan Reference Ltd, London.
124. Boleda, M. D., Pere, J., Moreno, A., and Pares, X. (1989). *Arch. Biochim. Biophys.*, **274**, 74.
125. Scopes, R. K. (1982). *Protein purification. Principles and practice*, p. 185. Springer-Verlag, New York.
126. Tubbs, P. K. (1962).*Biochem. J.*, **82**, 36.
127. Dalziel, K. (1962). *Biochem. J.*, **83**, 28P.
128. Houslay, M. D. and Tipton K. F. (1973). *Biochem. J.*, **135**, 735.
129. Tipton, K. F. (1977). *Biochem. Pharmacol.*, **26**, 1525.
130. Nimmo, H. G. and Tipton, K. F. (1982). In *Methods in enzymology* (ed. W. A. Wood), Vol. 90, p. 330. Academic Press Inc., New York.
131. Tipton, K. F. (1973). *Biochem. Pharmacol.*, **22**, 2933.

Chapter 2
Photometric assays

Robert A. John
School of Biomedical Sciences, University of Wales College of Cardiff,
PO 911, Cardiff CF1 3US, UK

1 Introduction

Photometric methods are the most frequently used of all kinds of enzyme assay. They are convenient and capable of rapidly providing accurate and reproducible results on large numbers of samples. Useful changes in the optical properties of the system under analysis frequently arise directly from the chemical transformations accompanying the enzyme-catalysed conversion of substrate to product. When the reaction catalysed by the enzyme under assay does not itself produce a useful change in optical properties, incorporation of appropriate additional reagents often allows the reaction to be monitored photometrically. Enzyme assays based on changes in the light *absorbed* by the solution as the reaction proceeds are more frequently used than other photometric methods. However, changes in *fluorescence* and changes in *turbidity* of the solution also provide the bases of useful optical methods for assaying enzymes.

2 Absorption

Visible light is electromagnetic radiation of wavelength between 400 and 750 nm. Compounds that absorb light in this wavelength range are coloured. Many colourless compounds also absorb light in the ultraviolet (UV) range, 200–400 nm. Although it is conventional to refer to visible and UV light by its wavelength, it is easier to understand the process of absorption in terms of frequency rather than the inversely related parameter, wavelength. (As with all propagating wave forms, wavelength, λ, and frequency, ν, are inversely related, $\lambda = c/\nu$. In this case c is the speed of light.)

Absorption of light occurs when electrons of irradiated absorbing molecules are promoted to a higher energy level, because the frequency of the electronic oscillation in the molecule coincides with the frequency of the irradiating light. The wavelength corresponding to this frequency (λ_{max}) is that at which the compound absorbs most light and is determined by the structure of the absorbing molecule. The amount of light absorbed depends upon the probability that the electronic transition occurs.

2.1 Terminology

Although, strictly speaking, the term *chromophore* describes a chemical group that brings colour by absorbing visible light, in this chapter the term will also be applied to groups that absorb in the UV part of the electromagnetic spectrum. Whereas *absorption* may be used to describe the process by which light is absorbed, the word *absorbance* refers to a quantity and has a precisely defined meaning. *Optical density* and *extinction* are synonymous with absorbance but these terms are no longer accepted by scientific journals.

2.2 Absorbance

Instruments designed to measure absorbance invariably make their measurements by determining the amount of light that is not absorbed, that is, the light which is transmitted by the solution. Naturally, the light transmitted by a solution of a chromophore decreases as the concentration of the chromophore increases. Furthermore, the proportionality between transmitted light and concentration of the chromophore is not linear but logarithmic. The quantity known as absorbance was deliberately defined so as to be directly proportional to the concentration of the chromophore but its value must always be derived from measurements of the transmitted light, no matter what the design of instrument. Thus any instrument that measures absorbance must, in some way, make a comparison of I_0, the light transmitted by a solution not containing the chromophore, with I, the light transmitted by a solution that does contain the chromophore. The fraction of incident light that is transmitted by the solution is I/I_0 and Equation 1 shows the relationship between absorbance (A) and this fraction.

$$A = -\log_{10}(I/I_0) \qquad (1)$$

Absorbance, being derived from a ratio, does not have units. However, the expressions 'absorbance units' or simply 'A' are often used.

2.2.1 Conversion of absorbance to concentration

In the absence of unusual complicating factors, such as association of the chromophore into polymers, and at the relatively low concentrations almost invariably used in biochemical experiments, absorbance is directly proportional to concentration. Absorbance is always directly proportional to path-length. These facts are combined in the Beer–Lambert relationship (Equation 2) relating absorbance (A) to concentration (c) and path-length (l).

$$A = \varepsilon c l \qquad (2)$$

Although absorbance is now the term recommended by scientific journals, the Greek letter ε (epsilon, for extinction) is still used as the proportionality constant relating absorbance to concentration. The expression extinction coefficient is still widely used in the scientific literature for absorbance coefficient and this is apparently acceptable.

The most widely used unit for absorbance coefficient in the biochemical literature is $1\ mol^{-1}\ cm^{-1}$. This is particularly convenient when, as is normal, solutions are contained in cuvettes of path-length 1 cm, because the concentration may be determined simply by dividing the measured absorbance by the relevant molar absorbance coefficient to give the concentration in $mol\ l^{-1}$ (Equation 3).

$$c = A/\varepsilon \qquad (3)$$

Another way of looking at absorbance coefficient expressed in these units is to consider it to be the absorbance of a 1 molar solution of the chromophore in a 1 cm path-length cuvette. The unit $cm^2\ mol^{-1}$ is used occasionally and, expressed in these units, the values are numerically 1000 times greater than those expressed in $1\ mol^{-1}\ cm^{-1}$, representing the absorbance that would be obtained for a $1000\ mol\ 1^{-1}$ solution of the chromophore.

2.2.2 Instrumentation

All instruments operate by the same basic principle, in that light from an appropriate source is passed through the solution containing the chromophore and thence to a photoelectric detector. However, because different chromophores absorb light in different parts of the spectrum, the instrument will incorporate some system for restricting the wavelength range of the light used in determining the absorbance. Until recently, restriction of the wavelength range was achieved before the light entered the cuvette and most available instruments still continue with this arrangement. A range of instruments has now become available in which the diffraction takes place after the light has passed through the cuvette. This has been made possible by the availability of the diode array and such instruments are known as diode-array spectrophotometers.

Figure 1 shows a schematic illustration of a conventional instrument designed to measure absorbance. This basic format is common to a very wide variety of instruments ranging from the simplest colorimeter to the highest precision spectrophotometer. However, the quality of the result obtained depends very much on the detailed configuration of the instrument used.

i. Light source

The main consideration here is the wavelength at which measurements are to be made. The most commonly used light sources are the glass-enveloped filament bulb, the quartz-enveloped tungsten halogen lamp, the deuterium lamp and, more recently, the xenon lamp. To cover the useful UV/visible range (200–700 nm) an instrument using both deuterium and quartz halogen lamps is usually necessary although diode array instruments rely on a single, low-pressure, deuterium lamp that emits sufficient light over the whole UV/visible range. A recently developed, rapid-scanning instrument uses a high-powered flashing xenon lamp to cover the whole wavelength range. With this instrument, photolytic damage is avoided by making the flashes very brief.

To make measurements in the visible and near UV, including the 340 nm required for NAD(P)/NAD(P)H-based measurements, an instrument using a quartz halogen lamp as sole light source is adequate. At 340 nm the energies of deuterium and quartz–halogen lamps are approximately equal and either may be used. The glass-enveloped filament lamp can only be used in the visible region and its value in enzyme assays is very limited.

ii. Wavelength selector

The system used to restrict the wavelength of light to that most appropriate to the relevant chromophore is important in determining the quality of the measurements made. The colorimeter simply uses a series of filters that have different absorbance characteristics. If, for example, the reading is to be made at 410 nm, the filter should be one that absorbs light at wavelengths higher and lower than 410 nm. However, given that the range of available filters

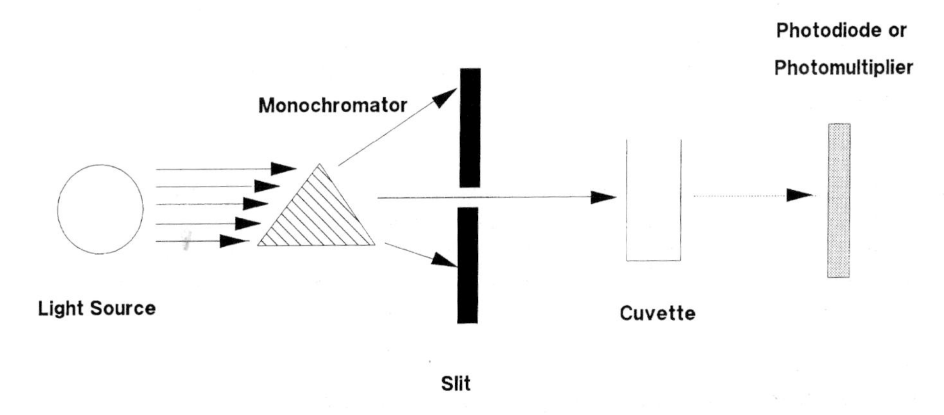

Figure 1 Essential features of an instrument designed to measure absorbance

is fairly small, the light incident on the cuvette will consist mainly of wavelengths well away from 410 nm, where the chromophore absorbs little or not at all. The absorbance readings obtained from such an instrument will therefore be less, sometimes very much less, than those predicted from the absorbance coefficient. The modern spectrophotometer uses a holographic diffraction grating to disperse the light from the source into a spectrum. Rotation of the diffraction grating varies the wavelength of emergent light passing through a slit, and the extent to which this approximates to monochromatic light depends on the width of the slit and is described in the specification of an instrument as the *bandwidth*. It is useful to be able to increase this slit width in situations where increased sensitivity of an assay is required, but on many instruments the bandwidth is predetermined. Photometric accuracy and precision can be increased by using a high bandwidth because the incident beam has more energy. However, this produces a beam that is less-nearly monochromatic, a fact that should be kept in mind when comparing instruments.

The diode array instrument uses a concave holographic grating that separates the light into its component wavelengths after it has passed through the sample. This diffracted light is reflected by the grating itself onto the fixed diode array so that each of the hundreds of diodes in the array receives light of different wavelength.

iii. Determination of absorption spectra

The most versatile spectrophotometers are capable of rapid determination of complete absorbance spectra. Despite the undoubted value of such instruments in the biochemical laboratory, the assay of enzyme activity *per se* does not require a wavelength-scanning instrument. Nevertheless, many problems experienced with photometric enzyme assays can be solved more rapidly if the absorbance spectrum of the solutions involved can be determined rapidly. Although this can only be achieved in an instrument that irradiates the sample with monochromatic light by the physical movement of the diffraction grating, modern instrumentation allows a full spectrum to be determined in less than 3 seconds. Thus, unless a very rapid reaction is being observed, the time difference between the two ends of the spectrum is not significant. With a diode array instrument, all the points on the spectrum are taken at the same time and spectra can be measured at 1 second intervals. However, it should be remembered that problems may arise because the sample is being irradiated with light at all wavelengths.

iv. Cuvette holder

For continuous assays it is essential that the temperature of the cuvette holder can be controlled. Electrical heating is superior to circulating water because of the danger of flooding the instrument with the latter. Peltier constant-temperature systems, which allow the temperature to be maintained below as well as above ambient, are available for some instruments and are undoubtedly useful in the investigation of enzyme kinetics. Many instruments have multiple cuvette holders so that several continuous assays can be conducted simultaneously by automatic mechanical movement of the carriage between measurements. This is a considerable advantage when multiple slow reactions have to be followed. Such additional compartments are also useful for holding cuvettes containing assay solution at the right temperature while a measurement on another cuvette is in progress. There is some occasional advantage in having a cuvette holder that can accept cells of path-length other than 1 cm.

2.3 Limitations and sources of error

It is well worth considering the factors that determine the quality of an absorbance measurement. The absorbance measurement reported by a spectrophotometer will differ from the true value due to three main causes.

2.3.1 Non-linearity arising from stray light

This problem arises from light that originates within the instrument but reaches the photodetector without passing through the cuvette. Although the Beer–Lambert law predicts linearity between concentration and absorbance, the relationship frequently appears not to be obeyed. This is almost always due to inadequacy of the instrument rather than to complex behaviour of the chromophore. The underlying reason for the non-linearity lies in the instrument's determination of values I_0 and I for the transmitted light. Despite all efforts, it is impossible to prevent some light from reaching the photodetector by routes other than that intended. This 'stray light' is a small but constant component of the light detected. Thus, whereas the true absorbance value is given by Equation 1, the instrument has to operate with an additional term I_s for stray light, which increases the apparent values of I and I_0 by the same amount (Equation 4).

$$A_{app} = -\log_{10}[(I + I_s) / (I_0 + I_s)] \quad (4)$$

It is easy to see that as I becomes smaller, I_s becomes more and more significant and the apparent value of the absorbance (A_{app}) falls below the true value. The result is an apparent deviation from the Beer–Lambert relationship of the type shown in *Figure 2*. It is due, of course, to a limitation of the instrument rather than a deviation from the Beer–Lambert relationship.

Stray light is normally given as part of the instrument's specification. It can be measured by first setting zero absorbance with nothing in the cuvette holder and then reading the absorbance reported by the instrument after putting a solid, opaque block, shaped like a cuvette, into the holder thus preventing any light being transmitted. The measurement is then based entirely on stray light. The percentage stray light is given by 100×10^{-A} where A is the reading reported by the instrument. However, many digital instruments will not report a reading under these circumstances because they are able to recognize that such high readings are meaningless.

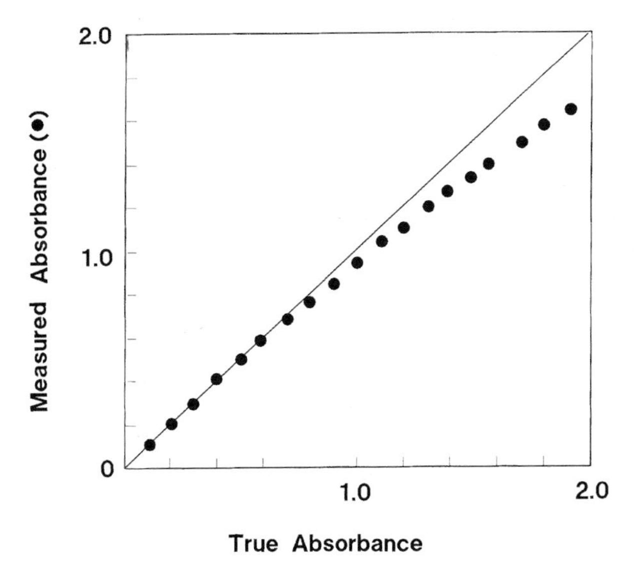

Figure 2 Effect of stray light on the relationship between true and measured absorbance. The experimental points show a deviation from linearity which is produced by approximately 1% stray light.

2.3.2 Instrumental noise

This is a random fluctuation in the output from the photodetector which originates within the instrument and does not arise from the solution.

2.3.3 Zero drift

This is a slow, steady change in apparent absorbance observed in single-beam instruments. It is caused by time-dependent changes in the photodetector and in the lamp so that the true value of I_0 changes, whereas the I_0 value used to determine absorbance is that measured when absorbance zero was set. Changes are large and rapid for the few minutes immediately after switching the instrument on (*Figure 3*). However, slow changes continue over a very much longer period and 'zero drift' never disappears completely with a single beam instrument. This problem may be avoided when using a diode-array spectrophotometer by internal referencing, that is, by subtracting the absorbance reading taken at a wavelength where there is no absorbance change due to the reaction. A split-beam instrument that determines absorbance from I and I_0 values measured simultaneously also avoids this problem.

2.3.4 Photochemical reactions

While the use of a diode-array spectrophotometer has some advantages over a conventional instrument, the fact that the sample is irradiated with light of all wavelengths means that the likelihood of significant changes caused by the irradiation itself is much greater. For example, NADH, one of the most commonly used chromophores in photometric assays, is photosensitive. *Figure 4* shows the changes that occur when a solution of NADH is observed in a diode array instrument in the absence of any enzyme or oxidizing substrate. The severity of the problem is reduced by taking measurements less frequently and by including an appropriate filter between the light source and the sample. With these precautions, the absorbance change due to photolysis is very small.

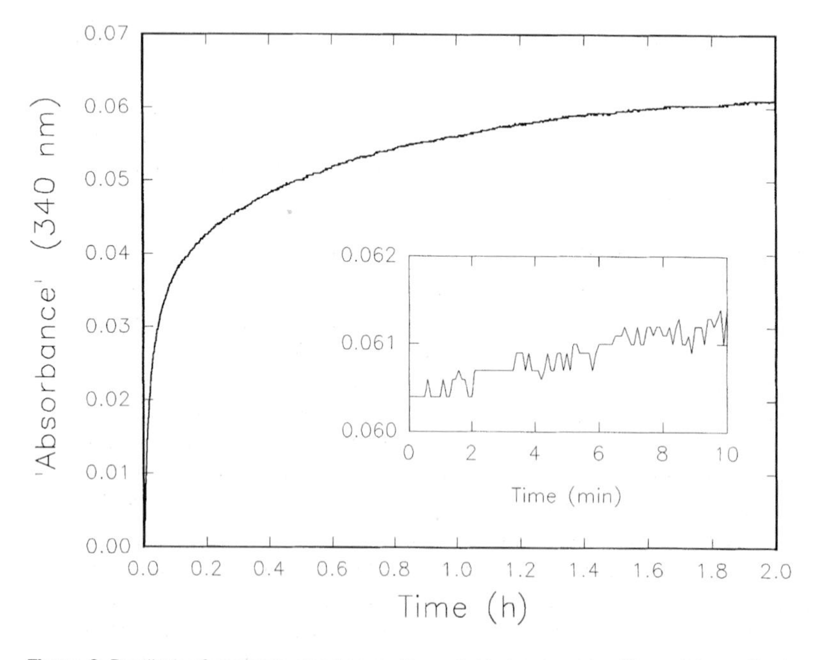

Figure 3 Readings of apparent absorbance after switching on tungsten filament lamp. The inset shows the slow continuous increase that persists after the instrument has been turned on for nearly 2 h.

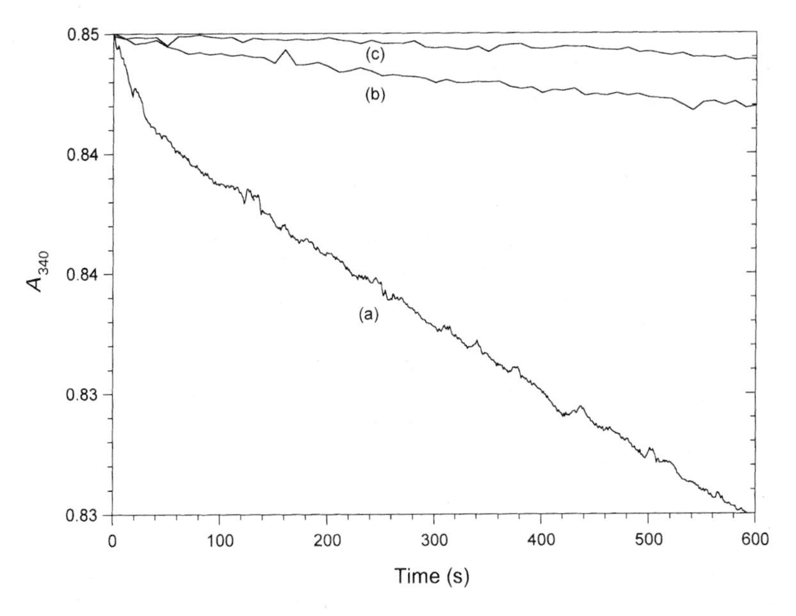

Figure 4 Photolysis of NADH. (a) A fall in absorbance observed when observations were made at 1 s intervals. Under these circumstances the solution is illuminated continuously by the xenon lamp of a diode array spectrophotometer. (b) Readings at 10 s intervals so that the solution is not continuously irradiated. (c) Changes occurring with a filter that cuts out light below 320 nm readings taken at 10 s intervals.

2.4 Absorbance range

The fact that determination of absorbance depends upon measurements of transmitted light means that the effects of these sources of error become increasingly important at high absorbance, as illustrated in *Figure 5*. Very low absorbance readings will be difficult to estimate because there is very little difference between the amounts of incident and transmitted light. The range of absorbance values that can be realistically measured varies greatly according to the quality of the instrument and, in the absence of the relevant information for a particular instrument, readings should be kept well within the range 0.01–1.0.

2.5 Measurement of low rates of absorbance change

Continuous improvements in instrumentation mean that the lower limit of detectability attainable by absorbance assay has fallen progressively. Because of wavelength-dependent variations in both the energy output from the source and the response of the detector, this limit depends on the wavelength at which the assay is conducted. The least favourable wavelength from this point of view is close to 340 nm, which coincides with the most useful of all the chromophores used in the assay of enzymes, namely NADPH and NADH. The lower limit of detectability depends very much upon the absolute value of the absorbance at which the assay takes place, small changes in absorbance being less easy to detect when the absolute absorbance is high. Spectrophotometers intended for making routine continuous enzyme assays typically have absorbance noise values in the range 0.0001 to 0.001 when the absolute value of absorbance is close to zero. This figure rises with increasing absorbance. In addition, the apparent absorbance reported by single-beam instruments changes continuously by about 0.003 h^{-1} because of time-dependent changes in the lamp and photodetector (*Figure 3*). The lower limit of detection is determined by these factors and is illustrated in *Figure 6*.

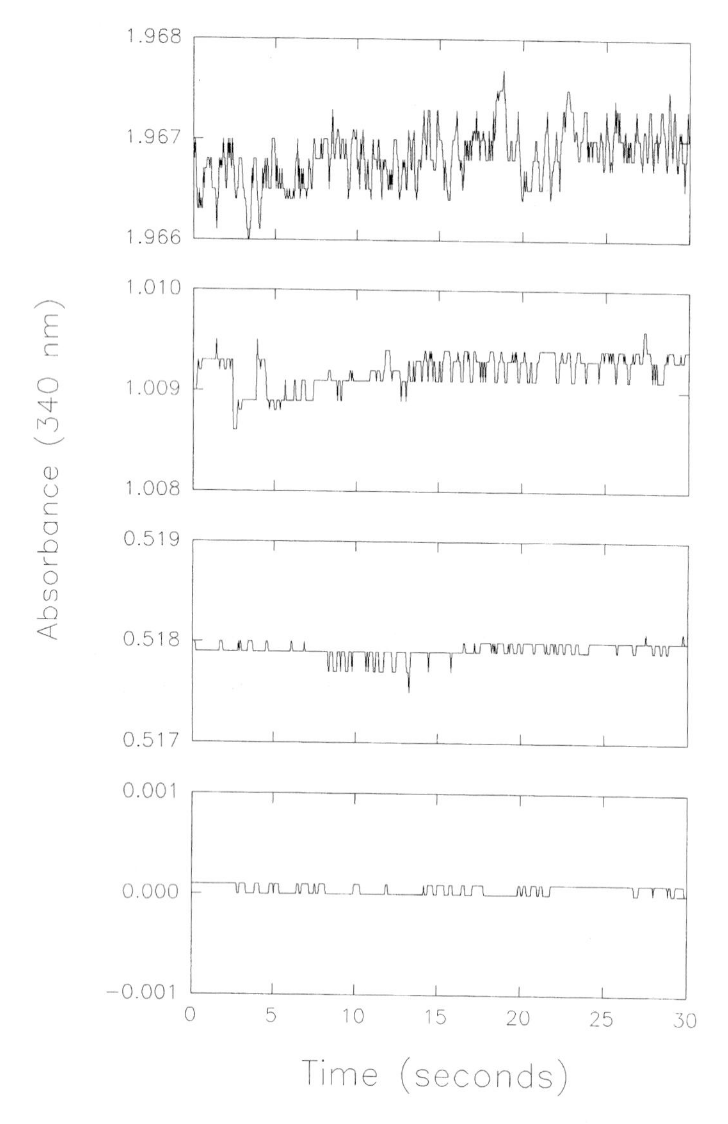

Figure 5 Instrument noise as a function of absorbance value.The traces show the level of instrument noise in the region of 0, 0.5, 1.0, and 2.0. The absorbance value of each solution was constant and observations were made at 340 nm.

2.6 Use of absorbance coefficient

So long as the instrument is used within its capabilities, the Beer–Lambert relationship will apply and rates of concentration change can be determined directly using the absorbance coefficient (extinction coefficient) for the compound. It should be remembered that published values for absorbance coefficients are determined with narrow bandwidths, in other words, with very nearly true monochromatic light. Errors will enter the determination of concentration if bandwidths are broad relative to the absorbance peak of the chromophore being determined. *Figure 7* illustrates this point. The problem is particularly acute when a

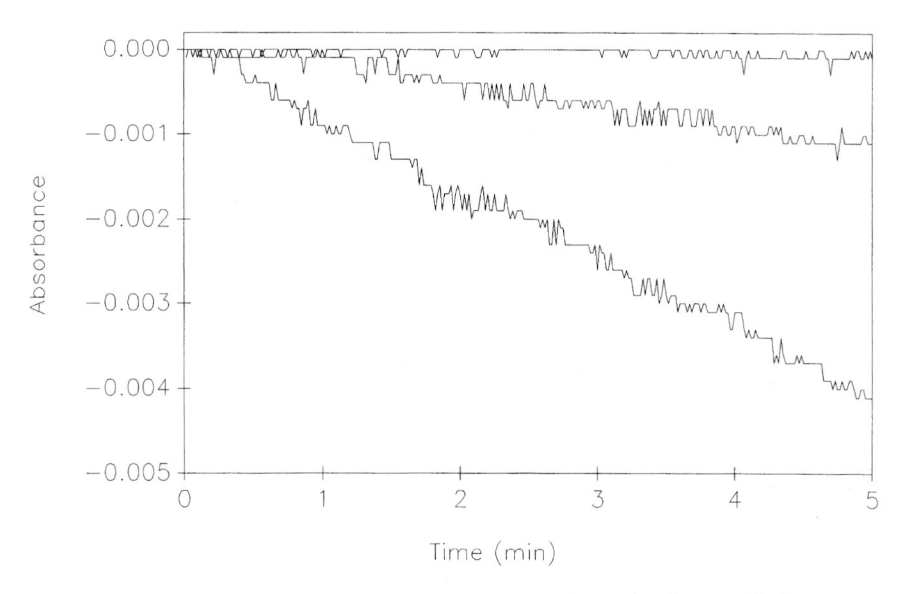

Figure 6 Continuous absorbance assays conducted near limits of detection. Lactate dehydrogenase was assayed using pyruvate and NADH as substrates as described in Section 2.10.1, using two concentrations of the enzyme to produce rates of absorbance change of 0.0003 min^{-1} and 0.001 min^{-1}, respectively. The line observed with no added enzyme is included for comparison and the initial absorbance was set to zero in each case.

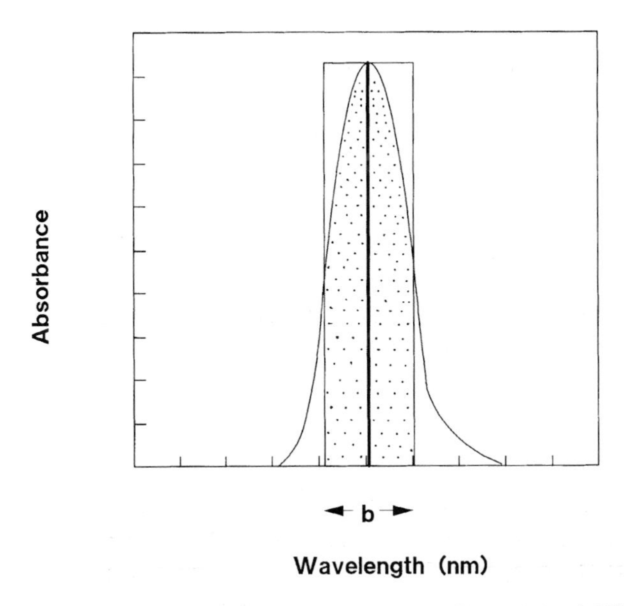

Figure 7 Dependence of apparent absorbance coefficient on bandwidth. Absorbances measured at broad bandwidth **b** will be lower than those predicted by an absorbance coefficient determined at the narrow bandwidth indicated by the solid line.The discrepancy will be in the ratio of the shaded and unshaded portions of the rectangle.

colorimeter is used to measure a chromophore that is not well matched by any of the available filters. In such a case, it is not exceptional to obtain readings an order of magnitude lower than those expected from the published absorbance coefficient.

2.7 Continuous assays

As has been explained in Chapter 1, a continuous assay, in which the enzyme-catalysed reaction is monitored as it proceeds, is very much to be preferred over one in which the enzyme reaction is run for a fixed time and then stopped before measuring products formed. Because the output of the spectrophotometer can readily be coupled to a computer, chart recorder, *x,y*-plotter, or printer, photometric methods are well suited to continuous assay. Some elementary but important practical points are worth considering.

2.7.1 Choice of solution for setting zero absorbance

Regardless of the instrument used and the system employed, at some stage the operator must make a choice as to what is in the cuvette holder when zero absorbance is set. Amongst various possibilities, this 'blank' may be a cuvette containing all of the ingredients except one, or a cuvette containing just water. Alternatively, the cuvette holder can be left empty, in which case the 'blank' is air. Absolute values of absorbance reported by the instrument during subsequent measurements will be reduced by the absorbance of the blank but rates of absorbance change will be unaffected. If the instrument used sets zero absorbance by attenuation of the light beam, the choice of 'blank' influences the noise level of the measurement. There should be an improvement in noise levels if the high-absorbing solution is used to set zero. Clearly, for continuous methods producing a fall in absorbance, a solution containing the relevant chromophore can only be used to set zero absorbance if the instrument is capable of reading negative values. *Figure* 8 shows the assay of LDH conducted (a) by setting the zero with an assay solution containing no NADH and (b) by setting the zero with a cuvette containing NADH at its starting concentration.

2.7.2 Temperature control

Photometric assays of an enzyme by a continuous method can only be achieved satisfactorily at a constant known temperature. It is particularly important to ensure that the contents of the cuvette are at the required temperature. The time taken for the contents of a cuvette to reach temperature depends on the nature and volume of contents. *Figure 9* (see page 61) shows how a solution approaches the required temperature when quartz or plastic cuvettes are used. Clearly, there is a need for the cuvettes to be brought to the correct temperature before the reaction is started. The inconvenience of waiting several minutes for temperature to be reached may be overcome in several ways. If the assay is rapid, that is complete within a few minutes, then the unused positions in a multiple cuvette holder provide convenient repositories in which successive cuvettes may be held before assay. For this system to work, the enzyme, or whatever solution is used to start the reaction, must be only a small part of the total volume. If addition of a relatively large volume to start the reaction is inescapable, then this must be brought separately to the assay temperature before it is added or allowance must be made as described in Chapter 1, Section 2.4.1. For continuous assays lasting significantly longer than the warming-up period it is more convenient to use all positions in the cuvette holder for assays but to delay starting the reaction for several minutes until assay temperature has been reached.

2.7.3 Starting the reaction

The operations involved in starting the reaction, that is, adding the final 'starter' ingredient, mixing adequately, replacing the cuvette in its holder, closing the lid, and beginning the

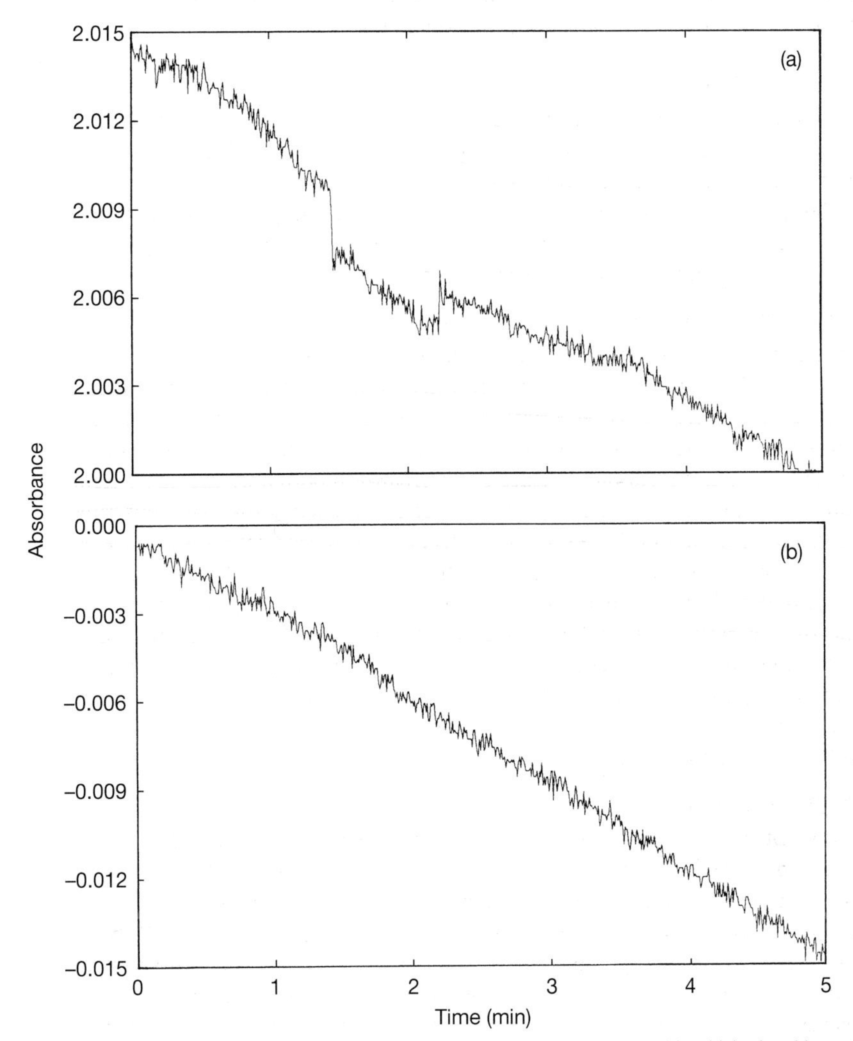

Figure 8 Continuous absorbance measurement obtained after setting zero with a high absorbing solution. The NADH concentration in the assay was set so that the absorbance was 2.0 and the course of reaction followed after (a) setting zero on a control solution containing no NADH, and (b) setting zero with the high-absorbing NADH solution.

recording, necessarily take time. Even the most skilled practitioner is unlikely to take less than 5 s. It is important to arrange things so that the amount of enzyme added does not produce such a rapid reaction that a significant part of the reaction is over before recording begins. It is not unusual to hear experimentalists unfamiliar with this sort of system reporting that there is no enzyme in a sample, whereas there is so much, that the reaction is complete by the time the recording has begun.

A convenient method for starting the reaction is as follows:

Protocol 1

Starting the reaction

Equipment and reagents

- Spectrophotometer
- Cuvette
- Parafilm
- Enzyme assay ingredients

1 Add all the ingredients except the 'starter' to the cuvette and allow it to attain the correct temperature. In the case of photometric assays, as in other enzyme assays, it is always best, if possible, to mix all the ingredients except one, in a stock solution and to start the reaction by adding a small (but not too small) volume of the omitted ingredient. (See Chapter 1 on the use of assay 'cocktails'.)

2 Add the 'starter' ingredient in a volume of 10–50 μl per ml of final assay mixture. This may be added directly into the solution. However, the time between mixing and measurement can be shortened if it is added as a 'hanging drop' on the side of the cuvette, so that it does not mix until the cuvette is inverted in the next step.

3 Use Parafilm and your thumb to close the cuvette and mix the contents by gently inverting the cuvette twice. Ensure that there is enough space at the top of the cuvette for the solution to mix adequately and that the solution really does mix by falling into this space (with semi-micro cuvettes the solution may remain in place even though the cuvette is inverted). The contents of the cuvette should not be mixed by shaking vigorously as this will introduce bubbles, which will interfere with the absorbance measurement and may possibly denature the enzyme.

4 Place the cuvette in the instrument and start the recording.

An alternative and equally good method of starting the reaction is to stir the contents of the solution with a commercially available plastic spatula shaped like a ladle or by drawing the solution in and out of a Pasteur pipette a few times. The operations of adding and mixing may be combined if the solution used to start the reaction is pipetted on to a spatula rather than directly into the cuvette (Chapter 1). One of these latter methods should always be used with a diode array instrument because the cuvette should be left in its holder after the zero absorbance measurement has been made.

If the rate of reaction is so fast that the process of mixing has to be hurried the enzyme sample must be diluted to slow the reaction down.

2.7.4 Volume needed in cuvette

When reagents and enzyme sample are not in short supply, it is better to use full-size (3 ml) cuvettes rather than semi-micro cuvettes. Pipetting errors are reduced, especially because a larger volume of enzyme can be used to start the reaction without lowering the temperature of the solution in the pre-incubated cuvette. Also, with very many spectrophotometers, problems arise because the beam width is greater than the optical face of the semi-micro cuvette so that some of the light does not pass through the solution. This may give unacceptable non-linearity at absorbance values well below 1. However, economy of reagents and enzyme frequently requires that the volume used be as small as possible. In this case it is worth looking at the dimensions of the beam at the point where it passes through the cuvette. This can be done as follows.

(a) Set the wavelength to something clearly visible (~ 500 nm, apple green) and increase the bandwidth to its maximum. This will make the beam sufficiently intense to be visible in a darkened room or under a black cloth.

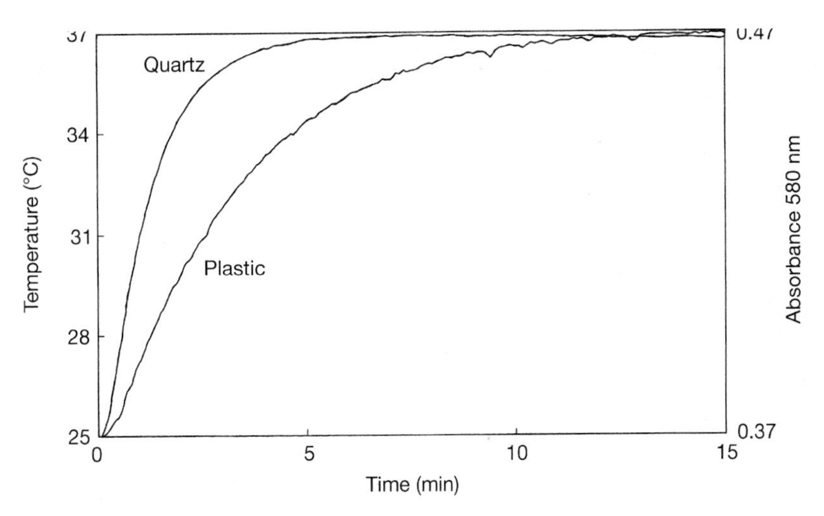

Figure 9 Temperature equilibration of cuvettes. Plastic and quartz cuvettes containing phenol red (6 mg l^{-1}) in 0.1 M Tris–HCl, pH 7.8, were transferred from room temperature to a cuvette holder at 37°C, and absorbance at 560 nm was monitored continuously.

(b) Insert a piece of white paper into the cuvette holder. The illumination of the paper will give a clear indication of the beam dimensions.

None of the light should get to the photodetector without passing through the solution, either through the sides or bottom of the cuvette or over the top of the liquid. Any light which gets through without going through the solution will have the same effect as stray light; absorbance values will be underestimated and the extent of the underestimation will increase with increasing absorbance. Light can be prevented from passing through the walls by masking the cuvette with black paint or tape. It can be prevented from passing through the air above the solution by adding enough solution to make the solution deep enough. It may be that the cuvette need not be pushed all the way into the holder for the beam to be completely contained within the solution. *Figure 10* shows the effect on the absorbance value reported by the spectrophotometer, of altering the volume in a full-size (3 ml) cuvette. With this particular instrument 1.5 ml is a 'safe' volume to work with. A similar effect is observed with semi-micro cuvettes. Frequently 0.4 ml or even less of a solution in a semi-micro cuvette is adequate to contain the beam completely. However, when working close to the limit in this way, it is important to avoid bubbles trapped in the surface layer as these may place a variable air–liquid interface within the beam.

2.7.5 Calculation of enzyme activity

If a chart recorder is being used to record the results, its speed should be set so as to produce a line with a slope that is neither too steep nor too shallow (the most accurate determination of gradient comes from a line of slope 45°). The line should be used to express the rate of change of absorbance with time (dA/dt)

Division by the molar absorbance coefficient, ε, gives the rate of change of concentration (dc/dt) in mol l^{-1} min^{-1}. The number of units (μmol/min) of enzyme present in the cuvette depends on the final volume, v_f (ml) it contains. Almost invariably the assay uses a very small part of a much larger sample. Suppose that v_s ml of enzyme was taken from a sample of total volume v_t ml and added to make a final volume of v_f ml in the cuvette. All three of these volumes must be used in determining the total number of units in the sample.

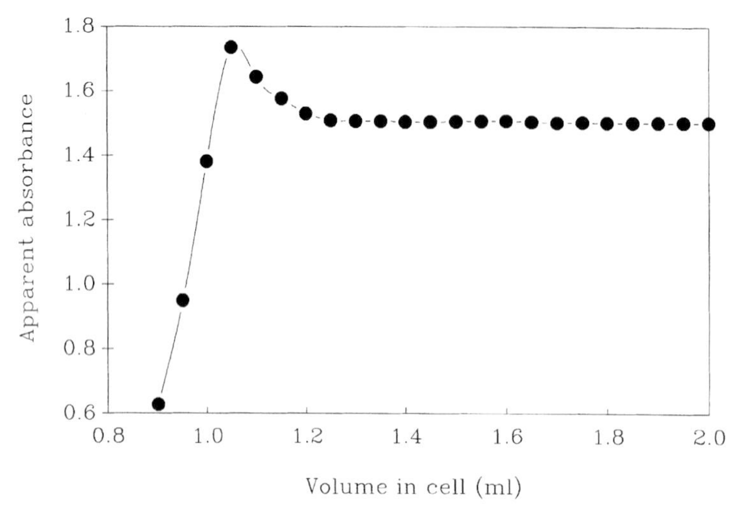

Figure 10 Effect of decreasing the volume of solution in cuvette. In this instrument, a constant reading is observed until the volume is reduced to about 1.3 ml. The apparent absorbance then rises as the region of the meniscus comes to lie in the beam. Thereafter apparent absorbance falls sharply. These measurements were obtained by removing successive 50 μl volumes from a cuvette with a total capacity of 3 ml.

The overall calculation is:

$$\text{Units of activity in sample} = [(dA/dt)/\varepsilon] \times v_f \times 1000 \times (v_t/v_s)$$

(The factor of 1000 arises because the units of activity are in μmol min^{-1}, use of the absorbance coefficient gives the concentration in mol l^{-1} and the cuvette volume is in ml.)

However, no self-respecting scientist should have to rely on formulae such as this for calculating enzyme units.

2.7.6 Causes of artefactual non-linearity

The best continuous assay methods should give linear initial-reaction rates from the time that recording begins until sufficient reaction has been recorded to establish the slope of the line clearly. However, even when the reactant solutions are correctly made, deviations from linearity may be observed because of errors in the use of the spectrophotometer or in the instrument itself.

i. Inadequate temperature equilibration

Upward curvature of the line as in *Figure 11a* indicates an accelerating reaction and suggests inadequate temperature pre-equilibration. The cure for this is simple and obvious. *Figure 9* shows how long it takes for temperature equilibration.

ii. Inadequate mixing

This is a common problem, particularly when the experimentalist is trying to catch the early stages of the reaction and hurries the mixing. The result is a random wavy line *(Figure 11b)*. The solution (obviously) is to mix the solutions thoroughly.

iii. Particles in solutions

This is also a common problem, exacerbated by the tendency of poorly trained experimentalists to use open beakers rather than narrow-necked stoppered flasks to store their solutions. The result is again a random, wavy line (*Figure 11c*). If the problem is simply due to extraneous particles like dust, it can be overcome by filtration. If the particles are an

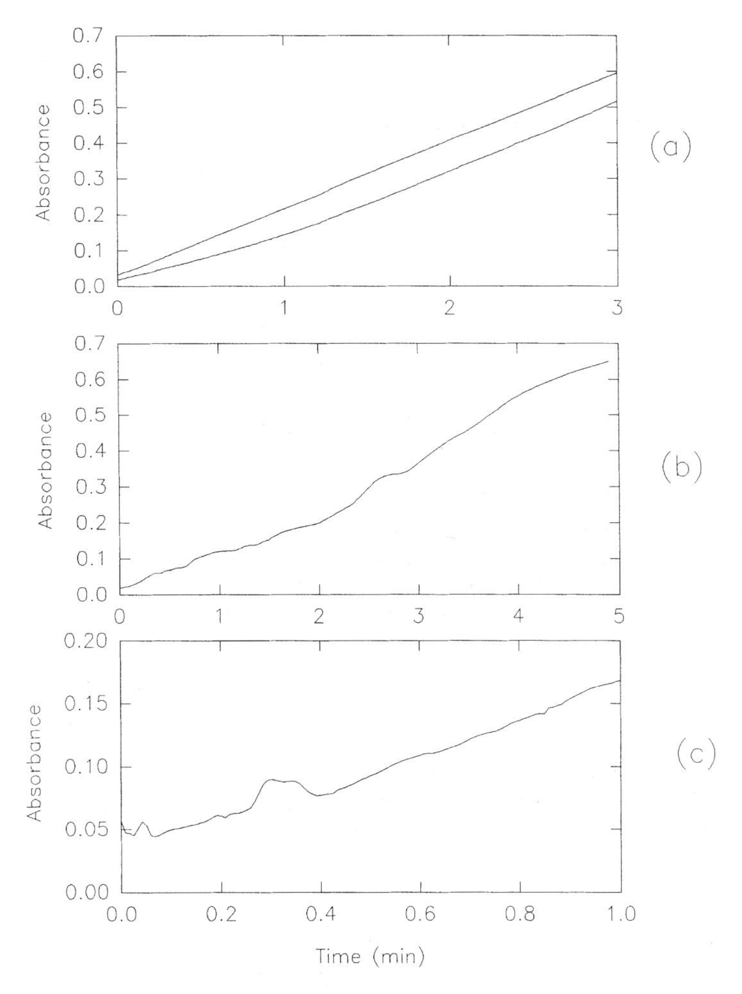

Figure 11 Some artefacts produced in continuous absorbance measurement. (a) Inadequate temperature equilibration (lower line) compared with adequate equilibration (upper line); (b) inadequate mixing; (c) dust particles in assay solution.

unavoidable component of the solution the problem can be neatly corrected using a diode array spectrophotometer and subtracting readings made at a wavelength where no absorbance change occurs from the readings taken at the wavelength of the assay. This technique is known as internal referencing. It works because changes in transmitted light, resulting from scattering by the particles, occur at both wavelengths.

iv. Stray light

Non-linearity will also be seen if the spectrophotometer does not give a linear response to concentration as discussed above in the section on stray light.

2.8 Discontinuous assays

The great value of absorbance measurement in the assay of enzymes is that the instrumentation is ideal for continuous assay. However, despite the fact that the eventual measurement is of absorbance, some systems cannot be assayed continuously. This may be because the conditions required for the enzyme-catalysed reaction are not the same as those for colour development. Alternatively, interfering background absorbance from, for example, crude tissue samples may be so high that deproteinization must precede absorbance measurement. In these circumstances the reaction is stopped at various intervals and the colour developed in a separate reaction.

2.8.1 Sources of error

As with any discontinuous assay there is a temptation to equate the concentration of product formed in a fixed time with the rate of the reaction. This is only true if the reaction runs at a constant rate over the period of the assay, an essential requirement that cannot be established with measurements made after only one time interval. The most satisfactory way of overcoming this problem is to make measurements at a minimum of two time intervals in addition to one for zero time. Alternatively, measurements at fixed time can be made using different volumes of enzyme sample (in the same total volume of assay). If the fixed-time absorbance value is linearly related to the volume added, then this provides good evidence that linearity of absorbance change with time is maintained for the duration of the assay. The necessity to make measurements at multiple time-intervals may justifiably be avoided in routine measurements if the conditions of the assay system, such as substrate concentration, are kept constant and the amounts of enzyme measured are well within already established bounds of linearity. For a detailed discussion of this point, see Chapter 1.

2.8.2 Use of microtitre-plate readers

The availability of plastic microtitre plates and associated plate readers considerably reduces the effort involved in obtaining absorbance measurements for fixed-point assays. The reproducibility of measurements made in this way is perfectly adequate for many types of analysis in which large numbers of samples need to be processed. Linearity is maintained to high absorbance readings (*Figure 12a*) and standard deviations in absorbance readings are constant at about 0.01 over this range (*Figure 12b*). Thus, if one is not prepared to accept errors greater than 10% from readings made in duplicate, minimum absorbance values of 0.2 should be used. The variance in the measurement appears to be associated with the fact that one optical face is the meniscus at the air–liquid interface. Consequently, to maximize the absorbance reading, it is best to fill the wells close to the top using a total volume of about 0.25 ml, which gives a pathlength of about 1 cm. Because the instruments use filters, the absorbances measured will be lower than those expected on the basis of published absorbance coefficients, and concentrations should be determined by using an appropriate standard.

2.9 Examples of enzymes assayed by absorbance change

The number of enzymes that can be assayed spectrophotometrically is so large that only a limited number of examples can be presented here in detail. For experimental detail of enzymes not mentioned here the reader is referred to two major collections of such assay methods, *Methods in Enzymology* (1) and *Methods in Enzymatic Analysis* (2). Much ingenuity has been applied to devising photometric assays for enzymes that catalyse reactions which do not themselves give rise to any direct absorbance change. The following examples are chosen, in part, to illustrate this experimental ingenuity in the hope that consideration of these successful methods will aid the design of new assays for other enzymes.

2.9.1 Direct observation of the reaction using the natural substrate

A limited number of enzyme-catalysed reactions are themselves accompanied by a useful change in absorbance. Table I shows the absorbance properties of some substrates that can be employed directly in the assay of the enzymes indicated.

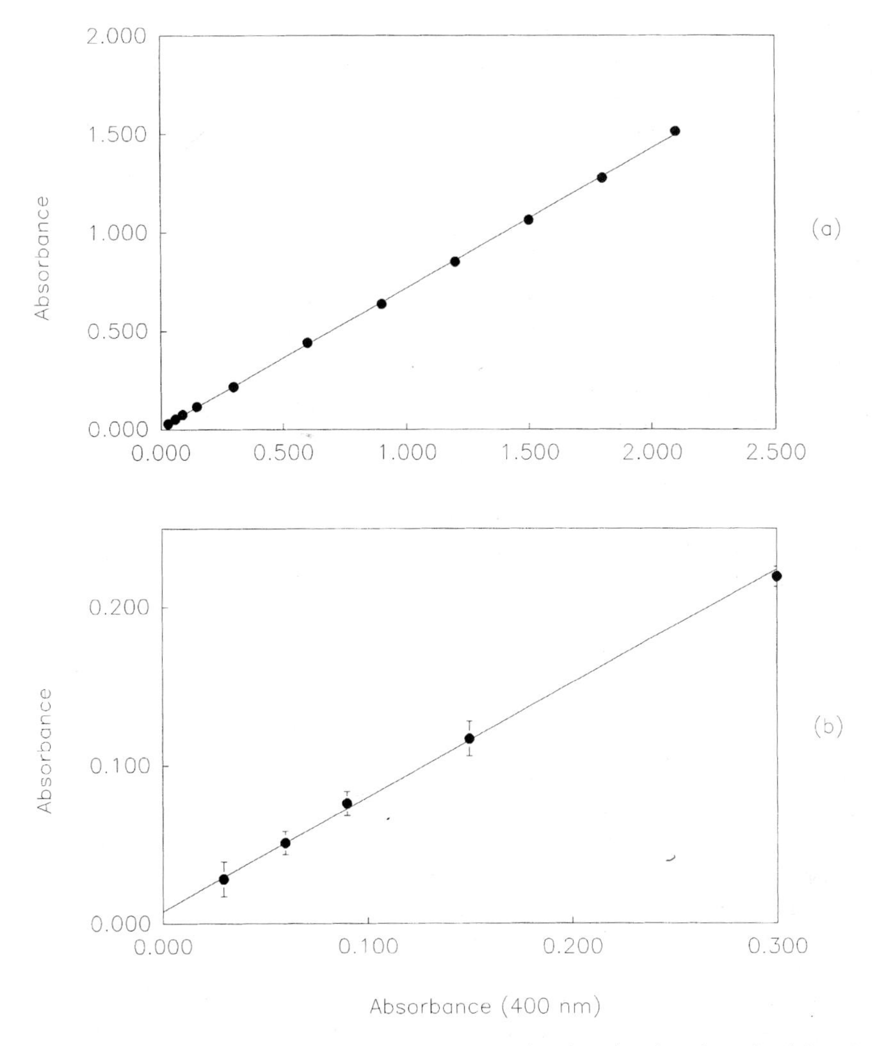

Figure 12 Errors associated with absorbance readings made using micro-titre plates. A solution of nitrophenol in 0.1 M NaOH was used to compare readings of true absorbance with values determined by a micro-titre plate reader. Wells were filled to a depth of 1 cm by adding 0.27 ml of solution. (a) Linearity is maintained to high absorbance values. (b) Some of the same data as in (a) but on an expanded scale so that the standard deviation of the measurements can be appreciated.

Oxidation of NADH (and NADPH) as well as the reaction in the opposite direction is accompanied by a large change in absorbance at 340 nm ($\varepsilon_{340} = 6200$ 1 $mol^{-1} cm^{-1}$). This makes the direct, continuous absorbance assay of a large and important group of enzymes, the dehydrogenases, very simple.

i. Lactate dehydrogenase

Lactate dehydrogenase (LDH) is one of the most frequently assayed of all enzymes because its presence in serum after tissue damage provides an important aid to clinical diagnosis. It is worth considering the factors that have led to the choice of assay conditions. Because the reaction catalysed is freely reversible, the assay can be made either in the direction of oxidation of lactate by NAD^+ or in the reverse direction in which pyruvate is reduced by NADH.

Table 1 Absorbance characteristics of some naturally occurring, UV-absorbing substrates. The wavelengths given are those which maximize the difference between substrate and product absorbances

Compound	Absorbance characteristics	Enzymes
Acetyl coenzyme A 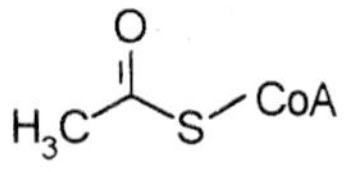	232 nm ε = 4500 l $mol^{-1}cm^{-1}$	Choline acetyl transferase and other acetyltransferases
Fumarate 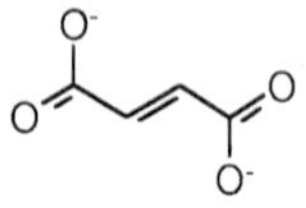	240 nm ε = 2440 l $mol^{-1}cm^{-1}$	Fumarase Argininosuccinase
Oxaloacetate 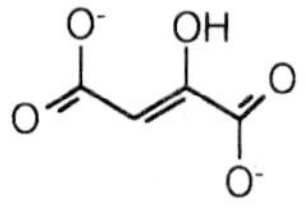	265 nm ε = 950 l $mol^{-1}cm^{-1}$	Oxaloacetate decarboxylase
Inosine 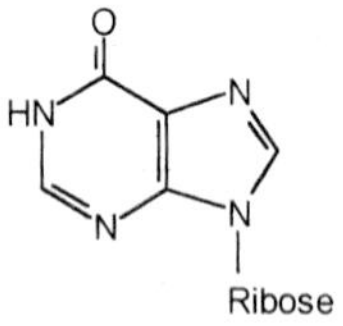	265 nm ε = 8100 l $mol^{-1}cm^{-1}$	Adenosine deaminase
Uric acid 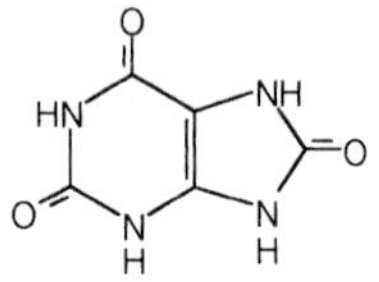	293 nm ε = 12000 l $mol^{-1}cm^{-1}$	Xanthine oxidase

Methods for assaying the enzyme based on both reaction directions are available but that which is almost invariably used begins with pyruvate and NADH and measures the fall in absorbance at 340 nm that accompanies the reaction.

$$\underset{\text{Pyruvate}}{CH_3COCOO^-} + NADH + H^+ \rightleftharpoons \underset{\text{Lactate}}{CH_3CHOHCOO^-} + NAD^+$$

Several reasons combine to make reduction of pyruvate the preferred direction for the assay:

(a) The equilibrium lies very much in this direction. Thus, only an insignificant fraction of the complete reaction need occur to give sufficient absorbance change for the measurement.

(b) The maximal velocity is higher in this direction. This makes the assay more sensitive.

(c) NADH can be kept in a more stable condition at the pH of the assay than can NAD^+.

The enzyme may be satisfactorily assayed at 30°C, pH 7.2 in 50 mM Tris, with substrate concentrations of 0.15 mM NADH and 1.2 mM sodium pyruvate. The reasoning behind the choice of concentrations is as follows:

(a) Pyruvate. Besides being the substrate, this compound is also an inhibitor because it binds to enzyme–NAD^+ to make an 'abortive complex'. Thus the concentration is deliberately kept low so that inhibition is avoided. The conditions given are intended for assay of the enzyme in serum, which is likely to contain two isoenzymes of LDH differing in their sensitivity to inhibition by pyruvate.

(b) NADH. The initial concentration gives a convenient absorbance measurement of 0.9 when the solution is used in a cuvette of pathlength 1 cm. A significant fall in this absorbance is essential for the assay but the low K_m for NADH ensures that the velocity does not fall significantly because the enzyme remains virtually saturated for most of the reaction.

2.9.2 Indirect assay by coupling with a dehydrogenase

The convenience of the change in absorbance when the nicotinamide coenzymes undergo oxidation or reduction, together with the large number of dehydrogenases having a wide range of specificities, mean that these enzymes frequently provide the final step of coupled enzyme assays (Chapter 1).

i. Aspartate aminotransferase

The clinical importance of measuring aspartate aminotransferase, which is released into the bloodstream in large quantities after heart or liver damage, means that, like lactate dehydrogenase, it is one of the most frequently measured of enzymes. Usually known in the clinical biochemistry laboratory as GOT (glutamate oxaloacetate transaminase) its measurement in serum is determined by a standardized method approved by the International Federation of Clinical Chemists, who recommend that the assay be conducted at 30°C, pH 7.8, in 80 mM Tris. Recommended substrate concentrations are 240 mM L-aspartate and 12 mM 2-oxoglutarate. To activate the significant amounts of apo-enzyme in serum, 0.1 mM pyridoxal phosphate is added. The coupling step and removal of interfering pyruvate require the addition of 0.18 mM NADH, 0.42 units of malate dehydrogenase/ml and 0.6 units of lactate dehydrogenase/ml.

$$\text{oxogluterate} + \text{aspartate} \xrightleftharpoons{\textit{aspartate aminotransferase}} \text{glutamate} + \text{oxaloacetate}$$

$$\text{oxaloacetate} + \text{NADH} + \text{H}^+ \xrightleftharpoons{\textit{malate dehydrogenase}} \text{malate} + \text{NAD}^+$$

When the assay method is used for measuring the activity of the pure enzyme, all the ingredients can be combined in one solution and the assay started by adding an appropriate amount of the enzyme in a small volume. However, when the enzyme is being measured in an unpurified sample such as serum or tissue homogenate, precautions must be taken to avoid artefacts arising from the presence of varying amounts of pyruvate and lactate dehydrogenase, which themselves oxidize NADH and give rise to a falsely high and non-linear initial rate. With such samples, the 2-oxoglutarate is left out of the assay mixture and the serum or other sample is added together with lactate dehydrogenase to remove pyruvate. The mixture is left until the absorbance is constant and the reaction is then started by adding 2-oxoglutarate.

ii. Triose phosphate isomerase

This enzyme catalyses the isomerization of glyceraldehyde-3-phosphate and dihydroxyacetone phosphate. It is normally assayed in the direction of dihydroxyacetone phosphate synthesis. Rabbit muscle α-glycerophosphate dehydrogenase is used to couple the reaction to the oxidation of NADH.

Conditions for the assay are 30°C, pH 7.9 in 0.1 M triethanolamine containing 0.14 mM NADH, 0.4 mM D,L-α-glyceraldehyde 3-phosphate, α-glycerophosphate dehydrogenase (2 μg/ml) and 5 mM EDTA.

$$\text{glyceraldehyde phosphate} \xrightleftharpoons{\textit{triose phosphate isomerase}} \text{dihydroxyacetone phosphate}$$

$$\text{dihydroxyacetone phosphate} + \text{NADH} + \text{H}^+ \xrightleftharpoons{\textit{glycerophosphate dehydrogenase}} \text{glycerophosphate} + \text{NAD}^+$$

iii. Enzymes catalysing production of CO_2

Photometric assay of enzymes producing CO_2 can be achieved by coupling with wheat phosphoenolpyruvate (PEP) carboxylase, an enzyme that produces oxaloacetate. A third enzyme, malate dehydrogenase, is necessary to complete the linkage, with oxidation of NADH producing a fall in 340 nm absorbance.

Decarboxylases functioning in the pH range 6–8 can be assayed in this way. An example is the assay of lysine decarboxylase (3).

$$\text{lysine} \xrightarrow{\textit{lysine decarboxylase}} \text{cadaverine} + \text{CO}_2$$

$$\text{CO2 1 PEP} \xrightarrow{\textit{PEP carboxylase}} \text{oxaloacetate}$$

$$\text{oxaloacetate} + \text{NADH} + \text{H}^+ \xrightarrow{\textit{malate dehydrogenase}} \text{malate} + \text{NAD}^+$$

Assay of lysine decarboxylase is conducted using a solution consisting of 0.1 M Tris–HCI pH 6.0, 8 mM $MgCl_2$, 10 μM pyridoxal phosphate, 10 Units malate dehydrogenase/ml, 1 Unit PEP carboxylase/ml, 45 mM PEP, 0.4 mM NADH, and 0.01% Nonidet detergent. The detergent is included to prevent aggregation of the enzymes. The solution is degassed by evacuation to remove dissolved CO_2

2.9.3 Problems with nicotinamide nucleotides

Preparations of these coenzymes have in the past been contaminated with inhibitors, and their storage in aqueous solution under the wrong conditions also allows inhibitors to form rapidly. Clearly, the use of such contaminated preparations is to be avoided. The clinical importance of assays based on these coenzymes means that commercial suppliers are very conscious of the need to provide high-quality reagents. After opening, the preparations should be stored dry at 0–4°C and protected from light. There is the additional problem of 'NADH oxidase' (Chapter 1) and that of photolysis in diode array spectrophotometers (Section 2.3.4).

2.9.4 Artificial chromogenic substrates

When the natural reaction is not accompanied by a useful absorbance change, it is commonplace to use a synthetic substrate. Ideally, but optimistically, the enzyme should have the same specificity for both synthetic and natural substrates. When synthetic substrates are used to measure the enzyme in a crude mixture, there is a very real risk of measuring an enzyme entirely different from that intended.

In the case of *p*-nitrophenol, one of the most commonly used chromogenic groups, the coloured species released by hydrolytic reactions is the nitrophenolate anion, which protonates to the colourless acid form with a p*K* of 7.15. The anion has an absorbance coefficient of 18000 $1\ mol^{-1}\ cm^{-1}$ (4). Care must be taken with such assays to ensure that the value used as absorbance coefficient takes pH into account. The absorbance coefficient that should be applied at different pH values takes the p*K* of the ionization into account according to Equation 5.

$$\varepsilon_{pH} = 18\,000 \times [10^{-7.15}/(10^{-pH} + 10^{-7.15})] \quad (5)$$

i. β-galactosidase

The 'normal' reaction for this enzyme is the hydrolysis of the naturally occurring β-galactoside, lactose, into galactose and glucose, a reaction which is not accompanied by an absorbance change. Assay of β-galactosidase is made simple by using the synthetic substrate nitrophenyl-galactoside. The simplicity of this assay greatly assisted the classical work that determined the control of gene expression via the *lac* operon. The enzyme is assayed in 0. 1 M Tris–HCl pH 7.6, at a nitrophenyl-galactoside concentration of 0.1 mM. At the pH of the assay the effective absorbance coefficient of the product nitrophenol is 13 300 $1\ mol^{-1}\ cm^{-1}$.

ii. α-amylase

A chromogenic substrate, namely 4-nitrophenylmaltoheptoside, provides the basis of an assay for this enzyme. The reaction is linked to the release of nitrophenol by α-glucosidase. Conditions for the assay are 30°C, 0.1 M sodium phosphate, pH 7.1, 0.05 M NaCl, 30 units of α-glucosidase/ml and 5 mM 4-nitrophenyl-maltoheptoside. The use of this concentration of the linking enzyme, α-glucosidase, results in a lag of about 3 min.

iii. Serine proteases

Considerable success has been achieved in assaying different serine proteases using synthetic substrates in which the acyl moiety is an oligopeptide, the sequence of which is intended to give specificity. Trypsin can be assayed with benzoyl-argininine-4-nitrophenylalanine, but this substrate is hydrolysed at similar rates by other serine proteases with a preference for basic amino acids as the acyl donor in the scissile bond. The synthetic tetrapeptide benzoyl-Ile-Glu-Gly-Arg-4-nitroaniline is a better substrate for the assay of trypsin in that sensitivity is increased 10-fold and specificity for trypsin is higher. Conditions for the assay are 40 mM Tris–HCl, pH 8.2, 16 mM Ca^{2+}, 0.1 mM benzoyl-Ile-Glu-Gly-Arg-4-nitroaniline.

iv. Carboxypeptidase A

A furoylacryloyl group incorporated at the amino terminus of a synthetic peptide provides an assay for this enzyme (5). Hydrolysis of the carboxyterminal residue from the substrate furanacryloyl-Phe-Phe (1 mM) gives a decrease in absorbance of the chromophore. In order to avoid high absorbance of the substrate, the assay is best conducted at a wavelength where the absorbance change is not at a maximum. At 350 nm the absorbance coefficient for the change is 800 $1\ mol^{-1} cm^{-1}$.

2.9.5 Chromogenic reagents

When an enzyme-catalysed reaction creates a product with a reactive grouping not present in the substrate, it may be possible to include a reagent which will react directly with the product to form a chromophore. For a satisfactory continuous assay, the chromogenic reactant should not interfere with the enzymic reaction. The rate of the second reaction will be proportional to the concentration of the reagent and this must be present at a concentration high enough to avoid an unacceptable lag.

i. Acetylcholinesterase

Dithio-bis(2-nitrobenzoic acid) (DTNB, Ellman's reagent) reacts with free thiols in an exchange reaction that produces the yellow 4-nitrothiolate anion. Acetylcholinesterase is conveniently assayed by replacing the normal substrate with acetylthiocholine. Hydrolysis releases a thioester and the rate of release is continuously measured by a reaction with DTNB included in the assay solution. Conditions are 30 °C, 0.1 M sodium phosphate, pH 7.5, 10 mM DTNB, 12.5 mM acetylthiocholine iodide.

2.9.6 Use of pH indicators in enzyme assays

Many reactions produce or consume protons and are therefore capable of altering the pH of a solution. The ionization of an appropriately chosen pH indicator may be exploited to convert the system into a spectrophotometric assay and in such systems the pH must be allowed to vary. However, so long as the pH change is small, linear velocities are observed. A knowledge of the pH profile of the enzyme will help to decide the magnitude of the pH change that can be tolerated. The sensitivity of such a system decreases with increasing concentration of buffer and increases with increasing concentration of indicator. Both sensitivity and linearity are best at the p*K* of the indicator used (6). Clearly, if an enzyme is being assayed from different samples, it is important that the sample does not contribute to the buffering capacity of the system or alter the pH. Quantification of rates of change must be accomplished by titration of the system using small volumes of standard acid or base and measuring the resulting absorbance changes. Amongst the enzymes that have been assayed in this way are the amino acid decarboxylases (6).

i. Arginine decarboxylase

This enzyme is very simply assayed in 0.05 M sodium acetate, pH 5.0, 0.025 M arginine, and including 10 μM bromocresol green as indicator. The rate of absorbance change is determined at 615 nm.

3 Turbidimetry

Enzymes that act upon insoluble polymers will frequently clarify a turbid solution and this property may be used to quantify the amount of enzyme present. The process involved is light scattering and not absorbance but it may be measured with a standard spectrophotometer. Such turbidimetric measurements are less easily standardized than absorbance measurements, partly because it is not easy to provide reproducible suspensions of insoluble polymeric substrate but also for reasons based upon instrumentation.

When an absorbance photometer is used to make measurements of turbidity, the reading that it gives is determined from the light transmitted by the solution in the same way that the instrument makes genuine absorbance measurements. In the case of turbidimetric measurements, however, some of the scattered light reaches the photodetector and the proportion depends on the distance of the detector from the cuvette (*Figure 13*). Clearly, an instrument (a), in which the cuvette is close to the detector, will receive more scattered light than (b), in which detector and cuvette are further apart. For the same solution, instrument (a) will give a lower apparent absorbance reading than instrument (b). As an illustration, a turbidimetric assay for lysozyme is described.

The enzyme lysozyme has the function of hydrolysing bacterial cell walls and is conveniently assayed by observing the change in turbidity that occurs when it is added to a suspension of dried bacterial cells. This decrease in turbidity is clearly the result of a complex

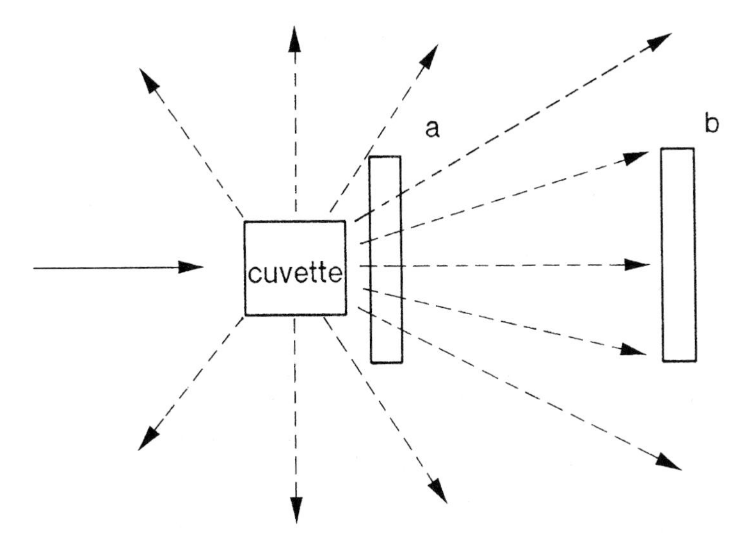

Figure 13 Use of a spectrophotometer for turbidimetric measurements. An instrument with photodetector in position **a** will detect more scattered light than one with the detector in position **b.**

process of progressive random hydrolysis and it is therefore not possible to express the rate in molar terms. The unit is defined in terms of the rate of decrease in turbidity. The wavelength chosen for these turbidity measurements is arbitrarily set at 450 nm and one unit of activity is defined as that which produces an initial rate of change in 'absorbance' of 0.001 per min when the volume in the cuvette is 2.6 ml (other conditions being pH 6.24 and 25 °C). It is important to note that in this case the volume in the cuvette must be specified. The same number of units added to a smaller volume of the same suspension would produce a proportionately higher change in turbidity.

4 Fluorescence

The phenomenon of fluorescence is, like that of absorption, the result of an electronic transition which converts the absorbing molecule to an excited state. Thus excitation and absorption are two words describing the same physical process. The difference between fluorescent and non-fluorescent compounds is determined by what happens when the excited state returns to the ground state. Whereas in non-fluorescent molecules the energy of the excited molecule is lost as heat, a fluorescent compound emits part of the energy as light. During the period ($\sim 10^{-9}$ s) between absorption and emission, the molecule loses some of its energy by vibrational relaxation so that the emitted light is of lower energy and consequently higher wavelength than the exciting light.

Fluorescence-based enzyme assays are potentially capable of much greater sensitivity than absorbance assays. This is because of a fundamental difference in the way that the measurements are made. Measurements of low values of absorbance are intrinsically difficult to make because they are based on the determination of two values of transmitted light that are high and nearly equal. The small amounts of light emitted by low concentrations of fluorescent material are intrinsically more readily measured because comparison is being made with the complete absence of emitted light when no fluor is present. It is by no means always true, however, that a fluorescence-based assay will be more sensitive than an absorbance assay based upon observation of the formation of the same compound.

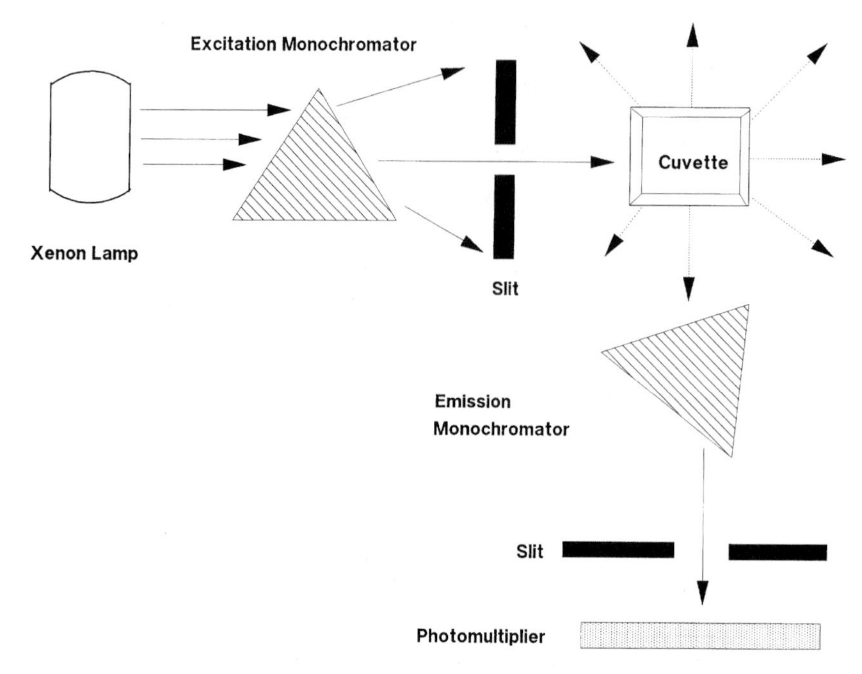

Figure 14 Essential features of a fluorimeter.

4.1 The fluorescence spectrometer

Figure 14 shows a schematic diagram of a spectrofluorometer. Up to the point of emergence of light from the cuvette the system is essentially the same as for an absorbance spectrophotometer. However, because fluorescence emission occurs in all directions, the detector is normally set to collect light emitted at 90° to the incident beam. A second monochromator is included for selecting the emission wavelength.

4.2 Quantitation of fluorescence

Although conditions may be arranged so that the relationship between measured fluorescence emission and concentration is virtually linear, the concentration of fluorophore in the sample cannot be calculated from the measured fluorescence emission by the application of a universal constant equivalent to an absorbance coefficient.

A fluorescent compound emits a constant fraction of the light energy that it absorbs. This fraction is known as the quantum yield (Q_f). Strongly fluorescent materials have both high absorbance coefficients and quantum yields close to 1, that is, they absorb a lot of light and re-emit most of it as light of higher wavelength. Although quantum yield is constant for a given set of experimental conditions, several features of the instrumentation prevent its use in the direct determination of concentration from fluorescence intensity measurements. The light incident on the cuvette varies from one instrument to another, and even with the same instrument from one occasion to another. Furthermore, although a given fraction of the absorbed light is emitted, only that part captured by the photodetector is registered, and this depends upon a variable feature of instrument design, namely the geometric relationship of cuvette to photodetector. Finally, although the photodetector response is proportional to the emitted light at a given wavelength, it does not provide an absolute measurement of light intensity, nor is its response the same at all wavelengths. Conversion of fluorescence emission values to units of concentration must therefore be achieved by the use of an appropriate standard.

The most straightforward standard to use is the same fluorescent compound that is generated by the assay. The concentration of this may be determined by weighing, or more simply and accurately by measuring its absorbance at the appropriate wavelength and calculating concentration using the absorbance coefficient. When running an assay for the first time, it is essential to construct a calibration line relating fluorescence to concentration over the range that is expected from the assay. Thereafter, so long as a linear relationship exists, calculations may be based upon the measured fluorescence of a single sample of the standard at known concentration. It is essential that these calibration measurements be made using the standard sample of fluorophore under the same conditions as the fluorophore produced in the assay. In cases where the fluorescent product is not readily available, an alternative secondary standard, having fluorescence excitation and emission in the same region as the test material, may be substituted. Alternatively, and very conveniently, manufacturers supply solid-state fluors cast in the shape and size of a cuvette and having a wide range of fluorescent properties.

4.3 Causes of non-linearity – the inner filter effect

All fluorescent compounds absorb light at the excitation wavelength. This means that the intensity of incident light (I_0) falls exponentially as the beam progresses through a solution of a fluorescing compound. The relationship between fluorescence emission (I_f) and concentration of fluor (c) is therefore non-linear (Equation 6).

$$I_f = I_0 Q_f (1 - 10^{ecl}) \tag{6}$$

In practice the relationship between *measured* fluorescence and concentration is more complex because of the way in which the instrument is constructed. When the absorbance of the solution at the exciting wavelength is high, the most intensely emitting part of the solution, that closest to the light source, is hidden behind the cuvette holder and the emitted light cannot be detected. A plot of measured fluorescence against concentration has the form of *Figure 15*, and it is quite possible therefore to conclude that a sample contains no fluorescent material when in fact it contains so much that the fluorescence cannot be detected. An experimentalist beginning to use techniques based on fluorescence will gain a real understanding of the processes involved (as well as a bit of fun) from looking at the colours produced, by observing the fluorescence in a cuvette directly with the lid of the fluorimeter open and adjusting the excitation wavelength, (*taking care not to expose the eyes and skin to wavelengths shorter than 320 nm.*).

From a practical point of view, a near-linear relationship between concentration and fluorescence emission is obtainable so long as absorbance values at both exciting and emitting wavelengths are sufficiently low (below 0.1).

4.4 Examples of fluorimetric enzyme assays

4.4.1 Direct observation of the natural reaction

Only a small proportion of the many naturally occurring compounds that absorb UV and visible light are sufficiently fluorescent to provide useful enzyme assays.

i NAD(P)H-dependent systems: glucose-6-phosphate dehydrogenase

This enzyme is chosen as an example of the many dehydrogenases that may be assayed fluorimetrically by making use of the fluorescence of the reduced nicotinamide nucleotide

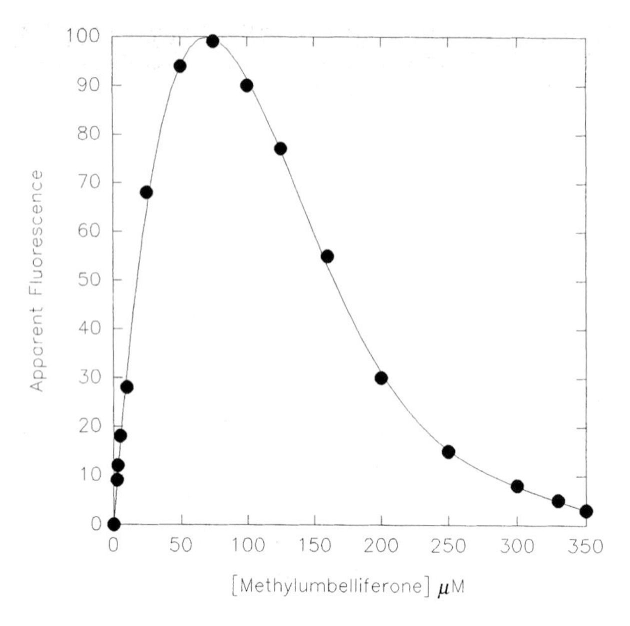

Figure 15 Relationship between measured fluorescence and concentration. The concentration dependence of the apparent fluorescence (λ_{ex} = 355 nm, λ_{em} = 460 nm) of methylumbelliferone was determined. The compound was dissolved in 0.1 M glycine, pH 10. At concentrations of fluor below about 10 μM the relationship is linear but becomes progressively more non-linear as light absorption becomes more significant. Eventually, at high concentrations, this inner-filter effect leads to a fall in measured fluorescence with increasing concentration.

coenzymes. The inner filter effect prevents the use of high concentrations of NAD(P)H and for this reason the systems most suited to fluorimetric assay are those run in the direction producing NAD(P)H. However, the very high sensitivity, often available from other fluorescence-based assays, is not obtainable because the fluorescence of NADH is strongly quenched in aqueous solutions, the quantum yield being only approximately 0.03. The lower limits of detection of the two methods obviously depend upon the instruments used. Comparisons made by continuously recording over 3 min using two popular instruments (Beckman DU7 Spectrophotometer and Perkin Elmer LS 5 Luminescence Spectrometer) showed the lower limit of rate measurement by absorbance to be in the 10^{-7} M min^{-1} range and that by fluorescence an order of magnitude more sensitive. At these levels of sensitivity both methods require the exclusion of dust from the solutions by filtration.

Glucose-6-phosphate dehydrogenase is assayed by adding 50 μl of enzyme sample to 3 ml of 0.1 M Tris-HCl, pH 7.8, containing 3 mM $MgCl_2$, 10 mM glucose-6-phosphate and 7 mM NAD^+ (λ_{ex} = 340 nm, λ_{em} = 465 nm).

ii. Porphobilinogen deaminase

Uroporphyrin I is a highly fluorescent compound formed by oxidation of uroporphyrinogen I, an intermediate that occurs in the later stages of the pathway leading to haem synthesis. Its determination by fluorometry is used as the basis of an assay for porphobilinogen deaminase and for other enzymes in the same pathway (7).

Protocol 2

Fluorometric assay of porphobilinogen deaminase

Equipment and reagents

- Spectrophotometer
- Centrifuge
- Long-wavelength UV light source
- 0.1 M Tris–HCl, pH 8.1
- 0.5 mM porphobilinogen
- Porphobilinogen deaminase
- 50% (w/v) trichloroacetic acid
- Commercial uroporphyrin

1. Mix 0.5 ml of Tris–HCl (0.1 M, pH 8.1) and 0.2 ml of 0.5 mM porphobilinogen.
2. Start the reaction by adding 50 μl of enzyme sample.
3. Keep solution in the dark at 37 °C for 30 min.
4. Stop the reaction by adding 0.25 ml of 50% (w/v) trichloroacetic acid.
5. Centrifuge to remove protein.
6. Expose to long-wavelength UV light to photo-oxidize the supernatant to uroporphyrin I.
7. Measure fluorescence (excitation wavelength 406 nm, emission wavelength 600 nm).
8. Standardize using a solution of commercial uroporphyrin, having determined its concentration by absorbance measurement ($\varepsilon_{406} = .05 \times 10^{-5}$ l mol^{-1} cm^{-1}).

iii. Anthranilate synthase

Anthranilate (*o*-aminobenzoate), a key intermediate of aromatic amino acid metabolism, is fluorescent. Its formation provides the basis of convenient linked assays of several enzymes as well as assay of anthranilate synthase itself (8).

$$\text{chorismate} + \text{glutamine} \xrightleftharpoons{\textit{anthranilate synthase}} \text{anthranilate} + \text{pyruvate}$$

Conditions for assay are 20 mM L-glutamine, 10 mM $MgCl_2$, 0.1 mM chorismate, 25 mM mercaptoethanol in 50 mM potassium phosphate, pH 7.4. The reaction is followed at excitation wavelength 325 nm and emission wavelength 400 nm. Anthranilate is used as standard.

iv. p-Aminobenzoate synthase

In this assay *p*-aminobenzoate (PABA), formed from the reaction of chorismate and glutamine, is extracted into ethyl acetate and its concentration determined by comparison with standard PABA extracted in the same way (9).

The following reagents are mixed in the volumes indicated: Hepes 0.6 M, pH 7.8, 50 μl; glutamine, 0.2 M, 100 μl; guanosine 0.2 M, 50 μl; chorismate 10 mM, 20 μl; $MgCl_2$, 0.2 M, 50 μl; solution containing 1.0 M EDTA, 60 mM mercaptoethanol, and 30% glycerol, 50 μl.

The reaction is started by adding enzyme and water in a total volume of 50 μl. The reaction is stopped by adding 100 ml of 1 M HCl, the PABA formed is extracted into 1.5 ml of ethyl acetate and its fluorescence measured (λ_{ex} = 290 nm, λ_{em} = 340 nm). Concentrations are determined using standard PABA solution.

4.4.2 Synthetic fluorigenic substrates

Just as nitrophenol and nitroaniline have been used extensively in the preparation of artificial chromogenic substrates for esterases and peptidases respectively, very many artificial substrates for hydrolases have been based upon the naturally occurring fluor umbelliferone

(7-hydroxycoumarin) and its amino analogue 7-amino-4-methylcoumarin. A consensus of the many papers describing these methods is that they are potentially at least two orders of magnitude more sensitive than the corresponding assays in which nitrophenol or nitroaniline are released. Methylumbelliferone can be reliably measured at concentrations of 10^{-10} M. Unfortunately, the useful fluorescent properties (λ_{ex} = 360 nm, λ_{em} = 455 nm) of methylumbelliferone reside in the anion, which is not formed at the acid pH values appropriate to the assay of many hydrolases. The enzyme-catalysed part of assays based on the release of this compound are therefore usually conducted at low pH. The reaction is then stopped and the fluorescent anion formed by the addition of a strong alkaline buffer. The p*K* for the conversion is 7.8 (10) so that this type of substrate can be used continuously for hydrolases such as alkaline phosphatase that act at alkaline pH values.

Many continuous fluorescent assays for peptidase substrates have been based on synthetic oligopeptides in which the scissile bond is an amide with 7-amino-4-methylcoumarin. There is sufficient difference between the fluorescence spectra of substrate and product that assays for peptidases can be made continuously over a range of pH values. However, there is significant overlap of the spectra and this means that the potential sensitivity of the fluorescence method is not fully realized. Despite conducting assays at excitation and emission wavelengths which are not maximal for the product, a significant fluorescence contribution from the substrate itself cannot be avoided. The assay of elastase with a fluorogenic peptide substrate of this type was found to be less sensitive than an absorbance assay based on thiopyridine (11). Attempts to overcome the overlap problem have been made by the introduction of fluors with different properties such as 6-aminoquinoline (12) and aminoacridone (13).

i. β-Glucuronidase

Assay of this enzyme using methylumbelliferyl glucuronide was first described in 1955 (14) and may be accomplished by adding the enzyme sample (10 μl) to 1 ml of 0.25 mM methylumbelliferyl glucuronide in 0.1 M sodium acetate, pH 4.6. After an appropriate period (1–30 min) the reaction is stopped by adding 4 ml of 0.1 M glycine /NaOH buffer, pH 10.4.

ii. Chitinase

Assay of this enzyme provides an example of the use of a more complex fluorogenic substrate, namely methylumbelliferyl-β-D-*N*,*N*′,*N*″-triacetylchitotriose. The enzyme is present in greatly increased quantities in the plasma of patients suffering from Gauchers disease (15). It may be assayed (15) at 37°C using the fluorogenic substrate (0.02 mM) dissolved in 100 μl 0.1 M citrate/0.2 M phosphate pH 5.2. After time intervals appropriate to the amount of enzyme used, the reaction is stopped by adding 2 ml of 0.3 M glycine/NaOH at pH 10.6. Any precipitate that has formed should be removed by centrifugation and the fluorescence determined (λ_{ex} = 360 nm, λ_{em} = 455 nm).

iii. Elastase

The synthetic peptide substrate methoxysuccinyl-Ala-Ala-Pro-Val-aminomethylcoumarin is reported to provide a sensitive assay for elastase (11). By appropriate choice of excitation and emission wavelengths (370 nm and 460 nm) more than 70% of the maximum fluorescence of aminomethylcoumarin can be used while the contribution of the unhydrolysed substrate contributes only 0.5% of its maximum fluorescence. The system is sensitive enough to detect 10 nM product and is linear to about 5 μM. Elastase may be assayed at pH 7.6 and 25°C in 50 mM Tris–HCl, 0.5 M NaCl, 0.1 M $CaCl_2$.

4.4.3 Relief of quenching

An ingenious method for assaying hydrolases relies upon a process known as radiationless energy transfer in which the energy from a fluorescent group on one part of a polymeric substrate is transferred without emission of light to a chromophore nearby in the same molecule. Hydrolysis of the susceptible bond interrupts the process by separating the two interacting groups. In one adaptation of this system, energy absorbed as light by excitation of the donor fluorophore at its characteristic excitation wavelength is transferred directly to the acceptor fluorophore and then emitted as light at a wavelength characteristic of the acceptor. The rate of hydrolysis is monitored either by measuring the decrease in fluorescence of the acceptor or the increasing fluorescence of the donor.

i. Carboxypeptidase A

The fluorescence of tryptophan is quenched by the dansylation of the amino terminus of the carboxypeptidase substrate glycyl-tryptophan to give dansyl-L-glycyl-L-tryptophan. Quenching is relieved by hydrolysis of the peptide bond. Enzyme is added to the synthetic substrate dissolved in 50 mM Hepes buffer, pH 7.5, containing 1 M NaCl. Fluorescence excitation is at 290 nm and emission is measured at 350 nm. At the start of the reaction the fluorescence at this wavelength is only 1% of that which results upon complete hydrolysis of the substrate (16).

ii. Aminopeptidase P

This enzyme hydrolyses the amino-terminal residue from peptides in which the next residue is proline. In the fluorogenic substrate (17) the fluorescent 2-aminobenzoyl group is linked by ethylene diamine to the carboxy terminus of the tripeptide NH_2-Lys-Pro-Pro-COOH. The quenching dinitrophenyl group is linked to the ε-amino group of the N-terminal lysine. A 160-fold increase in fluorescence emission is observed when the substrate is cleaved at the DNP-Lys-Pro bond. In the assay, 20 μl of enzyme sample is added to 0. 1 ml of 5 mM substrate in 0.2 M Tris–HCl, pH 8, containing 2.5 mM manganese sulphate and 10 mM trisodium citrate. The reaction is stopped after 20 min by adding 4.3 ml of a solution containing 1 mM dithiothreitol and 50 mM EDTA. Concentrations are determined by comparing increase in fluorescence (λ_{ex} = 320 nm, λ_{em} = 410 nm) with aminobenzoxyglycine standard.

iii. Phospholipase

Fluorescence is frequently quenched by a change in polarity of the environment of the fluorophore. The strong fluorescence of 6-carboxy fluorescein is quenched when it is incorporated into lecithin liposomes. A method for the assay of phospholipase is based on the disruption of the liposomes brought about by the hydrolysis of the phosphatidylcholine units from which the liposomes are composed (18).

4.4.4 Fluorogenic reagents

i. Amine oxidases

Enzymes generating hydrogen peroxide can be adapted to sensitive fluorometric assays using a continuous system in which homovanillic acid is converted to a fluorophore in the presence of horse radish peroxidase. As an example (19), diamine oxidase is assayed in 0.1 M phosphate, pH 7.8, containing 0.5 mM homovanillic acid and 10 μg/ml horseradish peroxidase. Enzyme in the form of tissue samples is first mixed and shaken at 37°C for 10 min with buffer and peroxidase before starting the reaction by adding homovanillic acid and 0. 1 mM putrescine as substrate. The reaction is followed continuously (λ_{ex} = 315 nm, λ_{em} = 425 nm).

References

1. *Methods in enzymology.* Academic Press, New York.
2. Bergmeyer, H. U. (1986). *Methods in enzymatic analysis,* Vol. 1–4. Verlag Chernie, Weinheim, Germany.
3. Burns, D. H. and Aberhart, D. J. (1988). *Anal. Biochem.,* **171**, 339.

4. Khalifah, R. G. (1971). *J. Biol. Chem.*, 246, 2561.
5. Plummer, T. H. and Kimmel, M. T. (1980). *Anal. Biochem.*, **108**, 348.
6. Rosenberg, R. M., Herreid, R. M., Piazza, G. J., and O'Leary, M. H. (1989). *Anal. Biochem.*, **181**, 59.
7. Bishop, D. F. and Desnick, R. J. (1986). *Methods Enzymol.*, **123**, 339.
8. Gozo, Y., Zalkin, H., Kein, P. S., and Heinrisburg, R. L. (1976). *J. Biol. Chem.*, **251**, 941.
9. Zalkin, H. (1985). *Methods Enzymol.*, **113**, 293.
10. Yakatan, G. J., Juneau, R. J., and Schulman, S. G. (1972). *Anal. Chem.*, **44**, 1044.
11. Castillo, M. J., Kiichiro, N., Zimmerman, M., and Powers, J. C. (1979). *Anal. Biochem.*, **99**, 53.
12. Byrnes, P. J., Bevilaqua, P., and Green, A. (1981). *Anal. Biochem.*, **116**, 408.
13. Baustert, J. H., Wolfbeius, 0. S., Moser, R., and Koller, E. (1988). Anal. *Biochem.*, **171**, 393.
14. Mead, J. A. R., Smith, J. N., and Williams, R. T. (1955). *Biochem. J.*, **61**, 569.
15 Hollak, C. E. M., van Weely, S., van Oers, M. H. J. and Aerts, J. M. F. G. (1994) *J. Clin. Invest.*, **93**, 1288–1292.
16. S. Latt, S. A., Auld, D. S., and Vallec, B. L. (1972). *Anal. Biochem.*, **50**, 56.
17. Holtzman, F. Pittey, G., Rosenthal, T., and Vaner, A. (1987). *Anal. Biochem.*, **162**, 476.
18. Chen, R. F. (1977). *Anal. Lett.*, **10**, 787.
19. Snyder, S. H. and Hendley, E. D. (1971). *Methods Enzymol.*, **17B**, 741.

Chapter 3
Radiometric assays

Kelvin T. Hughes
Screening Systems Department, Research & Development,
Amersham Pharmacia Biotech, Cardiff Laboratories, Forest Farm, Whitchurch,
Cardiff CF14 7YT, UK

1 Introduction

The accurate measurement of enzymatic activity in biological samples is important in many fields of cell biology, not only in routine biochemistry and in fundamental research, but also in clinical and pharmacological studies as well as in drug discovery and development.

For an enzyme assay to be 'fit for purpose', it should be specific, sensitive, quantitative and, ideally, simple and rapid to perform. It should also allow the assay of enzymatic activity in both crude and purified enzyme preparations.

Assay techniques that employ radioactively labelled substrates, commonly referred to as radiometric or radioenzymatic assays, have been successfully employed over many years to measure a range of enzyme activities. First used in the 1950s, the application of radiometric enzyme assays was initially restricted by, amongst other things, the limited commercial availability of suitable labelled substrates and scintillation counting instrumentation. The disappearance of these constraints in the 1960s led to a rapid increase in the use of this technology, and publications involving its use are now numbered in thousands.

Enzyme substrates labelled with a variety of isotopes such as ^{3}H, ^{14}C, ^{32}P, ^{33}P, ^{35}S and ^{125}I have been used successfully in radiometric enzyme assays. In most cases, the configuration of a radiometric assay will depend on the availability of a satisfactory quantitative method of separating labelled product from unreacted substrate. This is no longer a consideration with homogeneous radiometric techniques such as scintillation proximity assay (SPA) (Section 2.6).

Despite the fact that non-radioactive technologies such as fluorescence and luminescence for measurement of enzyme activities are now widely available, radiometric methods continue to enjoy widespread acceptance and utility for many types of enzyme measurements.

Because radioenzymatic assays have been so widely used for so long, it is not possible in the limited space available here to cover all assays currently in use, nor give detailed assay protocols for all such applications. I would strongly recommend that the reader refers to the original comprehensive review on radiometric assays written by my late ex-colleague Dr Ken Oldham in the first edition of this series (1), much of which remains highly relevant today.

In recent years, there has been widespread use of radiometric assays within the pharmaceutical industry, where there is a large and growing demand for screening assays in the early stages of drug development. The high sensitivity, specificity and freedom from interference of radiometric assays makes them ideally suited for such high throughput screening applications.

Chapter 11 reviews developments in the use of technologies such as SPA that have enabled drug screeners to achieve a 'quantum leap' in the number of enzyme assays that can be conducted in an automated/semi-automated fashion. The concomitant reduction in the 'hands-on' time required to run automated assays, as well as the reduced costs of reagents and consumables resulting from assay miniaturization, are very important considerations in such a high sample throughput environment.

2 Techniques

Over the years, many different radiometric assay methods have been employed for measuring enzymatic activities. In general, methods of monitoring enzyme activity are of two main types: continuous and sampling. In both methods either product formation or substrate utilization may be measured. Generally, unless the primary product is rapidly metabolized by other enzymes, enzymatic activity is monitored by estimating product formation. Radiometric enzyme assays are typically based on the conversion of radiolabelled substrates to labelled products.

Apart from homogeneous techniques such as SPA (see later section), the two major requirements of radiometric enzyme assays are the availability of a suitable labelled substrate and of a simple and rapid method of quantitatively separating product from unreacted substrate.

Published bibliographies are available (2) covering a plethora of different types of radiometric assays. Researchers wishing to configure their own assays are advised to carry out a literature search to identify related radiometric assays set up by others.

2.1 Ion-exchange methods

Both ion-exchange mini-columns and paper discs have been employed successfully for a wide range of applications. Commercially available columns containing a variety of sorbents, such as Amprep™ (from Amersham Pharmacia Biotech) and Sep-Pak™ (from Waters) are available together with detailed protocols covering a number of common separation methods. Vacuum manifolds, which permit the simultaneous and rapid processing of multiple samples, are also available from the above-mentioned suppliers. Researchers interested in using such devices should contact the manufacturers for details of published procedures and relevant application support literature.

As far as the use of ion-exchange papers is concerned, these have also been used extensively in radiometric assays of a wide range of enzymes. There are a number of suppliers who offer various papers/disks as well as 96-well microplates containing fitted filtration disks with a variety of membrane compositions (for example, those available from Millipore Corporation, Polyfiltronics, Inc., and Pall Corporation), suitable for handling large numbers of samples when used in conjunction with commercially available vacuum manifolds.

With the advent of appropriate instrumentation, radioactivity on filters can be readily counted using phosphor storage plate readers, such as the PhosphorImager (Amersham Pharmacia Biotech). Filtration membrane sheets can be mounted into a 96-well dot-blot manifold (Minifold II: Schleicher and Schuell, Inc.) for filtration and subsequently removed and exposed to a phosphor storage plate. For all these 96-well assay formats, care should be taken when using energetic β-emitters such as ^{32}P, due to the possibility of 'cross-talk' between adjacent samples. Non-specific binding of the labelled substrate can result in high blank readings, but there are a number of relatively straightforward ways of minimizing the contribution caused by this effect.

For an overview of the use of ion-exchange papers in radiometric enzyme assays, I would recommend contacting the manufacturers for relevant technical support information.

One particular application area that is of considerable research interest at the present time is protein phosphorylation/dephosphorylation. A whole range of tyrosine, serine and threonine kinase enzymes (and the corresponding phosphatases) are being isolated and investigated, due in part to the key roles they appear to play in many cellular processes such as cell growth and division. The practicalities of performing kinase assays are covered in more detail in Section 2.6.

2.2 Precipitation of macromolecules

Many enzymes involved in the synthesis of macromolecules such as DNA, RNA, amino-acyl tRNA, polypeptides, polysaccharides and related kinases have been assayed using predominantly ^{14}C and ^{3}H-labelled substrates or $^{32}P/^{33}P$-γ-ATP (for kinases) and measuring the radioactivity incorporated into a form insoluble in acidic or organic solvents.

Precipitation methods using the above and other isotopically labelled substrates (for example, employing ^{35}S and ^{125}I) have also proved useful in assays of a wide range of macromolecule-degrading enzymes.

An important family of enzymes currently the focus of much attention that are analysed by precipitation methods, following degradation of macromolecular radiolabelled substrates, are matrix metalloproteinases (MMPs). For example, collagenase activity is typically measured using ^{14}C- or ^{3}H-labelled collagen (3). A range of different acids and solvent systems has been effectively employed in such precipitation assays. Commonly used reagents include trichloroacetic acid (TCA), ethanol and ammonium sulphate. When synthetic peptide substrates are used in preference to higher molecular weight protein substrates, then TCA or similar precipitation methods are substituted by methods in which the modified peptides are captured onto, for example, ion-exchange discs.

A more detailed synopsis of the issues to be aware of when considering precipitation approaches in radiometric assays can be found in the chapter written by K. G. Oldham in the first edition of this book (1).

2.3 Solvent extraction methods

Solvent extraction methods have found widespread utility over the years for assaying a range of enzyme activities. Examples of important enzymes that are conventionally measured using solvent extraction include lipases and phospholipases such as phospholipase A_2, D and C, all of which cleave different membrane phospholipids to generate key cell signalling mediators such as eicosanoids and inositol phosphates.

2.4 Paper and thin-layer chromatographic (TLC) methods

This approach has been extensively applied to separate substrates and products in radiometric enzyme assays. An example of a class of enzymes that have traditionally been measured using this approach are phospholipases (4). Although TLC methods are not ideally suited to high sample throughput, this approach does allow easy detection of any degradation of substrate or product during the assay by other enzymes present in the enzyme preparation being used or by chemical reaction.

2.5 Electrophoretic methods

Radioactive tracer methods are commonly employed to assay polymerases and are also very amenable to nucleic acid processing enzymes. In general, polymerases are assayed by monitoring the incorporation of either ^{3}H or $^{33}P/^{32}P$-labelled mononucleotides into oligonucleotide products, or by the extension of 5′-end labelled primers (5, 6). Separations can be performed using denaturing gel electrophoresis. Similarly, radioactive endonuclease assays typically involve cleavage of end-labelled substrates, with separation of substrate and products by gel electrophoresis. Gel electrophoretic mobility shift assays are also widely employed for monitoring radiometric DNA/RNA modifying enzyme assays.

2.6 Scintillation Proximity Assay (SPA)

SPA is a proprietary, versatile, homogeneous technology that has been developed primarily for use within the pharmaceutical industry for screening large compound libraries for new drug candidates (7, 8). Most of the major pharmaceutical companies have access under licence to the technology from Amersham Pharmacia Biotech and employ this enabling high throughput screening technology in a wide range of assays including radiometric enzyme measurements.

The principle of SPA is that when a radioactive atom decays it releases sub-atomic particles such as electrons, and depending upon the isotope, other particles and various forms of energy such as γ-rays. The distance these particles will travel through aqueous solution is limited and is dependent upon the energy of the particle, normally expressed in MeV. SPA relies upon this limitation.

For example, when a tritium atom decays it releases a β-particle. If the tritium atom is within 1.5 μm of a suitable scintillant molecule, the energy of the β-particle will be sufficient to reach the scintillant and excite it to emit light. If the distance between the scintillant and the tritium atom is greater than 1.5 μm, then the β-particles will not have sufficient energy to travel the required distance. In an aqueous solution collisions with water molecules dissipate the β-particle energy and it therefore cannot stimulate the scintillant. Normally the addition of scintillation cocktail to samples containing radioactivity ensures that the majority of tritium emissions are captured and converted to light. In SPA the scintillant is incorporated into small fluomicrospheres. These microspheres or 'beads' are constructed in such a way as to bind specific molecules. If a radioactive molecule is bound to the bead it is brought in close enough proximity that it can stimulate the scintillant to emit light as depicted in *Figure 1*. The unbound radioactivity is too distant from the scintillant and the energy released is dissipated before reaching the bead and therefore these disintegrations are not detected.

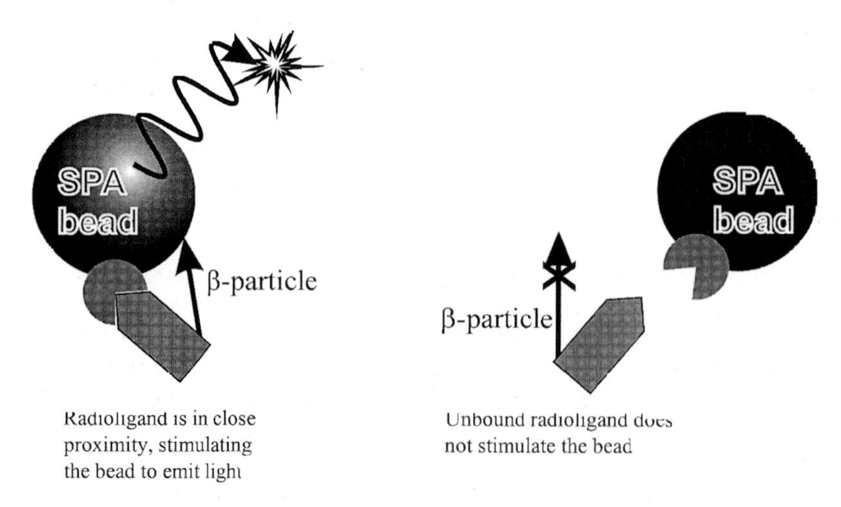

Figure 1 Diagrammatic representation of SPA (not to scale).

SPA microspheres have been developed from inorganic scintillators such as yttrium silicate (9) and hydrophobic polymers such as polyvinyltoluene (PVT). An optimized microsphere has been developed, consisting of a solid scintillant-containing PVT core coated with a polyhydroxy film that renders the bead more compatible with aqueous buffers. Coupling molecules, such as antibodies, are covalently attached to the coating, allowing generic links for assay design. These coupling formats are high-affinity biological molecule mediated linkages. No complex chemistry is required to achieve attachment of assay components to the SPA particles.

In terms of the type of radioisotopes employed, tritium is ideally suited for SPA assays as its β-particle has the extremely short path length through water of only 1.5 μm. This means that the background obtained from unbound tritium molecules is normally low, even when relatively large amounts of activity are used. Iodine-125 is another isotope that displays excellent properties for use in SPA. The ^{125}I atom decays by a process termed 'electron capture'. This type of decay gives rise to particles named Auger electrons and these electrons may be detected by SPA. Higher energy isotopes such as ^{35}S and ^{33}P have been used successfully with SPA provided certain relatively straightforward steps are taken to minimize the higher background counts that result from the so-called 'non-proximity effect'.

2.6.1 The application of SPA to enzyme assays

The catalytic action of enzymes can be determined by SPA. The basis of the majority of SPA enzyme assays involve the use of biotinylated substrates, which may either be immobilized on, or subsequently captured by, streptavidin-coated SPA beads. The biotin–streptavidin system is renowned for the strength of binding involved and, therefore, gives a reliable, reproducible and high-affinity capture system for use in SPA enzyme assays.

SPA enzyme assays have been developed for a number of enzyme classes including hydrolases, transferases, polymerases and kinases. The technique is applicable to ^{3}H-, ^{125}I-, ^{35}S- and ^{33}P-labelled substrates and, as with other assays, optimization is required.

The conversion of the substrate to product is monitored by designing the assay to either remove or add radioisotope with respect to the component that is captured on the SPA bead. The process can involve either the removal of radioactivity by the enzyme, resulting in a decrease in the SPA signal, or, conversely, the reaction may involve the addition of radioisotope causing an increase in the SPA signal. In all cases the discrimination of product from substrate does not require the components to be separated, as SPA is a homogeneous technique. This has the advantage in some instances that the incomplete recovery or detection of the product is not an issue. As the entire reaction takes place in one tube or well of a microtitre plate, there are no errors incurred by transfer and separation steps, which are traditionally employed for enzyme assays. SPA enzyme assays, therefore, show high precision and reproducibility when compared to methods such as precipitation and filtration. SPA is a powerful technique when large numbers of samples need to be assayed in a limited time frame. In this instance SPA may be considered an enabling technology for many enzyme assays. The removal of a laborious or cumbersome separation step means that SPA enzyme assays are fast, simple, precise and easy to automate. The enzyme reaction may be terminated by techniques such as a pH shift or the addition of a chelator of essential cofactors such as EDTA. In most instances the 'stop' reagent may be formulated with the SPA beads present. Therefore, the reaction can be stopped and the signal generated by a single pipetting step.

SPA is a solid-phase technology and the binding capacity of the bead surface is finite; therefore, the quantity of substrate that can be used is also finite. It is important to balance the quantity of bead required with the quantity of substrate, to obtain an adequate signal with a concentration of substrate which gives a kinetically competent assay. As in other assay techniques, this will invariably involve 'trade-offs' in assay volume, substrate concentration, signal obtained, and sensitivity, all of which will be specific for each individual assay.

When designing an SPA enzyme assay, two options are available: a solid-phase ('on'-the-bead) format or a solution phase ('off'-the-bead) assay format. The format selected depends to a large extent on the assay being developed and the intended application of that assay. For example, a solution-phase assay lends itself more to kinetic analysis compared with the solid-phase format.

The effect of SPA signal quenching caused by coloured samples, which will not be separated before the assay is counted, can be readily corrected for when using either conventional single-tube or microplate-based scintillation counting instrumentation.

2.6.2 Assay design

The fundamental aspects of designing an SPA enzyme assay are similar to those involved in traditional methods. It is, therefore, useful to consult the literature available for the enzyme of interest to ascertain the requirements for pH, ionic strength, cofactors and substrate specificity. In addition, to design the appropriate SPA assay, the substrate or product must be able to bind effectively to the SPA bead, so the inclusion of a biotinylation site may be necessary. This must not interfere with the activity of the enzyme. Another aspect to consider is a route to terminate the enzyme reaction.

In general, SPA enzyme assays are designed using the streptavidin-SPA bead. The biotin–streptavidin reaction is stable and rapid over a wide range of conditions and, therefore, provides the ideal capture system for application in SPA enzyme assays. Another strategy is to use SPA antibody-specific or protein-A beads to capture a reaction product using a specific antibody.

The key component in the design of an SPA enzyme assay is the substrate. If structure–activity studies have been performed on the enzyme of interest this information can be extremely valuable in designing the substrate. It is important to ascertain whether the biotinylation or the radiolabelling interfere with the kinetics of the enzyme action. This is normally determined by direct comparison of rates of activity in an SPA format and a traditional method such as HPLC. Biotinylation of substrates may be effected by a number of reagents, depending upon the moiety to be coupled.

The concept for a typical hydrolase-type signal decrease SPA enzyme assay is shown in *Figure 2* and enzyme dependence data from a model assay designed in this format (sphingomyelinase) are given in *Figure 3*.

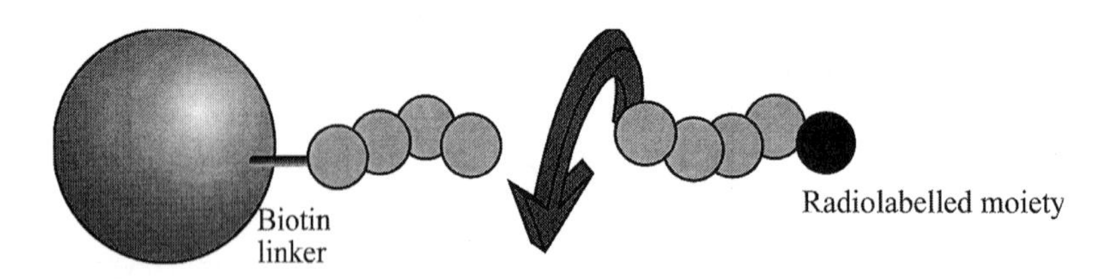

Figure 2 Diagrammatic representation of a signal-decrease SPA enzyme assay.

Similarly, an SPA signal increase enzyme assay format is shown in *Figure 4* and typical data from a model assay (farnesyl transferase) are illustrated in *Figure 5*.

An alternative approach is to use a product capture assay, whereby the radiolabelled product of an enzyme reaction is captured on an SPA bead, for example by a second antibody interaction. If an antibody is used for product capture it must be able to discriminate adequately between a small amount of product generated and the excess substrate present.

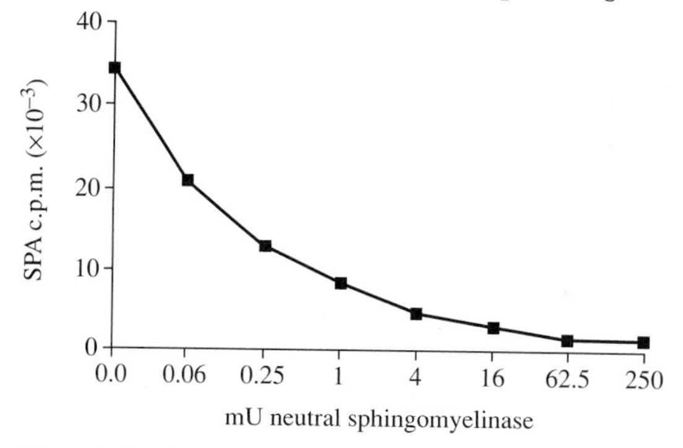

Figure 3 The determination of substrate cleavage by sphingomyelinase (human placenta) as a function of enzyme concentration. In this experiment, 0.06–250 mU of enzyme was added to 200 μg streptavidin-coated YSi beads and 50 nCi/0.6 pM substrate, and incubated for 30 min at 37°C. Results are ± SEM (n = 3).

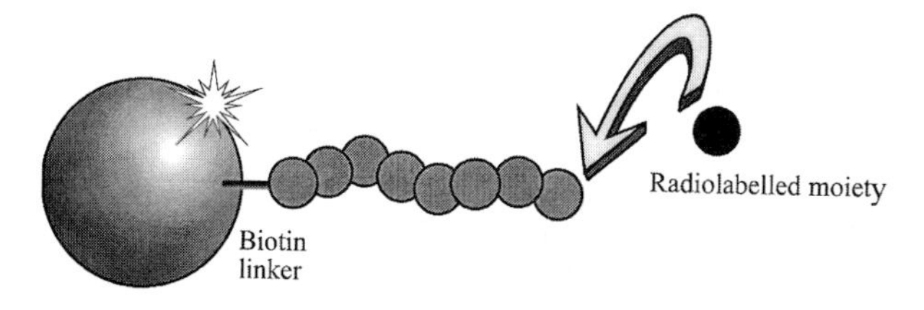

Figure 4 Diagrammatic representation of a signal-increase SPA enzyme assay.

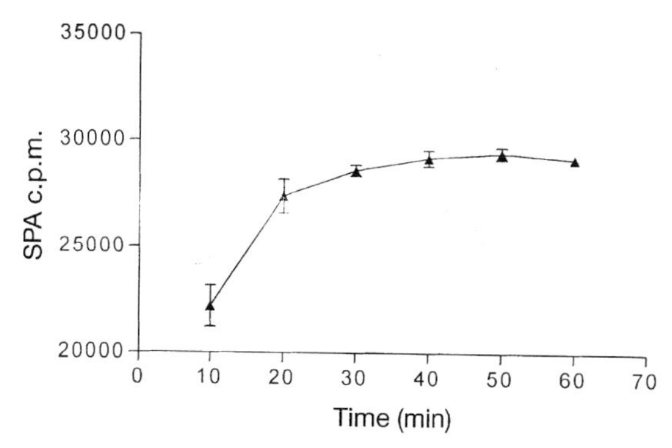

Figure 5 Time-course analysis of a farnesyl transferase SPA assay. This time-course analysis was performed using farnesyl transferase enzyme incubated with 4 μCi [^{3}H]farnesyl pyrophosphate and biotinylated human lamin B peptide substrate at 37°C. The reaction was stopped at time intervals by the addition of 0.13 mg of streptavidin-coated PVT SPA beads in 440 mM $MgCl_2$, 60 mM orthophosphoric acid, 0.03% (w/v) sodium azide, and 0.3% (w/v) BSA (final concentrations) Results are means ± SEM (n = 3).

The advantage of product capture assays is that they can be used to give signal increase assay formats for hydrolytic enzyme activities. However, the fact that there are multiple interactions (product–antibody, antibody–second antibody, etc.) involved in the assay may be an issue in screening applications. *Figure 6* illustrates the concept for such an assay.

In the case of a phosphodiesterase (PDE) SPA assay, the surface properties of the yttrium silicate particles are such that, under optimal buffer conditions, they are able to selectively capture the product of the enzyme reaction ([^{3}H]AMP). (See *Figure 7* for a typical time-course for measuring this important enzyme.

For most assays there is the option to design the assay in solution phase or as a solid phase assay.

2.6.3 Solid-phase versus solution-phase SPA enzyme assays

i Solid-phase SPA enzyme assays

In a solid-phase enzyme assay the substrate is immobilized onto a surface, in this case the SPA beads. The enzyme is then allowed to act upon this immobilized substrate. This format for assay design raises a number of points.

First, the process can no longer be assumed to behave with Michaelis–Menten kinetics for a number of reasons, primarily because the enzyme and the substrate are in effect (but not in reality) compartmentalized from one another. The substrate is concentrated at the surface of the bead, which occupies a finite, unchanging and particularly small volume in the assay tube. The enzyme, however, is assumed to be homogeneously distributed throughout the assay tube. Therefore, a substantial quantity of the enzyme will not be used in the reaction.

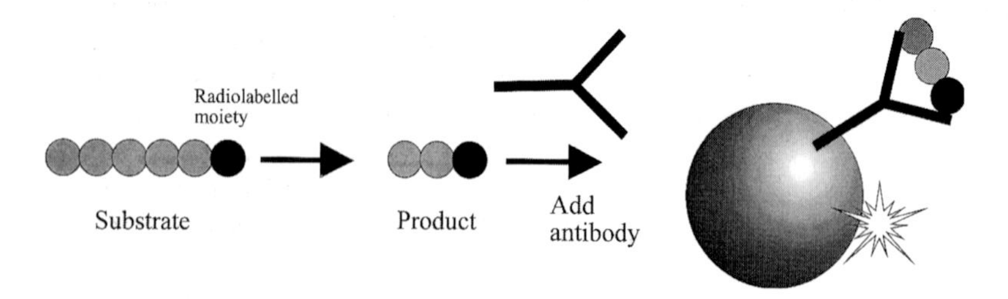

Figure 6 Diagrammatic representation of a product-capture SPA enzyme assay.

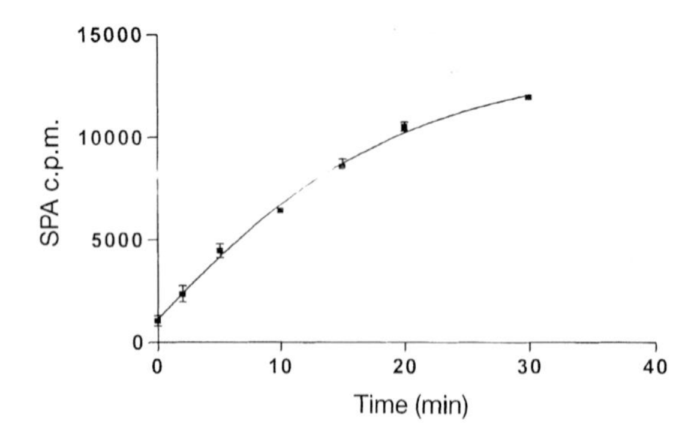

Figure 7 Time-course analysis of a phosphodiesterase SPA assay system. The time-course analysis was performed using human type IV phosphodiesterase incubated with 10 nM [^{3}H]cAMP at 30°C. The reaction was stopped at time intervals by the addition of 1 mg of underivatized YSi SPA beads in 6 mM zinc sulphate (final concentrations). Results are means ± SEM (n = 3).

Because the substrate is considered to be immobilized into a finite volume at the surface of the bead, it is not free to diffuse except 'en masse' with the movement of the bead. The movement of the bead cannot be considered to behave as the free substrate and, therefore, it is not possible to estimate the effective concentration of the substrate present in the reaction.

It is also important to consider whether the presence of the bead surface itself affects the kinetics. Interaction of the bound substrate with the bead could reduce the rate of association of enzyme and substrate. Alternatively, the bead itself may attract or repel the enzyme, thereby affecting the apparent association. These effects are important if enzyme kinetics are to be studied and, therefore, kinetic measurements should not be made on assays in a solid-phase format unless the system has been adequately characterized. However, the solid-phase assay does have an *in vivo* parallel, as a soluble protease acting on a membrane-bound substrate is analogous to a solid phase assay.

Figure 8 demonstrates a comparison of a solid-phase and solution-phase assay for interleukin converting enzyme (ICE). The substrate concentration has been adjusted so that the concentration at the surface of the bead is effectively saturating. The solution-phase assay is also performed at saturating substrate concentration. In this instance the free-solution-phase assay has a fast rate as all the enzyme present is able to participate in the reaction. In the solid-phase assay only a proportion of the enzyme added can participate in the reaction, and so although the substrate is saturating, the rate of the enzymatic reaction is zero.

At non-saturating substrate concentrations the time-course of the solid-phase assay can be faster at a given quantity of substrate, as all the substrate is concentrated at the bead surface and, therefore, the localized concentration is relatively high. In the solution-phase assay the same mass of substrate is distributed evenly throughout the assay volume and the concentration, and hence the rate, is lower. As an assay for drug screening, enzyme measurement or purification, the solid-phase format gives a fast, simple and reliable assay using less substrate.

One aspect that should be considered with solid-phase SPA enzyme assays is the non-specific binding properties of the bead on the assay components. It is essential that the labelled substrate or product does not stick to the bead directly. In a signal-decrease assay where the labelled substrate is linked directly to the bead through biotin, a number of interfering effects may be encountered. If the substrate binds to the bead without the biotin-streptavidin link it is likely that this substrate will not be in a conformation whereby it can be used by the enzyme. This means that only a proportion of the substrate may be cleaved

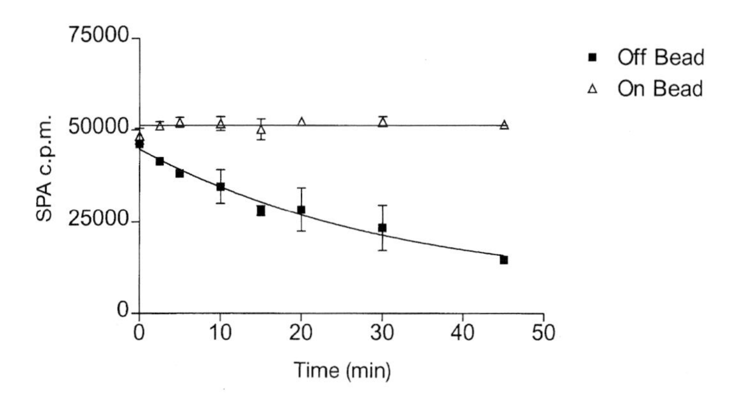

Figure 8 Time-course analysis of an interleukin-converting-enzyme (ICE) assay. The assay was run in 'off' and 'on' bead formats using 1 mg streptavidin-coated PVT SPA beads. 0.27 μCi [^{3}H]ICE substrate was incubated with 12 ng ICE and stopped at time intervals by the addition of 0.8 M orthophosphoric acid. Results are means $\pm$ SEM (n = 3).

from the bead and an artificially high baseline is established, which effectively closes down the potential signal-to-noise window. In addition, substrate that is not linked to the bead through biotin may be subsequently displaced by samples added to the assay, giving a false apparent rate of reaction. If the cleavage product of the reaction sticks to the bead then again an artificially high background will be apparent and a small assay window will result. These effects can normally be overcome by the presence of blocking agents such as detergents or BSA. However, if blocking agents are required it is important to ensure that these do not interfere with the reaction under study.

In signal increase assays similar effects should be considered. The unlabelled acceptor substrate must be immobilized to the bead via the biotin–streptavidin link to ensure it is available to participate in the reaction and that it is not susceptible to displacement by complex samples such as microbial broths. In addition, the radiolabelled donor substrate must not stick to the bead as this will produce a high 'no-enzyme' blank. Again, these effects can normally be overcome by the appropriate blocking agents provided these do not interfere with the reaction being studied. One advantage of performing assays in the solid-phase format is that once optimized and validated the assay can be performed in 'real time' without disturbing the process. This allows the possibility of monitoring the progress of the complete reaction by repeatedly counting one sample.

In summary, solid-phase SPA assays offer a rapid and convenient method to develop high throughput assays which need, in general, less substrate per assay. Properly validated, these assays can be optimized to detect inhibitors or measure relative rates of enzyme activity. However, not all enzymes perform satisfactorily in solid-phase SPA assays. This may be due to the shape of the active site or the size of the substrate used.

ii Solution-phase SPA enzyme assays

In a solution-phase assay all the components of the reaction are present in solution and the reaction is allowed to proceed in the absence of SPA beads. When the reaction has progressed to the desired conversion of substrate, the assay is terminated in the presence of the SPA beads. The products are then captured and the reaction rate may be measured. This format eliminates the concerns associated with interference by the bead.

The beads have a finite capacity to bind biotin and, therefore, the assay must be designed either with an excess of bead or a known amount of bead to capture a constant proportion of the product. This can be estimated from the biotin-binding capacity of the beads but it is advisable to determine the binding capacity empirically for each assay for a number of reasons. First, there are some instances where the rate of association of a biotinylated component may be much slower than that for free biotin. In addition, the structure and size of the biotinylated component may also cause the apparent binding capacity of the beads to be lowered. It is, therefore, preferable to determine empirically the quantity of beads required.

2.6.4 Non-specific binding (NSB) and non-proximity effect (NPE)

If an excess of bead is added at the end of an assay to capture all of the components, a large amount of bead may be present. This may result in a high background owing to NSB and/or NPE. The latter effect occurs when using high-energy isotopes such as ^{33}P and ^{35}S. In this situation, electrons from labelled assay components not bound to SPA beads can also can give rise to a stable and reproducible light signal, which will form part of the observed background counts in the assay.

Non-specific binding to the bead may occur with some components and may be distinguished from NPE by dilution of the bead–substrate mixture: NPE shows a dilution dependence whereas NSB does not. NSB is a quantitative phenomenon dependent upon the surface area of the bead, and therefore the effects are exacerbated if there is a need for large quantities of bead in the assay. It can often be overcome by the use of blocking agents such as BSA. NSB effects should be investigated using the unbiotinylated components to ensure capture by the bead is via the biotin–streptavidin system. Non-specific capture by the bead may result in variable performance, particularly if crude samples such as microbial broths are used, as these mixtures may displace labelled components, not linked through biotin, in a variable manner. It is also important to investigate the non-specific binding of the biotinylated products. The reaction product may vary substantially in its non-specific binding to the bead and as for the substrate it is important to ensure that all the product is captured by the biotin link and not through a non-specific interaction. Non-specific interaction of either substrate or product may cause apparent variability in the assay performance in either the blank or signal, or both, depending upon the assay format.

Both NPE and NSB effects can normally be overcome in the formulation of the 'stop' reagent. Designing the assay so that a substantial dilution occurs can reduce NPE. Formulating the beads in the 'stop' reagent with blocking agents can reduce the NSB.

2.6.5 Stopping the reaction

The assay termination buffer or 'stop' reagent can be a complex mixture depending upon the mode of termination. Normally a pH shift is an appropriate means of terminating the reaction rapidly; the reaction can be stopped immediately and does not drift appreciably with time.

Streptavidin SPA beads can be formulated into 'stop' reagents over a wide pH range without substantially affecting their biotin-binding function. Streptavidin SPA beads are capable of binding [^{3}H]biotin when formulated in buffers of pH 1.0–11.0. The stability of any other components in the 'stop' buffer mixture will need to be assessed.

Other methods of assay termination could involve sequestering of essential cofactors such as divalent cations. In general, the method of assay termination will depend on the type of enzyme being studied and the formulation will vary depending upon the assay format and the properties of the assay components.

2.6.6 Coupling strategies

In SPA-based enzyme assays, the coupling strategy of the substrate or product to the SPA bead is of paramount importance as this allows the measurement of the activity of the enzyme under investigation. The majority of enzyme assays developed by Amersham Pharmacia Biotech have used the biotin–streptavidin system as a robust and reliable coupling system. This uses streptavidin-coated SPA beads and a biotinylated substrate. In our laboratories we have also successfully employed the use of a specific antibody to enable product or fusion protein capture to the bead. Amersham Pharmacia Biotech's Bio Labelling Service is available to customers for the preparation and analysis of radiolabelled and biotinylated substrates.

There are many different ways of incorporating biotin into proteins, peptides, sugars and oligonucleotides.The required chemical derivatives are commercially available (e.g. Pierce & Warriner and Amersham Pharmacia Biotech). The choice of biotinylation reagent depends on the interaction of the substrate with the enzyme, as the biotinylated moiety must not interfere with the enzyme function. The suppliers of such reagents provide detailed protocols.

Whether the substrate for biotinylation is a protein or a nucleic acid, it is important that it is in a pure form with as few interfering substances as possible present in the reaction. If impurities in the substrate also become biotinylated, they will bind to the bead, which will result in a poor assay signal being obtained. For SPA enzyme assays, biotin insertion levels of one molecule of biotin per molecule of substrate are ideal. This allows for a minimal amount of bead to be used in the assay to capture the biotinylated substrate, giving a low background and a good signal-to-noise ratio.Examples of known enzyme assays that have been developed using SPA are given in *Table 1*.

2.6.7 Summary of SPA enzyme assay design procedure

As with the development of any enzyme assay, careful consideration of a number of key factors prior to commencement of practical work can save a considerable amount of time and effort. Some of these are highlighted below:

i Source of enzyme

The source of enzyme can be recombinant, purified from a suitable cell or tissue source, etc. It is important to ensure that the enzyme is free from contaminating activities and is in the correct molecular configuration for optimal catalytic activity.

ii Design of substrate

In terms of the substrate used in SPA enzyme assays, the researcher will need to consider and decide on factors such as the appropriate radiolabel, the preferred method of binding to the bead and whether to use a synthetic or naturally occurring substrate. Depending on the particular application, the substrate must also have the appropriate K_m, specificity, solubility and stability.

iii Signal increase or decrease assay

The choice of format is usually dictated by the nature of the enzymatic activity under investigation; for example, hydrolytic-type activities lend themselves to a signal decrease approach, whereas transferases typically can be formatted as signal-increase SPA assays. The preference for a specific format will also depend on the level of complexity of the assay system and the availability of reagents for capture to bead.

iv On or off bead

Benefits of an on-bead SPA assay approach include the use of less radiolabelled substrate, the requirement for less pipetting steps and it also allows 'real time' measurement of enzymatic activity. Off-bead assays are recommended if true kinetic analysis is required and if a direct comparison with other assay formats is needed. Unlike on-bead assays, which can suffer from steric hindrance problems, this is not an issue when SPA assays are configured in an off-bead format.

Table 1 Examples of enzyme assayed by the SPA technique

HIV integrase	Telomerase
HIV-1 protease	DNA polymerase
HIV-2 protease	Farnesyltransferase
CMV protease	DNA helicase
Endothelin converting enzyme	RNA helicase
Interleukin-converting enzyme	Geranyltransferase
Phospholipase A2	RNaseH
Phospholipase D	Cyclin-dependent kinase
Reverse transcriptase	Sphingomyelinase
MAP (mitogen-activated protein) kinase	Cholesterol ester transfer protein
Fucosyltransferase	Phosphodiesterase
Cathepsin G	Calcineurin phosphatase

See Sections 2.6.9 and 2.6.10 for further details.

v Optimization of incubation conditions

Again, as with any enzyme assay, careful optimization of buffer conditions, cofactor requirements, pH, time and temperature of incubation and concentration of components is important. With SPA being a bead-based technology, the requirement for agitation during the incubation period also needs to be evaluated, although this is seldom required in enzyme assays.

vi Selection of SPA bead type

As stated above, streptavidin beads are used predominantly in enzyme assays. However, other bead types such as Protein A, second-antibody-coated beads and those capable of capturing His-tagged substrates, have been used successfully to configure assays.

vii Optimization of the amount of SPA bead

The amount of bead employed in an assay will depend on a number of factors including whether total or proportional product capture is appropriate. It should always be borne in mind that background counts due to non-specific binding effects are likely to be greater the more bead is present in an assay.

viii Performance criteria

As SPA is predominantly employed in high throughput screening applications within the pharmaceutical industry, key performance parameters are assay drift, variation across and between microplates, and how the assay performs manually (hand pipetted) or robotically. Also, as SPA is a homogeneous technique and screening assays typically involve hundreds of microplates, the stability of signal is of paramount importance to ensure meaningful results.

ix Validation of an SPA assay

As with any type of radiometric assay, checking parameters such as K_{m}, K_{i}, and IC_{50} values and inhibitor profiles with known or published figures are good ways of ensuring an SPA enzyme assay is performing as expected.

2.6.8 Examples of signal decrease assays

The signal decrease assay format is suitable for hydrolytic enzymes that act at a single cleavage site on the chosen substrate. The substrate is designed with a site for bead binding and a radiolabelling site, and these are separated by a cleavage sequence for the enzyme. Most signal decrease assays can be designed as either solution-phase or solid-phase assays but not all are amenable to the solid phase format.

A simple concept for hydrolase activities was first demonstrated using HIV-1 proteinase (10). A peptide substrate was designed which was labelled with ^{125}I at the N-terminal tyrosine and biotinylated at the C-terminus. The 12 amino acid sequence used had a single cleavage site. The action of the proteinase separates the radiolabel from the biotinylated portion of the peptide. On termination of the assay, the biotinylated peptide is captured by streptavidin-SPA beads and the amount of cleavage determined. This assay format results in a signal decrease, which is generally perceived as a less sensitive assay strategy because the signal window may be small relative to the high background. Despite this limitation, Fehrentz and co-workers used this assay to determine comparative IC_{50} values for a series of statin-based peptides as inhibitors of the HIV-1 proteinase (11,12). Wilkinson *et al.* (13) further adapted the assay to perform high-volume screening of samples of serum for proteinase inhibitor levels in pharmacological studies on rats. The signal-decrease SPA format was compared with an HPLC method, a renin bioassay method and a protease assay linked to immunoassay of angiotensin I. All methods showed comparable performance in measuring Ditekiren (a peptide that exhibits dual renin and protease inhibitory activity) levels extracted from rat serum (13).

The signal decrease format has subsequently been applied to a number of proteinases for both high throughput screening and research assays. Brown *et al* used this approach with a tritiated β-amyloid peptide sequence to screen cathepsin G. These authors also used the SPA methodology in a signal decrease format to determine K_m and V_{max} values for the β-amyloid substrate (14). Furthermore, SPA has been employed in the development of an assay for the human cytomegalovirus UL80 proteinase (15). Other hydrolase activity assays such as RNase H, phospholipase A_2, phospholipase D and sphingomyelinase have been developed using biotinylated substrates that have been strategically radiolabelled to release the radioactive moiety due to the action of the enzyme of interest.

An SPA that allows Ca^{2+} /calmodulin-dependent serine/threonine phosphoprotein phosphatase 2B (calcineurin) activity to be analysed has been developed employing a biotinylated peptide containing a partial sequence of the regulatory subunit (RII) of the cyclic adenosine monophosphate (cAMP)-dependent protein kinase (16).

A signal-decrease SPA has been configured by Nare *et al* for the measurement of the important eukaryotic gene expression regulatory enzyme histone deacetylase, employing a biotinylated [^{3}H]acetyl-histone H4 peptide (17). The authors demonstrated that the SPA faithfully reproduced results obtained with the more traditional non-homogeneous organic extraction method.

2.6.9 Examples of signal-increase SPA assays

Signal-increase SPA formats have also used biotinylated acceptor molecules as substrates for ^{125}I- or ^{3}H-labelled donor substrates. By using biotinylated DNA primers to DNA or RNA templates and tritiated nucleotides, a number of assays have been constructed for DNA and RNA polymerases. In view of their attraction as therapeutic targets, the various assay methodological approaches for high-volume sample screening of viral polymerases and related proteins were recently reviewed by Cole (5).

Reverse transcriptase (RT) activity has been measured routinely using SPA (18). The assay has been reported to be useful in evaluating both chemical and natural product non-nucleoside inhibitors (19). Taylor *et al* (20) used an SPA RT assay to characterize the complete kinetic profile of a new and novel class of RT inhibitors, the inophyllums, while Cannon *et al* used SPA RT assays to measure the replication rate of mutant HIV-1 virus in cultured T-cell lines (21).

Another SPA transferase assay concept for farnesyl transferase was described by Santos *et al.* (22). In this assay, [^{3}H]farnesyl pyrophosphate was used as a donor substrate to a biotinylated human lamin B peptide sequence containing a single farnesylation site. The SPA assay has been reported to offer a rapid and more sensitive alternative to traditional TCA precipitation methods (23), and the simplicity of the assay technology enhances the automation capability for HTS. The assay was also used to perform detailed kinetic analyses on both the enzyme substrates and for short peptide inhibitors. In the same way, using SPA, the four ras proteins that are critically involved in cell signalling and differentiation have been examined as substrates for human farnesyl and geranylgeranyl protein transferases (24).

The measurement of cholesteryl ester transfer protein (CETP) has been assayed by SPA in two ways. Amersham Pharmacia Biotech has produced a commercially available kit which uses HDL-containing [^{3}H]cholesteryl ester as a donor substrate and biotinylated LDL as an acceptor molecule. Following incubation with purified preparations of CETP, the biotinylated LDL is captured with streptavidin-SPA beads and the transfer of [^{3}H]cholesteryl ester from HDL to LDL may be monitored. Coval *et al* used this system in HTS to identify the natural product wiedendiol as an inhibitor of CETP (25, 26). Lagros *et al* achieved a similar result using an anti-apolipoprotein B antibody coupled to a donkey anti-sheep SPA bead (27). This assay format allowed the measurement of CETP activity in total human serum. In addition, by using an anti-apo B antibody the transfer to both LDL and VLDL could be determined.

Sensitive, rapid screening assays have also been performed in signal addition formats by using product capture assay strategies. Takahashi *et al.* (28, 29) adapted a porcine lung endothelin-1 receptor assay (30) to capture and quantitate levels of endothelin for studies on the endothelin-converting enzyme (ECE). Endothelin production from [^{125}I]big endothelin was monitored by capturing the product using the receptor bound to wheat germ agglutinin (WGA) SPA beads. This novel approach was successfully applied during the purification of ECE from rat lung. Kinetic parameters and pH dependence were also determined. The assay was subsequently used to monitor phosphoramidon-sensitive activity during cloning and expression of the enzyme from rat endothelial cells (29, 31).

Norey *et al.* used a biotinylated polynucleotide probe to capture complementary ^{3}H-labelled oligonucleotides. This concept was successfully applied to configure a DNA helicase assay (32). A radiolabelled oligonucleotide was bound to an M13 plasmid and the action of bacterial or viral helicases resulted in an ATP-dependent unwinding of the duplex. The unwound [^{3}H]oligonucleotide was subsequently captured with the biotinylated probe and quantified by the addition of streptavidin-SPA beads. A similar approach has been employed to configure an assay for detecting hepatitis C virus RNA helicase activity (33).

Other DNA-modifying enzymes that have been assayed by SPA include telomerase (34, 35), HIV integrase (36, 37) and topoisomerase (38, 39); the last assay monitors the binding of ^{3}H-labelled supercoiled plasmid DNA to *E. coli* topoisomerase I immobilized on streptavidin SPA beads.

With the current high level of interest in protein kinases as therapeutic targets, many screening assays for these key enzymes are being developed using SPA (40–42). The practicalities of conducting an SPA for such enzymes compared with the more conventional filtration approach are given below (see *Protocols 1* and *2*).

2.6.10 Practical examples of SPA enzyme assays: protein kinases

Protein kinases are a class of enzymes that transfer phosphate groups onto specific recognition sequences on proteins, thereby regulating their function (43). These enzymes are involved in a wide variety of cellular responses including cell growth , cell differentiation and inflammation, and are classified according to their functional properties and their location within cells (44). These cellular mechanisms are important sites for therapeutic intervention (45).

Phosphorylation events are triggered by extracellular signals and lead to the enzymatic amplification of the initial signal via signalling cascade mechanisms involving many different kinases. This can be through second messenger systems such as cAMP, or by a variety of kinases that phosphorylate specific substrates; kinases can also undergo autophosphorylation. Kinases transfer the γ-phosphate group from ATP to a hydroxyl group of an acceptor amino acid, which can be serine, tyrosine or threonine, with the subsequent release of ADP. They can be studied *in vitro* and *in vivo* allowing detection, activity, and inhibitor studies and kinetic data to be generated by the inclusion of radiolabelled ATP. This allows the transfer of the labelled γ-phosphate group to be incorporated into the substrate. Phosphorylation is detected by methods using either ^{32}P- or ^{33}P-ATP labelled in the gamma phosphate group.^{32}P was the label of choice but it is a high-energy beta emitter (1.709 MeV) and has since been superseded by ^{33}P, which has a lower emission energy (0.249 MeV) and thus does not require such rigorous safety procedures.

To study kinases a simple approach is to use specific substrates made from consensus sequences recognized by the enzyme or proteins containing that sequence (for example, myelin basic protein); alternatively, generic or specific peptides can be used. Chemical modifications can also be applied to the substrate to assist in the detection of that substrate.

Once phosphorylated, the substrate can be captured using binding filter papers or SPA (see *Protocols 1* and *2* for experimental details).

In the filter binding method, following incubation the reactants are spotted onto the filter paper, which is usually phosphocellulose. Excess label is removed by various washing strategies, the filters are dried, placed into glass vials, scintillant is added and the vials are counted; any substrate that is labelled with ^{32}P or ^{33}P will produce a signal.

The substrate is captured onto the membrane through weak electrostatic forces, consequently the efficiency of the system can be improved by using a biotinylated substrate and filter papers coated with streptavidin (for example, the SAM^2 Biotin Capture Membranes available from Promega).

Additionally, solid phase kinase assays have been described using streptavidin-coated scintillating microplates (FlashPlates™) available from New England Nuclear (45) and ScintiStrips™ supplied by EG & G Wallac, Oy (46).

Protocol 1

Kinase assay using a filtration separation method

Equipment and reagents

- Binding papers (Whatman p81 code 3698915)
- Assay buffer (75 mM HEPES, 270 μM sodium orthovanadate, pH 7.4)[a]
- Diluent for label (1.2 mM ATP, 80 mM $MgCl_2$, 33 mM HEPES, pH 7.4)
- Stop solution (150 mM orthophosphoric acid)
- Wash solution (75 mM orthophosphoric acid)
- [γ-^{33}P]ATP (e.g. Amersham Pharmacia Biotech, codes AH9968/PB10132) at 200 μCi/ml[b]
- Substrate, e.g. peptide SLP 76 (sequence: biotin-SFEEDDYESPNDDQKKK), diluted if necessary in assay buffer
- Kinase enzyme (recombinant or derived from lysed cells or tissue preparation, e.g. ZAP 70 kinase involved in T-cell receptor upregulation)[c]
- Rackbeta™ liquid scintillation counter (Wallac)

Method

1. Pipette 10 μl assay buffer into each well or tube (buffer will depend on the kinase being assayed), inhibitor or inhibitor solvent as a control.
2. Pipette 10 μl substrate into each well or tube.
3. Pipette 5 μl recombinant enzyme or cell lysate into each appropriate well or tube (concentration will be dependent on the specific activity of the kinase).
4. Start the reaction by adding 5 μl magnesium [γ-^{32}P]ATP solution.
5. Incubate for 1 h at 30 °C. During the incubation period prepare a solution of 75 mM orthophosphoric acid to wash the papers.
6. Terminate the reaction by adding 10 μl 'stop' reagent.
7. Separate phosphorylated substrate as described below.
8. Aliquot up to 30 μl of the terminated reaction mixture onto numbered squares of binding paper. Allow the solution to soak completely into the paper.
9. Place the binding papers into 75 mM orthophosphoric acid, or 1% (v/v) acetic acid. Use at least 10 ml of this wash reagent per paper. Leave for 10 min with intermittent gentle mixing.

Protocol 1 continued

10 Decant the wash solution and dispose of the liquid waste (use an appropriate disposal route). Add a similar volume of wash reagent and leave the papers for a further 10 min with gentle mixing.

11 Wash the papers twice using distilled water, then air dry.

12 When dry, place the papers into individual scintillation vials. Add 10 ml scintillant and count using a scintillation counter.

[a] Typical assay buffer (note the buffer is dependent on the particular kinase being assayed; e.g. some kinases have an absolute requirement for Mg^{2+} or Mn^{2+} ions).

[b] Magnesium [γ-^{32}P]ATP (^{33}P can also be used). Dilute [γ-^{32}P]ATP to 200 μCi ml^{-1} using 1.2 mM ATP, 80 mM $MgCl_2$, 33 mM HEPES, pH 7.4.

[c] Cell samples from tissue culture should be lysed and homogenized in a buffer containing protease and phosphatase inhibitors. Cells may be lysed in 10 mM Tris, 150 mM NaCl, 2 mM EGTA, 2 mM DTT, 1 mM orthovanadate, 1 mM PMSF, 10 μg ml^{-1} leupeptin, 10 μg ml^{-1} aprotinin, pH 7.4, measured at 4 °C. Cellular debris should be pelleted at 25 000 **g** for 20 min and the supernatant retained.

Calculation of results

The ^{32}P incorporated into the peptide is quantitatively measured by the binding papers. In the presence of enzyme, the ^{32}P counted on the papers is the sum of non-specific [^{32}P]ATP binding, specific binding of phosphorylated peptide and binding of phosphorylated proteins in the cellular extract (A).In the absence of enzyme the ^{32}P counted on the papers is the non-specific binding of [^{32}P]ATP or its radiolytic decomposition products (B). Kinase activity is therefore obtained from (A − B)

Calculation of specific activity (*R*) of 1.2 mM Mg[^{32}P]ATP

5 μl of 1.2 mM Mg[^{32}P]ATP contains 6×10^{-9} moles ATP

$$R = \frac{\text{c.p.m. per 5 } \mu\text{l Mg[}^{32}\text{P]ATP}}{6} \text{ c.p.m. nmole}^{-1}$$

Calculation of total phosphate (*T*) transferred to peptide and endogenous proteins

30 μl spotted on to binding paper

Total terminated volume 40 μl

$T = [(A) - (B)] \times 1.33$

Calculation of pmoles phosphate (*P*) transferred per minute

$$P = \frac{T \times 1000 \text{ pmoles min}^{-1}}{I \times R}$$

where I = incubation time (min)

Calculation of % inhibition

$$\frac{(\text{non-inhibited control c.p.m.} - \text{sample c.p.m.}) \times 100}{\text{non-inhibited control c.p.m.}} = \%\text{ inhibition}$$

Protocol 2

Kinase assay using SPA

Alternatively, the phosphorylated kinase reaction product can be captured using streptavidin-coated SPA beads.

Equipment and reagents

- Streptavidin-coated SPA beads (Amersham Pharmacia Biotech code RPNQ0006/7) reconstituted in phosphate-buffered saline (PBS) to 50 mg ml^{-1}
- ×10 Kinase reaction buffer; 500 mM MOPS pH 7.2, 10μM ATP, 50 mM $MgCl_2$ and 25μM biotinylated myelin basic protein substrate
- 'Stop' solution (×10) 500 μM ATP, 10 μM EDTA,1% (v/v) Triton™ X100 in phosphate-buffered saline
- $[^{33}P]/[^{32}P]\gamma$-ATP (as *Protocol 1*)[a]
- Kinase enzyme, e.g. extracellular signal-related kinase (Erk-1) available from Upstate Biotechnology (0.25–1.0 μg per well or tube)[b]
- Substrate, e.g. biotinylated myelin basic protein
- 96-well microplates
- Microplate or single-tube scintillation counter (MicroBeta™ Trilux and RackBeta™ Wallac Oy, TopCount™, Packard Instruments)

Method

1. To each well or tube add: 5 μl 10× kinase reaction buffer (the choice of buffer will depend on the kinase used); *x* μl diluted $[\gamma\text{-}^{33}P]$ATP (equal to 0.5 μCi, as calculated below); 10 μl test compound inhibitor sample or suitable control solvent; Analytical grade water to a final volume of 40 μl.
2. Dispense enzyme into wells or tubes in a total volume of 10 μl.
3. Dispense 40 μl assay mix into each well or tube. If using a microplate, cover with a tight-fitting lid, but do not seal.
4. Incubate for 30 min at 37 °C.
5. Stop the reaction by adding 200 μl of 'stop' buffer/bead mix (containing a mixture of 20 μl 'stop' solution; 10 μl bead suspension and 170 μl PBS) per well or tube if the beads are to be settled by gravity or by centrifugation, or 75 μl (20 μl 'stop'; 10 μl bead suspension and 45 μl PBS) if the beads are to be reverse-settled by caesium chloride (i.e. floated). Add 125 μl caesium chloride solution (recommend 80% w/v) to each well or tube to be reverse settled. Microplates may be sealed at this point.
6. Incubate at room temperature for at least 10 min.
7. If microplates are being used, allow beads to settle for at least 10 h prior to counting or pellet the beads using appropriate centrifuge equipment(10 min at 120 **g**).
8. If plates are reverse-settled, incubate at room temperature for 1 h.
9. If tubes are being used, pellet the beads in a microfuge at 1000 **g** for 5 min. If beads are reverse-settled, incubate at room temperature for 1 h.
10. Count using an appropriate microplate or single-tube counter (see *Protocol 1*).

[a] The quantity of $[\gamma\text{-}^{33}P]$ATP needed must be calculated. The example below is sufficient for one 96-well plate or 96 tubes. Scale-up for larger numbers, if required. The quantity of $[\gamma\text{-}^{33}P]$ATP required for each well or tube is 0.5 μCi. Total quantity for one plate or 96 tubes = 50 μCi

[b] For other kinases the concentration of enzyme used should be determined empirically.

The isotope is normally supplied at a radioactive concentration (RAC) of 10 mCi/ml (10μCi/μl) at the reference date. From the accompanying data sheet, calculate the actual volume needed for 50 μCi on the date of use. For example, at 4 days past the reference date, the actual RAC is 89.7% of the reference RAC.

Reference RAC = 10 μCi/μl

Reference (+ 4 days) RAC = 8.97 μCi/μl

Therefore for 50 μCi, 5.57 μl of the stock $[\gamma\text{-}^{33}P]$ATP is required.

Dilute 5.57 μl of the stock $[\gamma\text{-}^{33}P]$ATP 100× in analytical grade water. The new RAC is then 50 μCi in 557 μl. Dispense 0.5 μCi, 95.57 μl per well or tube.

2.6.11 Imaging of SPA enzyme assays

The move towards ultra-high throughput SPA assays within the pharmaceutical industry has increased the pressure on conventional plate-counting technology and reagents. As plates have increased in well density, so counting times have increased. Typically, scintillation counters can only read up to 12 wells of a 384-well plate at any one time, giving count times of up to 40 minutes per plate. Imaging devices such as the LEADseeker™ developed jointly by Amersham Pharmacia Biotech and Imaging Research Inc. comprise a charge-coupled device (CCD) camera and a set of new SPA-bead reagents. This system is capable of imaging all wells of a plate in a single pass and thus has the potential to reduce count times. In addition, the new bead types have emission characteristics that greatly reduce the effect of colour quenching on signal (48).

2.6.12 LEADseeker bead types

As standard photomultiplier tubes (PMTs) are most sensitive to light in the blue region of the emission spectrum, SPA beads were developed to emit light in this region, at 400 to 450 nm. However, CCD chips are more sensitive to light in the red region of the spectrum than the blue, and Amersham Pharmacia Biotech, using proprietary technology, has therefore developed new bead types which emit light in the ~ 600 nm region and are optimized for use with the LEADseeker. Another advantage of these beads is that the quenching effect of yellow and red compounds is eliminated.

There are two core bead types, one based on polystyrene (PS), the other on yttrium oxide (YOx). Both types have an emission spectrum with a peak at 615 nm and exhibit a much higher light output than SPA beads. Each may be derivatized by covalently linking molecules such as streptavidin to the bead in the same way as with SPA beads.

2.6.13 Well miniaturization

Since the early 1990s, SPA has been successfully applied to many drug-screening applications. A range of enzyme activities have been measured using this enabling technology. Assays are routinely performed in 96-well microplates, but for greater throughput, plates with 384 wells and also 1536 wells are increasingly being used.

There are a number of advantages of assay miniaturization, including reduction in both plate and reagent usage and increased throughput, particularly if the process is automated. Imaging allows throughput to be further increased by capturing the signal from all wells of a plate in a single exposure of not more than 10 minutes. A number of SPA assays have therefore been miniaturized to 384-well format and subsequently optimized for use with the LEADseeker.

2.6.14 Imaging assay for mitogen-activated protein kinase

The mitogen-activated protein (MAP) family of kinases are important components in signal transduction pathways (49, 50). The imaging assay relies on the enzyme-catalyzed incorporation of ^{33}P from [γ-^{33}P]ATP to biotinylated myelin basic protein (bMBP), which is subsequently captured by streptavidin-coated PS or YOx imaging beads. The close proximity of bound ^{33}P to the scintillant-containing beads causes the emission of light, which is detected by the LEADseeker imaging system. The inhibitory effect of staurosporine (a non-specific inhibitor of protein kinases) on the [^{33}P]ATP phosphorylation of bMBP substrate by Erk-1 MAP kinase was determined using 384-well microplates using the conditions shown in *Table 2* (see *Protocol 3*).

Protocol 3
Imaging assay for MAP kinase

Equipment and reagents

- Biotinylated myelin basic protein substrate (bMBP)
- [γ-^{33}P]ATP (see *Protocol 1*)
- Assay buffer: 50 mM MOPS, pH 7.2 containing 25 mM $MgCl_2$
- 25 pM ATP
- LEADseeker CCD imager
- 'Stop' buffer: 500 μM ATP, 5 mM EDTA, 0.1% (v/v) Triton X-100 and 0.25 mg streptavidin-coated imaging beads in phosphate buffered saline (PBS)
- Erk-1 kinase enzyme
- Staurosporine
- 384-well microplates

Method

1. The reaction mixture contained 50 pmol biotinylated MBP, 0.1 μCi [γ-^{33}P]ATP, 25 pmol unlabelled ATP and 0.5 μg Erk-1 kinase in 50 mM MOPS, pH 7.2, containing 250 nmol $MgCl_2$. Staurosporine dissolved in DMSO was added to the assays to cover the range 0.001–100 μM. Assays were performed in 384-well plates and the reaction volume was 30 μl.
2. All assays were incubated at 37 °C for 30 min prior to the addition of a 'stop' buffer containing 500 μM unlabelled ATP, 5 mM EDTA, 0.1% (v/v) Triton™ X-100 and 0.25 mg LEADseeker beads in PBS. The final volume was 75 μl.
3. Plates were centrifuged at 1000 **g** for 10 min and then imaged for 10 min.

Table 2 MAP kinase assay reagents

Assay component	96-well (volume)	96-well (amount/conc.)	384-well (volume)	384-well (amount/conc.)
bMBP[a]	5 μl	125 pmol/50 pM	5 μl	50 pmol/25 pmol
[^{33}P]-γ-ATP	5 μl	0.5 μCi	5 μl	0.1 μCi
Enzyme	10 μl	1 μg	10 μl	0.5 μg
Assay buffer	20 μl	–	5 μl	–
Staurosporine	10 μl	0.001–100 μM	10 μl	0.01–100 μM
SPA bead	200 μl	0.5 mg	20 μl	0.25 mg

[a] biotinylated myelin basic protein

The effect of staurosporine on Erk-1 kinase activity was shown by a signal decrease as the concentration of staurosporine was increased. Images were obtained from assays using PS and YOx LEADseeker beads (*Colour Plate*).

Typical inhibition curves were obtained for staurosporine (*Figure 9*). IC_{50} values of 4.95 μM and 4.76 μM were obtained for PS and YOx imaging beads respectively.

3 Experimental design

The robustness of any assay depends to a large extent on the amount of effort spent in the design and optimization of the key parameters that influence enzyme rates and their measurement. Factors such as substrate concentration, purity, reaction volume and, in the

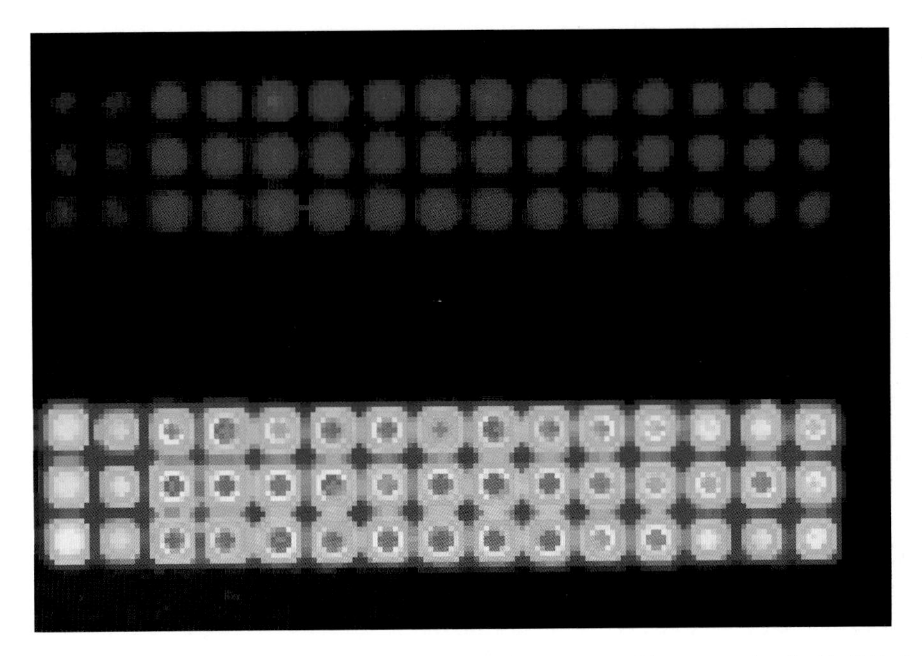

Plate 1 Inhibition of Erk-1 kinase activity by staurosporine. The assay conditions are as described in *Table 2* and *Protocol 3*, using polystyrene (PS) and yttrium oxide (Yox) LEADseeker beads. Assays were performed in triplicate. The image from the CCD camera represents the amount of light emitted from reagents contained within each well of the microplate. The varying light intensities are represented by different colours in the pseudo-colour image shown above, where blue depicts wells containing the least active components, and red those with the most activity. (NEC, no enzyme control, ± 10% DMSO; NIC, no inhibitor control, ±10% DMSO).

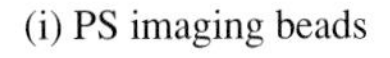

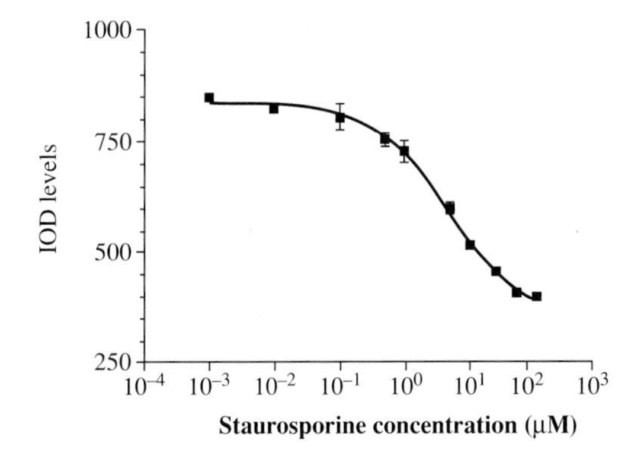

(ii) YOx imaging beads

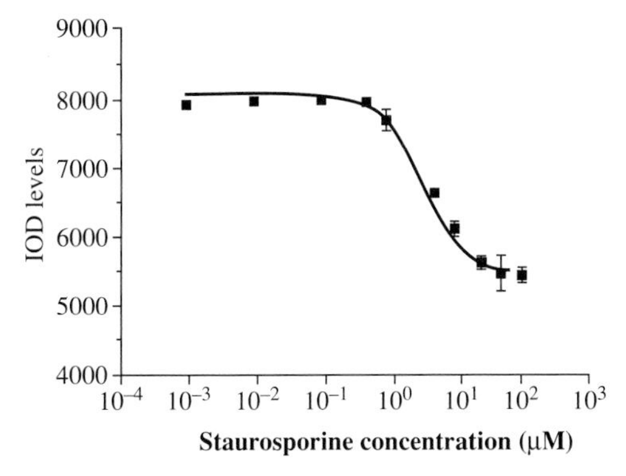

Figure 9 Typical inhibition curve for Erk-1 kinase by staurosporine. Assays were carried out using (i) streptavidin-coated polystyrene (PS) beads and streptavidin-coated yttrium oxide (YOx) beads, 0.5 μg Erk-1 kinase, 50 pM biotinylated myelin basic protein (bMBP), staurosporine in the range 0.001–100 μM, 25 pM ATP, and 0.1 μCi [^{33}P]ATP, as described in *Table 2* and *Protocol 3*. Values are means ± SEM (n = 3). Integrated optical density (IOD) levels equate to the light intensity from each well of the microplate.

case of radiometric assays, the choice of radiolabel and the optimal specific radioactivity of the substrate, all need careful consideration.

The various aspects of radiometric enzyme assay experimental design are covered in considerable detail in the first edition of this book (1).

4 Microplate technology

There is no doubt that micro-well plate devices have helped considerably in increasing sample throughput in enzyme assay measurements. During the 1990s, 96-well microplates have gained widespread utility in the semi and total automation of many radiometric enzyme assays. A host of vendors now supply a wide range of 96-well microplates that conform to a standard design 'footprint' (for example, Corning, Nunc , Dynatech), for microplate scintillation counters.

Due to the properties of some radiolabelled substrates, non-specific binding to microplates can give rise to high background assay counts. In order to minimize this effect, microplates are now available with chemically modified polymer surfaces. Corning's non-binding surface plates can significantly reduce non-specific binding and thus improve the observed signal-to-noise ratio in assays.

Within the pharmaceutical industry, advances in molecular biology, genomics and combinatorial chemistry have necessitated miniaturized assay formats for ultra-high throughput screening. In order to accommodate this, plates with more wells have become available that allow 384 and 1536 samples to be analysed. The parallel evolution of microplate scintillation counters and imaging devices has enabled radiometric assays to be readily configured in these formats.

5 Measurement of radioactivity

During the 1990s, the introduction of microplate scintillation counters helped facilitate the miniaturization of radiometric assays to 96- and 384-well formats. Both the MicroBeta (Wallac), which employs dual photomultiplier tubes (PMTs) and classical coincidence counting, and the TopCount (Packard), with a single PMT and a time-resolved counting approach, can be used to determine beta-particle labels such as ^{3}H, ^{35}S, ^{45}Ca, ^{14}C and ^{32}P or ^{33}P in conventional scintillation counting, as well as ^{125}I when used with SPA. Depending on the desired sample throughput, these instruments can be purchased with a variety of different detector versions (up to 12), which allow samples in a complete 384-well plate to be counted in approximately 45 minutes.

Imaging devices such as those mentioned in Section 2.6.12 are also becoming available. These allow even greater throughput of samples in radiometric enzyme assays.

6 Automation of assays

The need to assay more samples has increased considerably, particularly during the early stages of drug development within the pharmaceutical industry. It is not surprising, therefore, that in recent years much attention has focused on improving productivity and cost-efficiency by automating, or semi-automating, radiometric assays in 96 and 384-well microplate formats.

Homogeneous radiometric technologies such as SPA are amenable to automation using the wide variety of robotic systems that are currently commercially available (see Chapter 11 for an in-depth review of approaches to automating enzyme assays within the pharmaceutical sector).

Acknowledgements

This chapter is essentially a follow-on from the original review of this subject written by the late Dr Ken Oldham. I am therefore greatly indebted to him for this earlier treatise. As far as the work on SPA and LEADseeker technologies are concerned, I am extremely grateful to all my colleagues at Amersham Pharmacia Biotech, too numerous to mention here, who have developed the vast array of assay applications, some of which are covered in this report.

References

1. Oldham, K. G. (1992). In *Enzyme assays: a practical approach* (ed. R. Eisenthal and M. J.Danson), p. 93. Oxford University Press, Oxford.
2. Oldham, K. G. (1977). *Methods Biochem. Anal.*, **21**, 191.
3. Cawston, T. E. and Barrett, A. J. (1979). *Anal. Biochem.*, **183**, 340.
4. Blank, M. L. and Snyder, F. (1991). *Methods Enzymol.*, **197**, 158.
5. Cole, J. L. (1996). *Methods Enzymol.*, **275**, 310.
6. Kuchta, R. D. (1996). *Methods Enzymol.*, **275**, 241.
7. Bertogglio-Matte, J. H. (1986). US Patent No. 4,568,649.

8. Cook, N. D. (1996). *Drug Discovery Today*, **1** (7), 287.
9. Bosworth, N. and Towers, P. (1989). *Nature*, **341**, 167.
10. Cook, N. D., *et al.* (1991). In *Structure and function of the aspartic proteinases* (ed. B. M. Dunn), p. 525. Plenum Press.
11. Fehrentz, J. A., *et al.* (1992). *Biochem. Biophys. Res. Commun.*, **188** (2), 865.
12. Fehrentz, J. A., *et al.* (1992). *Biochem, Biophys. Res. Commun.*, **188** (2), 873.
13. Wilkinson, K. F., *et al.* (1993.) *Pharm.Res.*,**10** (4), 562.
14 Brown, A. M., *et al.* (1994). *Anal. Biochem.*, **217**, 139.
15. Baum, E. Z., *et al.* (1996). *Anal. Biochem.*, **237**, 129.
16. Sullivan, E., *et al.* (1997). *J. Biomol. Screening*, **2**, 19.
17. Nare, B., *et al.* (1999). *Anal. Biochem.*, **267**, 390.
18. Lemaitre, M., *et al.* (1992). *Antiviral Res.*, **17** (1), 48.
19. Kleim, J. P., *et al.* (1993). *Antimicrob. Agents Chemother.*, **37** (8), 1659.
20. Taylor, P. B., *et al.* (1994). *J. Biol. Chem.*, **269** (9), 6325.
21. Cannon, P. M., *et al.* (1994). *J. Virol.*, **68** (8), 4768.
22. Santos, A. F. and Cook, N. D. (1992) In *Proceedings from rapid functional screens for drug development*, p. 7. International Business Communication Southborough, MA.
23. Tahraoui, L., *et al.* (1993) In *Proceedings from 1st international conference on advanced pharmaceutical screening*, p. 5. Vienna, Austria.
24. Zhang, F. L., *et al.* (1997). *J. Biol. Chem.*, **272**, 10232.
25. Coval, S. J., *et al.* (1995). *Biorg. Med. Chem. Lett.*, **5** (6), 605.
26. Chackalamannil, S., *et al.* (1995) *Bioorg. Med. Chem. Lett.* **5** (17), 2005.
27. Lagros, L., *et al.* (1995). *Clin. Chem.*, **41** (6), 914.
28. Takahashi, M., *et al.* (1993). *J. Biol. Chem.*, **268** (28), 21394.
29. Shimada, K., Takahashi, M,. and Tanzawa, K. (1994). *J. Biol. Chem.*, **269** (28), 18275.
30. Berry, J. A., Burgess, A. J., and Towers, P. (1991.) *J. Cardiovasc. Pharmacol.* **17** (Suppl.7), S143.
31. Matsumura, Y., *et al.* (1992). *Life Sci.*, **51**, 1603.
32. Norey, C. G., *et al.* (1994) In *Proceedings of international forum on advances in screening technologies and data management*, p. 4.
33. Kyono, K., Miyashiro, M., and Taguchi, I. (1998). *Anal. Biochem.*, **257**, 120.
34. Savoysky, E., *et al.* (1996). *Nucl. Acid. Res.* **24** (6), 1175.
35. Savoysky, E., *et al.* (1996). *Biochim. Biophys. Res. Comm.*, **226**, 329.
36. Downes, M. J., *et al.* (1994). In *Proceedings of international forum on advances in screening technologies and data management*, p. 7.
37. Pernelle, C., *et al.* (1995) In *2nd European conference on high throughput screening*, Budapest.
38. Lerner, C. G., *et al.* (1996.) *J. Biomol. Screening*, **1**, 135.
39. Lerner, C. G. and Saiki, A. Y. C. (1996). *Anal. Biochem.*, **240**, 185.
40. Spencer-Fry, J. E., *et al.* (1997). *J. Biomol. Screening*, **2**, 25.
41. Antonsson, K., *et al.* (1999). *Anal. Biochem.*, **267**, 294.
42. Park, Y. W., *et al.* (1999). *Anal. Biochem.*, **269**, 94.
43. Woodgett, J. R. (ed.) (1994). *Protein kinases*. IRL Press, Oxford.
44. Krebs, E. G. (1994). *Trends Biochem Sci.*, **19**, 439.
45. Levitski, A and Gazit, A. (1995). *Science*. **267**, 1782.
46. Braunwalder, A. F., *et al.* (1996). *Anal. Biochem.*, **234**, 23.
47. Nakayama, G. R., Nova, M. P., and Parandoosh, Z. (1998). *J. Biomol. Screening*, **3**, 43.
48. Jessop, R. A. (1998). In *Systems and technologies for clinical diagnostics and drug discovery* (ed. G. E. Cohn), *Proc.SPIE*, **3259**, 228.
49. Davis, R. J. (1993). *J. Biol.Chem.*, **268**, 14553.
50. Nishida, E. and Gotoh, Y. (1993). *Trends Biochem. Sci.*, **18**, 128.

Chapter 4
High performance liquid chromatographic assays

Shabih E. H. Syed
Research Career Awards, MRC Head Office, 20 Park Crescent, London W1N 3BG, UK

1 Introduction

Liquid chromatography is a separation process in which a mixture is separated into its individual components followed by their detection with a suitable monitor. Optimization of the resolution of the separated components, as well as speed of separation, has been a major interest of many researchers for a very long time. It is well known that the above parameters are directly affected by the size and nature of stationary-phase particles. In conventional liquid chromatography, where gravity was usually used to pull the solvent or mobile phase through a column packed with a stationary phase, a lower limit on the size of particles was eventually reached beyond which flow under gravity completely diminished.

This raised the need for the development of pumps capable of generating high pressures. However, at the pressures generally operated in high pressure liquid chromatography (HPLC) (up to 5000 p.s.i.) conventional soft matrices, for example Sephadex and Sepharose, will collapse. These days, silica or polymer-based matrices are the most commonly used in HPLC; they are available in particle sizes as low as 3 μm and can additionally withstand high pressures. A typical HPLC separation may take between 5 and 30 min compared to several hours in the case of conventional liquid chromatography.

A whole range of stationary phases and instrumentation (types of detection, solvent delivery systems, and data processing) have now become available for enhancement of resolution, sensitivity, and rapid data analysis. Advancement in the geometry and volume of flow cells has considerably increased the sensitivity of detection.

It is the aim of this chapter to describe briefly the principles of the technique of HPLC, to give a detailed practical discussion of the instrumentation available, the various stationary phases, and the limitations of the technique. A major part of the chapter is then devoted to the application of HPLC specifically to enzymatic analysis. In this respect, examples of each class of enzymes will be given with complete practical details for carrying out the desired separation. It is hoped that this will provide a sufficient background for an intending user of HPLC to adapt the procedure to his/her particular enzymatic reaction.

2 Theory of HPLC

2.1 Introduction

The retention of a mixture of solutes to a stationary phase occurs as a result of different mechanisms, and thus it is a complex process. The elementary forces acting on the molecules are as follows:

- Van der Waals forces operate between molecules
- dipole interactions arise in molecules and result in electrostatic attraction
- hydrogen bonding interactions
- dielectric interactions resulting from electrostatic attraction between solute molecules and a solvent of high dielectric constant
- electrostatic interactions

2.2 Chromatographic parameters

2.2.1 Retention

Before considering the theory in more detail, the reader should become familiar with the basic parameters related to a chromatogram. *Figure 1* shows a typical HPLC chromatogram in which detector response versus time or elution volume is plotted. In *Figure 1*:

- t_0 is the time taken for unretarded solvent front or any components of the mixture to elute from the column
- V_0 is the void volume, which represents the sum of the interstitial volume between particles and the accessible volume within the particle pores
- t_R is retention time
- V_R is retention volume, i.e. volume passed during t_R

V_R and t_R can be related by the following equation:

$$V_R = t_R \times F \tag{1}$$

where F is the flow rate of mobile phase. The retention volume is directly related to the distribution or partition coefficient of solute (k) between stationary and mobile phases:

$$V_R = V_m + kV_s \tag{2}$$

where V_m is volume of mobile phase and V_s is the volume of stationary phase. In the process of moving through the column the molecules continually fluctuate between the mobile phase and stationary phase. t_R can therefore be divided into the time the molecules spend in the mobile phase (t_m) and in the stationary phase (t_s):

$$t_R = t_m + t_s \tag{3}$$

The capacity factor (k'), which is a common measure of the degree of retention, is given by the equation:

$$k' = (t_s/t_m) = (t_R - t_m)/t_m = (V_R - V_m)/V_m \tag{4}$$

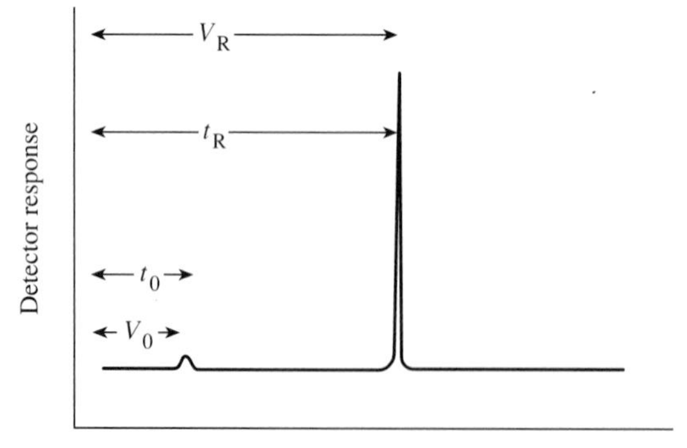

Figure 1 A typical HPLC profile showing various parameters defined in the text.

The capacity factor is related to the distribution coefficient by the following equation:

$$K = k' (V_m/V_s) \tag{5}$$

and

$$K = (C_s/C_m) \tag{6}$$

where C_s and C_m are the concentrations of solute in the stationary and mobile phases, respectively. The capacity factor is therefore the ratio of the mass of solute in stationary phase to the mass of solute in the mobile phase.

The ratio of capacity factors of two solutes 1 and 2 is called the separation factor, α, or sometimes the selectivity.

$$\alpha = (k'_2/k'_1) \text{ where } (k'_2 > k'_1) \tag{7}$$

2.2.2 Characterization of band broadening

On migration though a column, the individual zones of solute undergo dispersion or broadening as shown in *Figure* 2, which is a typical HPLC peak assumed to be Gaussian in shape.

W_2 is the width of the peak at its base, and W_1 is the width at half height. For any Gaussian peak,

$$W_2 = 2 \times W_1 = 4\sigma \tag{8}$$

where σ is the standard deviation of the HPLC peak. The efficiency of the chromatographic column can be determined using the theoretical plate by subdividing the column into many theoretical plates. In each plate, the solute molecules achieve an equilibrium distribution between the mobile phase and the stationary phase. The number of theoretical plates is both a measure of column efficiency and band broadening. N can be obtained from the equation:

$$N = (t_R/\sigma)^2, \text{ i.e. } N = (4t_R/W_2)^2 \tag{9}$$

The plate number can be easily derived from an HPLC chromatogram by measuring t_R and σ, as shown in *Figure* 2. The former is obtained with relative ease and good accuracy in the case of most modern machines. A second method, perhaps more accurate, involves the measurement of peak width at half height, W_1, given by:

$$N = 5.54 \times (t_R/W_1)^2 \tag{10}$$

The larger the value of N, the better the separation. N increases with smaller size of particulars, low flow rates, high temperatures, less viscous solvents, small solute molecules, and

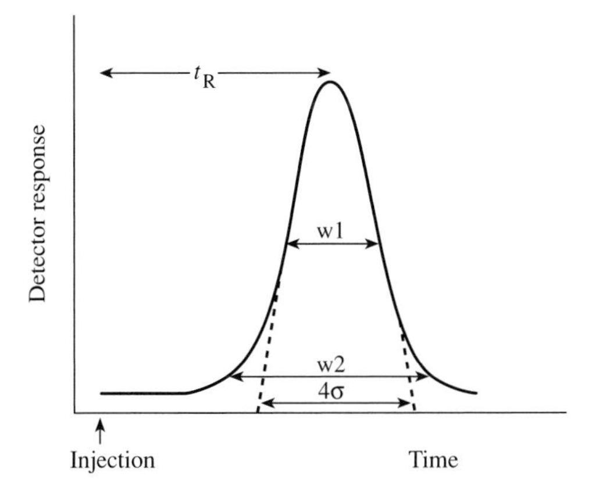

Figure 2 Terms describing the peak broadening.

good column packing. Since N is directly proportional to column length (L), it is more useful to measure height equivalent to a theoretical plate, H.

$$H = L/N \tag{11}$$

H is a measure of band broadening over a certain distance. It should be the aim of the user to use a large number of plates for a given length and, therefore, shorter plate height. HPLC columns are specified with the number of theoretical plates under the conditions of a given solute, temperature, flow rate, eluents, and so on. The column performance should be checked periodically by using the recommended solute or your own standard. The four main factors contributing to H, and therefore to band broadening, are eddy diffusion, longitudinal diffusion, variations in mass transfer, and excessive dead volume of the chromatographic system.

2.2.3 Resolution

The optimization process of chromatography involves balancing the maximisation of separation with the minimization of band broadening. The ratio of the two effects is expressed as resolution R_s, and is given by Equation 12:

$$R_s = 2(t_R{}^B - t_R{}^A) / W_2{}^A + W_2{}^B \tag{12}$$

Resolution can also be expressed in terms of the fundamental chromatographic factors, namely the selectivity, α, the capacity factor, k', and the plate number N. To obtain the best resolution, these factors can be optimized separately. It turns out that the selectivity factor is the most important in terms of resolution and can be modified by altering temperature or composition of the mobile phase or stationary phase. Change in the mobile phase can be most readily implemented in comparison to temperature and stationary phase.

3 Retention mechanism

3.1 Characteristics of silica

Nowadays, nearly all HPLC separations are carried out on chemically bonded phases where alkyl groups are covalently linked to the surface of silica, thereby overcoming the problem of phase stripping which occurs in the case of non-covalently-bonded phases. However, the presence of silica cannot be ignored. Commercially supplied silicas have differing physical properties including specific surface area, average pore diameter, specific pore volume and particle shape. *Table 1* lists these properties for some commonly used silicas.

Silicas used in HPLC have an average surface area of around 300 m^2/g, an average pore diameter of 10 nm, and therefore a specific pore volume of $\approx$1 ml/g. These properties allow the separation of solutes of molecular weight less than 5000 daltons. Whilst surface area affects the capacity and retentivity of the silica, pore volume affects its strength and robustness. The higher the surface area, the greater is its potential retentivity and capacity. The greater its pore volume the more brittle the material becomes. The particle size range is carefully controlled to avoid blocking of frits and meshes by fine particles and the disturbance of bed homogeniety by large particles. A tight distribution is essential to the production of a high performance column as determined by the number of theoretical plates/metre (*Table 1*). The spherical packing materials generally give columns of higher performance than those packed with irregular material due to improved particle uniformity. Generally, the silicas in aqueous suspension are acidic, having a pH value in the range of pH 4–5. The differences in surface pH of silicas influence the selectivity of the bonded phase in polar solvents and, in the case of non-polar solvents, the solute retention and column efficiency. As an example, the

Table 1 Properties of some commercial silicas

	Particle size (μm)	Surface area (m^2/g)	Pore diameter (Å)	Efficiency (1000 plates/meter)	Particle shape *
Exsil	3	200	100	100–130	S
Exsil	5	200	100	60–80	S
Exsil	10	200	100	30–40	S
Hypersil	3	170	120	100–130	S
Hypersil	5	170	120	60–80	S
Inertsil	5	320	150	60–80	S
Inertsil	5	350	100	60–80	S
Kromasil	5	340	100	70–90	S
Kromasil	7	340	100	45–60	S
Kromasil	10	340	100	30–45	S
LiChrosorb	5	300	100	50–70	I
LiChrosorb	10	300	100	20–30	I
LiChrospher	5	450	100	60–80	S
LiChrospher	5	650	60	60–80	S
Nucleosil	3	200	120	100–130	S
Nucleosil	5	200	120	60–80	S
Partisil	5	350	85	50–70	I
Partisil	10	350	85	30–40	I
Spherisorb	3	220	80	100–130	S
Spherisorb	5	220	80	60–80	S
Spherisorb	10	220	80	30–40	S

• I, irregular; S, spherical

separation of aromatic amines is significantly affected on using pH 7.2 and pH 9.0 silica with the same pore volume (1). Most silicas are stable in the pH range 2–8 only.

Silica on its own has been commonly used in normal phase chromatography where the stationary phase is either solid silica or a polar liquid phase coated on to silica particles. However, most biochemical analyses have not used this type of chromatography and, in any case, it suffers from the serious problem of phase stripping during gradient elution.

3.2 Polymeric packings

Polymeric matrices are based on cross-linked styrene-divinylbenzene or methacrylates. They do not have the same rigidity as the silica matrices, but are compatible with the typical pressures of HPLC. These resins are more hydrophobic than silica-based matrices and are therefore compatible for all mobile phases including the entire aqueous pH range. The compatibility with strongly acidic and basic eluents allows thorough cleaning of the packing. Due to the penetration of solvents and small molecules into the polymeric matrix, the mass transfer of small molecules is slower in polymeric supports than in silica-based packings. This leads to lower separation efficiencies for small molecules. For large molecules like proteins and synthetic polymers, the performance of polymeric matrices is comparable to silica-based supports. The polymeric packings are therefore extensivley used for the separation of natural and synthetic polymers using non-aqueous mobile phases. As with silica-based resins, a broad range of surface chemistries is available for reverse-phase, ion-exchange, hydrophobic interaction, or size-exclusion chromatography (*Tables 3, 5* and *6*).

3.3 Reverse phase chromatography (RPC)

In contrast to normal phase chromatography, in the case of reverse phase chromatography the stationary phase is non-polar while the mobile phase is polar and contains one or more non-polar organic modifiers. One of the huge advantages of RPC is the relative inertness of the silica beads where only the interactions between the solute and the covalently linked hydrophobic alkyl chain are possible. This allows the exploitation of a wide range of solvent effects through addition of salts or organic modifiers to the mobile phase, temperature and pH.

3.3.1 Preparation of reverse phases

The akyl groups are bonded by chemical reaction of alkylsilanes containing reactive mono-, di-, tri-chloro, or alkoxy groups with silanols on the surface of silica to give a new siloxane bond as shown in the diagram.

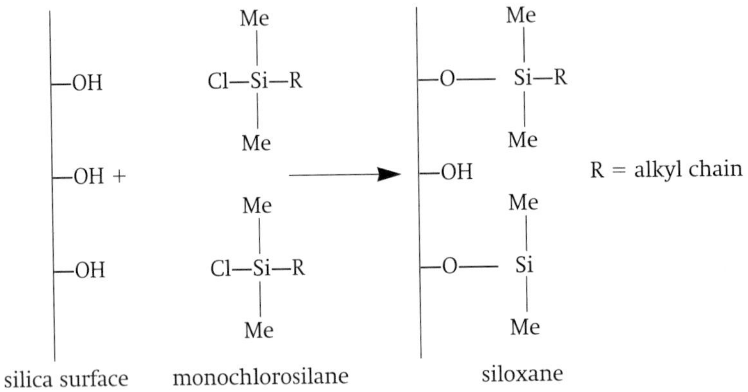

silica surface		monochlorosilane		siloxane	
–OH		Cl–Si(Me)(Me)–R		–O–Si(Me)(Me)–R	
–OH +			⟶	–OH	R = alkyl chain
–OH		Cl–Si(Me)(Me)–R		–O–Si(Me)(Me)	

The unreacted silane should be removed by reaction with trimethylochlorosilane after hydrolysis of the chloro group. This is called end-capping and is essential for good and reproducible separations. The surface coverage of silica can be calculated from the carbon content of the bonded phase determined by organic analysis (2, 3). Due to the colloidal properties of silica and the processes involved in its production, the concentration of unreacted silanols varies from batch to batch making an additional contribution to retention. This makes it difficult to compare separations carried out on columns purchased from different manufacturers.

In general the k' values of solutes increase with increasing carbon content/unit volume of column and with increasing chain length, provided the surface coverage is identical. Thus, a non-polar solute is retained to a greater extent on a C-18 than on a C-8 reverse phase. The former has a high selectivity for structurally homologous compounds, for example, ATP, ADP, CMP, and so on, while the separation of highly lipophilic compounds, for example, large peptides, is preferably carried out on C-8 reverse phase. *Tables 2* and *3* list some of the commonly used silica-based and polymeric packings, respectively. Phenyl and alkyl-nitrile groups are also included as reverse phases. However, it must be said that C-18 is by far the most commonly used reverse phase and the reader is referred to references (4–6) for a survey of its versatile applicability to separations of amino acids, peptides, proteins, vitamins, fats, steroids, antibiotics, nucleotides, and sugars.

3.3.2 Theory of reverse phase chromatography

Recent years have seen a rapid expansion in literature of separation mechanisms in RPC. The available experimental evidence suggests that the mechanism is not simple. However, it is

Table 2 Silica-based chemically bonded phases

Matrix	Chemical group	Pore diameter (Å)	Particle size (μm)	Preferred application
Hypersil SAS (C1)	Methyl	120	3, 5	Reverse phase chromatography. Highly lipophilic compounds
Hypersil C4	Butyl	100	5,10	Reverse phase chromatography. Moderately to highly lipophilic compounds
Nucleosil C8	Octyl	100 120	5 3, 5, 10	Reverse phase and ion-pair chromatography. Moderately to highly polar compounds such as small peptides and proteins,steroids, nucleosides, polar pharmaceuticals
Nucleosil C18	Octadecyl	100 120	3, 5, 10 3, 5, 10	Reverse phase and ion-pair chromatography. Non-polar to moderately polar compounds such as fatty acids, glycerides, fat-soluble vitamins, steroids, PTH amino acids, prostaglandins
Nucleosil C5H6	Phenyl	100 120	5 7	Reverse phase and ion-pairing chromatography. Moderately polar compounds. Retention characteristics are similar to C8 but with different selectivity for polycyclic aromatic polar and non-polar compounds, fatty acids
Nucleosil CN	Cyano (Nitrile)	100 120	5 7	Normal and reverse phase chromatography. In normal phase, the CN packing, due to its rapid equilibration, is much more suitable than unmodified silica for gradient separations of similar polar compounds using relatively non-polar solvents
Nucleosil NO2	Nitro	100	5	Separation of polycyclic aromatic compounds
Nucleosil NH3	Amino	100 120	5 7	Normal phase, weak anion exchange and reverse phase of polar compounds such as carbohydrates
Nucleosil OH	Diol	100	7	Normal and reverse phase chromatography. The diol packing is less polar than unmodified silica and very easily wettable with water. Separation of peptides and proteins

Hypersil, Kromasil, LiChrosorb, LiChrospher, Partisil, Spherisorb, μBondapak, Bondapak, Delta-Pak, Nova-Pak, Resolve, Symmetry300, Xterra RP and TSK-Gel also offer the range of matrices given in this table.

Table 3 Polymer-based reverse phase columns

Matrix	Features	Pore size (Å)	Particle size (μm)	Applications
Supelcogel ODP-50	C18 alkyl bonded to spherical beads of polyvinyl alcohol polymer Surface area: 150 m^2/g pH range: 2–13	250	5	High pH separations of basic drugs
TSKgel Octadecyl –2PW	C18 alkyl (monomeric) bonded to a methacrylate-based resin. pH range: 2–12. Exclusion limit: 100– 8000 Da	100–200	5	Pharmaceutical and environmental-research. Peptides and low MW oligomers
TSKgel Octadecyl –4PW	C18 alkyl (monomeric) bonded to a gel filtration packing, TSKgel G4000PW. pH range: 2–12; Exclusion limit: 1000–400 000 Da	500	7	Medium and high MW peptides and proteins and for peptides unstable at low pH
TSKgel Octadecyl –NPR	C18 alkyl (monomeric) bonded to a non-porous pelicular support. Able to withstand higher pressures than porous resins. pH range: 2–12 Exclusion limit: 1000–500 000 Da	NA	2.5	Rapid separations of medium and high MW peptides and proteins Micropreparative 10 ng–100 μg purifications and high pH applications
TSKgel Phenyl-5PW RP	C18 alkyl (monomeric, high coverage) bonded via an ether linkage to the gel filtration matrix, TSKgel G5000PW Exclusion limit: 10 000–1000 000 Da. pH range: 2–12	1000	10	High MW proteins. Use phenyl group to influence selectivity High pH applications

out of the scope of this Chapter to discuss these mechanisms in detail and the reader is referred to references (7–12) and references therein.

3.4 Influence of composition of mobile phase

In general, highest retention is obtained in pure water while water-miscible solvents such as methanol, acetonitrile, higher alcohols, dioxane, or tetrahydrofuran (THF) are used to enhance elution of a particular solute. The elution power increases in the order given due to decreasing polarity of the solvent and hence its increased retention on the reverse phase. The organic solvents have been arranged in order of increasing elution power in the so-called 'eluotropic series' given in *Table* 4. However, it must be pointed out that not all the solvents in *Table 4* are compatible with HPLC detectors.

η-Alkanes are generally used as mixtures in the more polar solvents such as methanol and acentonitrile. For water–methanol and water–acetonitrile mixtures, a linear decrease of log k' versus % of organic modifier is obtained.

As slight changes in eluent composition affect k' values significantly, caution should be exercised when preparing eluent mixtures. The composition of the mobile phase changes during the de-gassing procedure or on storage due to selective evaporation of volatile components. To exemplify the use of such mixtures to separate and elute solutes, *Figure* 3 shows

Table 4 Physicochemical properties of various solvents

Solvent	Density	BP	RI (n_D^{20})	Viscosity (cP, 20°C)	Polarity [e° (Al_2O_3)]	UV cut-off (nm)
Fluoroalkanes					–0.25	
n-Pentane	0.626	36.0	1.358	0.23	0.00	210
2,2,4-Trimethylpentane	0.692	98.5	1.392		0.01	210
Hexane	0.659	86.2	1.375	0.33	0.01	200
Cyclohexane	0.779	81.4	1.427	1.00	0.04	210
Carbon tetrachloride	1.590	76.8	1.466	0.97	0.18	265
Toluene	0.867	110.6	1.497	0.59	0.29	285
Benzene	0.874	80.0	1.501	0.65	0.32	280
Diethyl ether	0.713	34.6	1.353	0.23	0.38	220
Chloroform	1.500	61.2	1.443	0.57	0.40	245
Methylene chloride	1.336	40.1	1.424	0.44	0.42	240
Tetrahydrofuran	0.880	66.0	1.408	0.55	0.45	215
Methylethylketone	0.805	80.0	1.378		0.51	330
Acetone	0.818	56.5	1.359	0.32	0.56	330
Dioxane	1.033	101.3	1.422	1.54	0.56	220
Ethylacetate	0.901	77.2	1.370	0.46	0.58	260
Triethylamine	0.728	89.5	1.401	0.38	0.63	
Acetonitrile	0.782	82.0	1.344	0.37	0.65	200
Pyridine	0.987	115.0	1.510	0.94	0.71	305
n-Propanol	0.804	97.0	1.380	2.30	0.82	210
iso-Propanol	0.785	82.4	1.377	2.30	0.82	210
Ethanol	0.789	78.5	1.361	1.20	0.88	210
Methanol	0.796	64.7	1.329	0.60	0.95	205
Water	1.000	100.0	1.330	1.00	High	
Acetic acid	1.049	117.9	1.372	1,26	High	

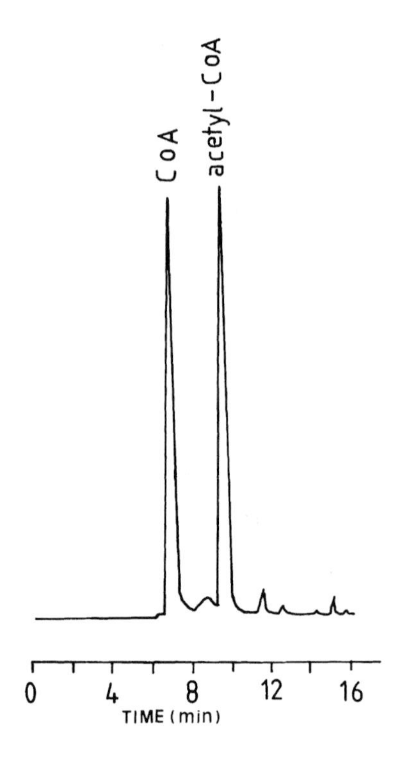

Figure 3 Separation of a mixture of acteyl-CoA and CoA on a Spherisorb C18, 5 μm column (1.5 × 250 mm). The column was eluted with a 0–20% methanol gradient in water containing 10 mM potassium phosphate, pH 6.7, at a flow rate of 1.5 ml/min. The detection wavelength was 254 nm.

the separation of a mixture containing acetyl-CoA and CoA, using a water–potassium phosphate–methanol mixture.

3.5 Effect of pH and salts

In the case of compounds that can undergo ionization, a change in pH can markedly affect their retention and selectivity. As an example, acidic solutes will be eluted before inert solutes at pHs near their p*K* values. The reasons for this behaviour are the presence of unreacted but dissociated silanols within the pores or the surface that can cause solute exclusion, and an increase in the electrostatic interactions with aqueous solvents. Both these effects can be minimized by lowering the pH by addition of acetic acid. The silanols can also interact with basic substances in an ion-exchange mechanism. If this extra mechanism proves to be undesirable, it can be suppressed by the addition of a base or an acid to form the salt. Of course, an appropriate concentration of NaCl could also be added to compete for ion-exchange sites. The addition of the latter also increases the polarity of solvent thereby causing the k' value of non-polar solutes to increase and the resulting difference in selectivity between solutes can only lead to their enhanced separation on the chromatographic column.

3.6 Influence of temperature

The speed can usually be enhanced by increasing column temperature (T) and a number of different column heaters are now commercially available. *Figure* 4 shows a Van't Hoff plot of 1n k' versus $1/T$ for a series of catecholamines. In general, a 10°C increase in temperature reduces the capacity factor by two-fold. However, in some cases, the increase in temperature leads to increased retention, for example, molecules that adopt a compact, near-spherical configuration are retained for a longer period. Furthermore, changes in temperature may also affect the ionization of either the buffers or the solute and this may result in an altered

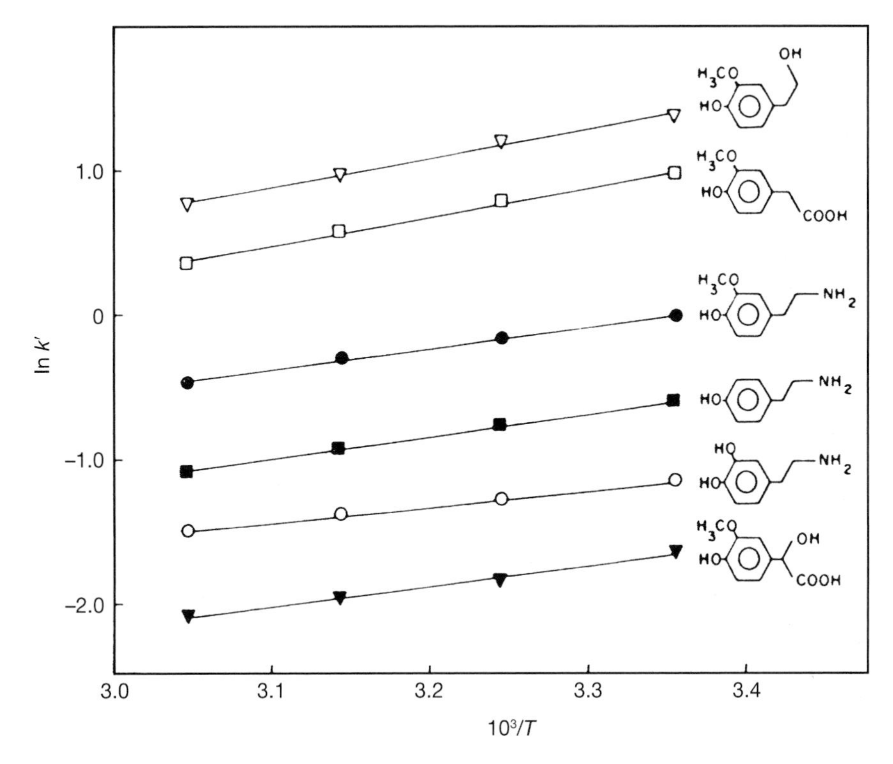

Figure 4 Van't Hoff plots of capacity factors (k'). Chromatographic conditions: column, Partisil 1025 ODS (250 mm × 4.5 mm ID); mobile phase 0.05 M KH_2PO_4; flow rate, 1 ml min^{-1}; detection, UV at 254 nm. T indicates the column temperature (K). Reproduced from (10) with permission.

k' value. It should also be noted that the increased temperature will lead to a reduction in viscosity of the mobile phase especially in the case of methanol–water mixtures, thereby allowing increased flow rates and a reduced separation time. However, caution must be exercised in that the decreased retention time must not be compensated by loss in selectivity, which may lead to loss in resolution.

3.7 Ion-pair chromatography

This technique, which may be used to separate both ionic and non-ionized solutes, is similar to RPC with similar factors affecting separation. Thus, in addition to a hydrophobic reverse phase such as C-18 and an aqueous mobile phase containing an organic modifier, a counter-ion is also added to form an ion-pair with the charged solute molecules. The retention properties of the solute are altered as a result of this. Generally, for separation of acidic solutes, a hydrophobic organic base is added, while in the case of basic solutes a hydrophobic organic acid is used.

As with RPC, there is some controversy over the mechanism of separation. Five basic models have been proposed to explain the influence of ion-pairing agents on separation of solutes. For a detailed discussion of retention mechanism, the reader should consult the review (13). The mechanisms of ion-pair chromatography can basically be described by two extremes. First, the counter-ions and solute ions form dissociated ion-pairs in the mobile phase and are selectively retarded by the stationary phase and separated. Second, the counter-ions are first absorbed by the stationary phase and then interact with solute ions. The extent to which a particular mechanism operates depends upon the counter-ions and their hydrophobicity, the amount of organic modifier in mobile phase, and the nature of solute.

The following equation represents the ion-pair formation:

$RCOO^-_{(aq)}$	+ $TBA^+_{(aq)}$	$= (RCOO^- \; TBA^+)_{org}$.
charged solute (less retarded)	(tertiarybutylammonium hydroxide)$^+$	retarded by the stationary phase

Generally, a tertiary amine such as trioctylamine or a quartenary amine such as TBA is a suitable choice initially. For basic compounds, an akyl sulphonate (e.g. heptane sulphonate) or an alkyl sulphate (e.g. sodium lauryl sulphate) are often used. It has been further shown that the k' factor increases in a non-linear fashion with increasing chain length and, therefore, the increasing hydrophobicity of the alkyl sulphate (14) (*Figure* 5).

3.8 Ion-exchange resins

Table 5 shows a list of commercially available ion-exchange stationary phases together with their properties.

The mobile phase usually consists of buffered solutions to which an organic modifier, such a methanol, acetonitrile, or dioxane, may be added for greater selectivity. The pH of the mobile phase is an important parameter in affecting retention in ion-exchange as this would affect the ionization of the ion-exchange group as well as the molecules to be separated. The ionic strength of the mobile phase may also be changed to alter $k'\alpha^1/(\text{ionic strength})^2$. This is the case when only the ion-exchange mechanism is operating. However, in the case of hydrophobic interaction, two mechanisms operate: the ion-exchange mechanism which

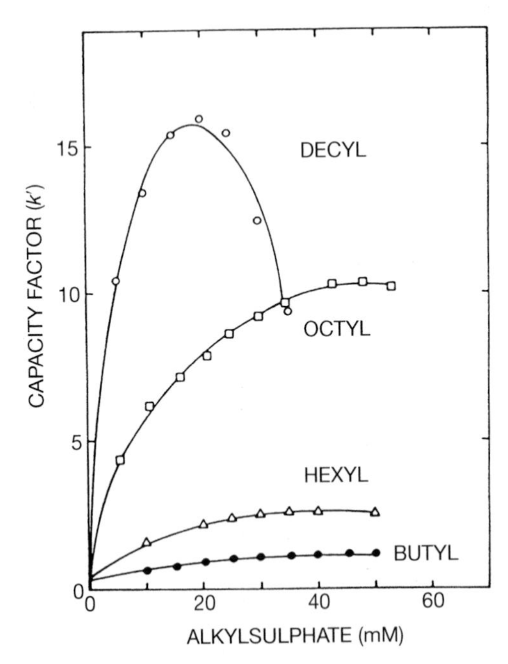

Figure 5 Plots of capacity factor (k') of adrenaline versus the counter-ion concentration for various *n*-alkylsulphates. Chromatographic conditions: column, Partisil ODS (10 μm); mobile phase, 500 mM potassium phosphate buffer (pH 2.55) containing various concentrations of counter-ion; flow rate, 2 ml/min; temperature, 40°C; inlet pressure, 400 p.s.i.; detection, UV at 254 nm. Reproduced from (14) with permission.

Table 5 Commercially available ion-exchange resins.

Matrix	Type	Capacity	Pore size (Å)	Particle size (μm)	Features and applications
TSKgel Q-5W	strong anion	100 mg BSA/ml	1000	10	Trimethylamino and sulphopropyl groups attached to a
TSKgel SP-5PW	strong cation	40 mg Hb/ml			hydrophilic methacrylate resin. Exclusion limit: 1000 000 Da; pH range: 2–12
TSKgel DEAE-5PW TSKgel CM-5PW	weak anion weak cation	30 mg BSA/ml 45 mg Hb/ml	1000	10, 13, 20	Diethylaminoethyl and carboxymethyl groups attached to polymeric methacrylate resin. Exclusion limit: 1000 000 Da; pH range: 2–12.
TSKgel DEAE-2SW TSKgel CM-2SW	weak anion weak cation	NA 110mg Hb/ml	125, 250	5, 10	Groups bonded to porous spherical silica Exclusion limit: 10 000 Da Useful for the separation of nucleotides, pharmaceuticals, catecholamines and small peptides
Hypersil APS And APS2	weak anion	NA	90–150	3, 5, 10	Amino groups bonded to porous silica pH range: 2–7.5. Analysis of sugars and vitamins
Hypersil SAX	strong anion	NA	90–150	5	Groups attached to silica. High stability to aqueous and low pH mobile phases. Suitable for the analysis of nucleotides and organic acids
Lichrosorb NH	weak anion	NA	100	5, 10	Amino groups bonded to porous irregular silica Similar application to Hypersil SAX
.Nucleosil SA	strong cation	NA	5, 10		Sulphonate groups bonded to porous silica pH range: 1–9. Nucleosil 100 matrix can withstand high pressures of up to 8500 psi
Nucleosil SB	strong anion	NA	NA		Quaternary ammonium groups attached to silica S Similar specifications to Nuceosil SA

Partisil, Spherisorb, Kromasil, and Exsil (silica-based) and Waters Protein-Pak HR (polymer-based) packings also offer a range of strong and weak exchangers.
NA, not available

causes the k' value to decrease with increasing ionic strength, and the salting-out mechanism which leads to increased k' values due to increased hydrophobic interaction of solute with the organic stationary phase.

3.9 Size exclusion chromatography (SEC)

Soft dextrans and agaroses, which lack mechanical stability, are not suitable for HPLC. A list of commercially available stationary phases is given in *Table 6*. The two most commonly used for aqueous SEC are cross-linked polyether or polyester and silica-based phases. Each has hydroxyl groups covalently attached to the surface. For aqueous SEC, TSK PW, TSK SW, and Zorbax GF series are quite suitable for proteins, peptides, and nucleic acids over very wide molecular weight ranges (e.g. TSK SW 30 000 → 500 000). For non-aqueous chromatography TSK HXL series can be used. The disadvantage of SEC over other forms of column

Table 6 Resins used for size-exclusion HPLC

Matrix	Pore size (Å)	Particle size (μm)	Comments
Protein-Pak 60 Protein-Pak 125 Protein-Pak 300SW	60 125 300	10 10 10	Diol-bonded silica packings for gel filtration. MW ranges for separation of proteins and peptides are 1000–20000 (Pak-60), 2000–80000 (Pak-125) and 10000–300000 Capacities of up to 1 mg for 7.8 mm × 300 mm columns
Shodex packings Protein KW-802.5 Protein KW-803 Protein KW-804	 NA NA NA	 7 7 7	MW ranges for the three matrices are 100–50000, 100–150000 and 500–600000 Da. Silica packings for gel filtration chromatography (GFC)
TSKgel SW and SWXL series Super SW2000	125	4	Separation ranges for globular proteins of 5000–7000000
Super SW3000 G2000–4000	250 125–450	4 5–17	All comprise rigid spherical silica gel chemically bonded with hydrophilic compounds for GFC. A 30 cm SWXL column provides a similar resolution to a 60 cm SW column, but the SWXL requires half as much time. Capacities of 5–100 mg
TSKgel PW and PWXL series G1000-G6000	100–1000	6–22	Hydrophilic, rigid, spherical, porous methacrylate beads used over a pH range of 2–12 with up to 50% organic solvent. Suitable for aqueous GFC of proteins, peptides polysaccharides, oligosaccharides, DNA, RNA, water-soluble organic polymers
TSKgel HHR and HXL series G1000-G7000	8 pore sizes	5,6,9	MW range of 2000–10000000, made from polystyrene divinylbenzene particles for non-aqueous gel permeation chromatography (GPC) using e.g. toluene, tetra-hydrofuran. Capacity of <5 mg
Styragel HMW 2–7	NA	20	Ultra-high MW analysis of shear-sensitive polymers. Ambient to 150°C. Exclusion limit of 100–10000 × 10^6
Styragel HT2-6	NA	10	Mid to high MW range, 10–5000 × 10^6 Suitable for ambient to high temperature analysis of polymers
Styragel HR 0.5-6	NA	5	Low MW analysis, up to 4000000. Ambient to 80°C All three ranges of matrices are based upon polystyrene Divinylbenzene particles for non-aqueous GPC

NA, not available

chromatography is its low capacity since total volumes must not exceed 1–2% of total column volume. Similar factors to those involved in conventional gel filtration also affect the separation in the case of HPLC matrices, for example, pore size distribution, pore volume, and pH.

4 Instrumentation

4.1 Essential components of an HPLC system

Figure 6 is a schematic representation of a typical HPLC system. The various components shown will be discussed briefly in turn in this section.

4.2 Pumps

The pump is the most important feature of an HPLC system (apart from the column). Manufacturers have sought to develop various operating principles with the main aim of producing pumps that could deliver pulseless, constant, and reproducible flow of solvent over short and long periods.

Amongst the early designs were gas pressure-driven and high pressure single- and twin-headed syringe pumps. These have largely been superseded by reciprocating single, dual, or triple pistons with or without diaphragm. The characteristics of each type of pump have been discussed by Snyder and Kirkland (15). Most common types of reciprocating pumps consist of a sapphire piston which displaces a volume of liquid (via a camdrive). The direction of flow is controlled by inlet and outlet check valves. This design suffers from a major drawback in that, as the piston moves through the whole length of the chamber, the pulse of solvent being generated leads to uneven flow. Some pump designs incorporated a hydro-pneumatic dampener consisting of a length of flattened tubing between pump and column leading to unwanted extra dead-volume. However, most recent pump designs have instead incorporated a fast refill cycle in the piston in conjunction with a dampener/transducer, thereby reducing pulsation considerably and leading to relatively smooth solvent delivery, for example, the Beckman System Gold *Nouveau* (single floating piston pump, models118 and 126) and Gilson (single reciprocating piston pumps, models 321, 322, 305–308). The twin-headed pumps where the pistons operate 180° out-of-phase (for example, Waters 600E, and the dual floating piston design of Hewlett-Packard 1100 series or Bio-Rad's Duo-Flow system with F10/F40 pump models) deliver essentially pulseless flow since the refill stroke of one piston is compensated for by the fill stroke of the other. It may be noted that the use of floating pistons has been employed in some designs to minimize the seal wear (for example, Beckman System Gold *Nouveau* 118 and 126). These pump designs have, in addition, incorporated mechanisms to compensate for the compressibility of different solvents. The delivery flow rate decreases

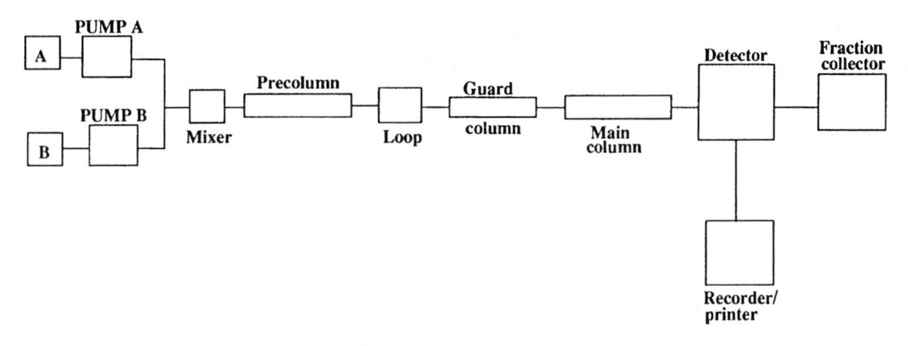

Figure 6 Essential components of an HPLC system.

when the pressure increases due to liquid compressibility. The most common solution has been the use of a pressure transducer which operates independently from the pump, measures flow rate and, via feedback circuitry, corrects pump drive when the measured flow-rate deviates from the setting. In the case of Beckman System Gold, the U-flow liquid heads together with compensation circuitry for solvent compressibility handle air bubbles with ease, eliminating the need for either external or internal degassing units. Waters have introduced the Alliance System (2690 Separations Module) encompassing a solvent management system with completely independent, digitally controlled piston drives under the control of two independent pressure transducers. In addition, the serial flow design of the 2690 module incorporates primary and accumulator piston chambers, with the advantage of maintaining only two check valves as opposed to four in the conventional reciprocating piston design with dynamic mixing for gradient operations. In the case of the 2690 module, the primary head receives the proportioned solvent via a gradient proportioning valve and delivers it to the accumulator head for thorough mixing during its fill cycle.

The usual flow rates attainable with a ± 0.1% error are in the range 0.1–10 ml/min for analytical applications. However, Beckman claim to have reproducible flow rates (± 0.1% in the range 0.001–10 ml/min, System Gold 118 and 126). For preparative HPLC, flow rates in the range 30 ml (Beckman system Gold) to 200 ml (Gilson model 307) are achievable with a coefficient of variation in the range 0.1–0.6% in the case of Gilson pumps. The latter offer the largest available range of pressures and flow rates (5 μl/min at 60 MPa to 200 ml/min at 3.5 MPa) with nine different pumps of different size and material which are suitable for microbore to preparative scales.

4.3 Biocompatibility

For particular applications of HPLC, such as purification of nucleic acids and proteins, biocompatibility of the system may be desirable. For example, Gilson offer a titanium-based system including pump heads, tubing, pressure dampeners/transducers, flow cells and solvent mixing chambers. Titanium is completely inert and of a high chemical purity and will not release any ions during separation of biological macromolecules. However, most HPLC systems now provide the option of using either stainless steel or PEEK flow path, for example, Bio-Rad's biocompatible Duo-Flow system in which all components, including the pumps, normally in contact with the solvent are made of PEEK.

4.4 Sample injection

A precise, quantitative result of HPLC analysis requires injection of well-defined sample volumes in a highly reproducible manner. Commercially available injector types allow the variation of volume in the range 0.5–20 μl for analytical separations and 0.1–10 ml in the case of preparative runs. Early designs were influenced by those used in gas chromatography. These have now been largely superseded by the high-pressure injection valves (the loop type, *Figure* 7).

Beckman model 210A injection valve (four-ports), for example, is able to accommodate 5–2000 μl sample volumes and can withstand pressures of up to 10000 p.s.i. (689 MPa). To reduce sample dilution, the diameter of the ports has been reduced to only 0.255 mm. Alternative methods of, for example, stopped-flow septum injections have been found to be inferior due to generation of particulate matter from the septum after repeated injections. The injector type shown above can also be operated automatically when a large number of samples are to be analysed. For example, such a system could be run overnight for automated methods development. Beckman System Gold 508 Autosampler provides a spectrum of injection modes from full loop to partial loop with a variety of pre-injection capabilities such as

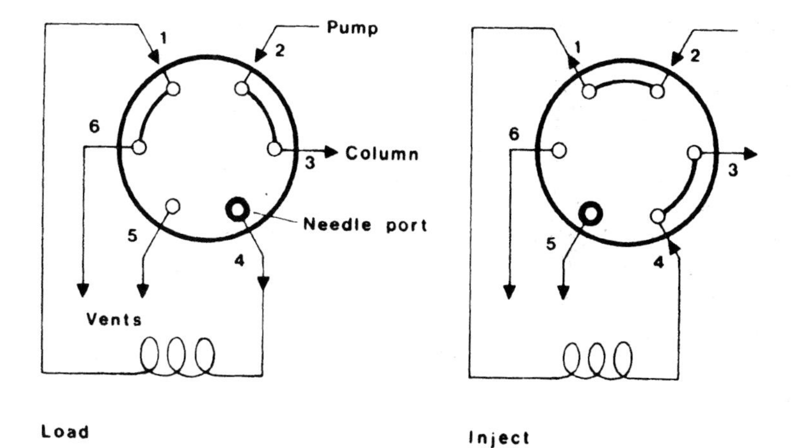

Figure 7 Illustration of the arrangement of a high-pressure injection valve. In the load position, the sample is introduced via the sample loop (positions 1, 4). Upon injection the solvent flow from the pump (2) is directed through the sample loop to the column (3).

serial dilution, reagent addition and standard addition; all can be programmed through Gold software.

5 Detectors

Much progress has been made in the design of specific (depends on some property of solute) and non-specific (depends on bulk property of mobile phase) detectors with regard to addressing the need for achieving high sensitivity and satisfactory lower limit of detection, in addition to obtaining low levels of noise and drift. It is outside the scope of this Chapter to discuss each type of detector in detail and the reader is referred to the literature produced by the various manufacturers for their respective detection systems. A brief survey of the most commonly used systems is therefore given below.

5.1 UV/visible detectors

The UV/visible type detectors are the most widely used detectors due to their versatility, high sensitivity, wide dynamic range, and relative insensitivity to temperature and flow variations. The detectors can be classified as follows.

5.1.1 Fixed-wavelength detectors

Fixed-wavelength detectors utilize lamps that emit light at discrete wavelengths (the pressure mercury lamp is in common use emitting at wavelengths of 254, 313, 365, 405, 436, 546, and 578 nm). By use of suitable cut-off filters, a particular wavelength can be selected. Light sources are also available for monitoring at lower wavelengths e.g. zinc at 214 nm (for example, Bio-Rad's 280/254 nm detector as part of the Duo-Flow System with an optional 214 nm detection kit and a series of filters in the 313–546nm range) and cadmium at 229 and 326nm. The advantages of using fixed-wavelengths detectors include low noise level, low operating cost, and simplicity of operation. It should be noted that the majority of HPLC detection is carried out at 254 nm. The sensitivity is in the range 0.0001–2.0 absorbance units (AU) with a noise level of less than 2×10^{-5} AU at 254 nm. However, most users of HPLC opt for the variable wavelength detection systems described below.

5.1.2 Variable wavelength detectors

Variable wavelength detectors use light sources that give a continuous emission spectrum in the range 190–700 nm. A continuously adjustable monochromator is used for wavelength selection. The light travels from the lamp (deuterium or tungsten) to focusing and steering mirrors, then to the diffraction grating, through the flow cell and, finally, to the photodiodes. The entrance slit, mirrors, diffraction grating, and the grating drive mechanisms are all part of the monochromator assembly. On selecting a new wavelength, the grating is moved by a stepper motor at a speed of 70 nm/s. Multi-wavelength monitoring is achieved by rapidly adjusting the grating between any two selected wavelengths. These data can be useful in peak identification. The Beckman System Gold model 166, Gilson's 151/155 and Hewlett Packard's 1100 uv/visible detector series offer programmable detector modules. Noise levels are less than 2×10^{-5} AU and detector response is linear up to 2 AU.

The latest development in this field is the Diode Array Detector (DAD). Photodiode array technology positions multiple detectors side-by-side on a silicon crystal, with a dedicated capacitor to convert light to electric charge. Polychromic light from the grating can now be detected in the same time it takes to measure a single wavelength with a conventional spectrophotometer. The image of the absorbance spectrum of the eluting substance is formed on the diode array. A diode array, unlike stepper-motor driven wavelength changes, is not subject to mechanical irreproducibility. The Hewlett Packard 1100 series DAD has an extended wavelength range from 190 to 950 nm, with simultaneous illumination using combined deuterium and tungsten lamps, 1024 diodes and a 1 nm slit for a high spectral resolution. Gilson 170 and Waters 996 DAD models also provide a 1 nm spectral resolution using the same number of diodes, albeit with a smaller wavelength range (1–700 nm). Beckman model 168, however, has an array of 512 diodes with a spectroscopic resolution of 1 nm/diode over the range of 190–700 nm. Linearity of detector response, noise levels and sensitivity are similar to those mentioned above for variable wavelength detectors. Records of spectra at different positions of a single peak can indicate the purity of the eluting peak. Two major advantages of the DAD over motor-driven monochromators are that they possess fewer moving parts and that they are less prone to peak skewing at high flow rates.

The sensitivity of UV detectors is generally in the low nanogram range. It is also affected by flow cell geometry and volume. A well-designed cell should minimize the refractive index effects, contain no unswept areas, and be able to withstand high pressures. Most cells have volumes of 8–10 μl and an optical pathlength of 10 mm. Cells with lower volumes are also available, for example, Gilson offer a range of 0.3–13 μl with variable path lengths and able to withstand different pressures (500 to 10000 p.s.i.). The choice of the cell is dependent upon the particular application, for example, analytical versus preparative.

5.2 Fluorescence detectors

Fluorometric detection possesses inherently higher sensitivity than absorption since the intensity of emitted light is directly proportional to power of exciting radiation. This method can be used to detect any compound that cannot be conveniently detected by any other method (for example, UV/refractive index) and has intrinsic fluorescence or can be derivatized by reacting with a fluorescent compound, for example, fluorescamine, dansyl chloride. This high sensitivity and selectivity have been extensively applied in biochemical systems since many biologically important compounds strongly fluoresce, such as porphyrins, riboflavins, vitamins, certain drugs, nucleotides, and so on.

The conventional detector consists of:

(a) a light source (xenon, mercury arc, quartz halogen)

(b) a wavelength reflector using any monochromator assembly suitable for a UV–visible spectrophotometer. Thus, excitation spectra can be obtained for the compound of interest. The Hewlett Packard 1100 series detector, for example, uses concave holographic gratings for measuring excitation (200–700 nm) and emission (280–900 nm) spectra of a given solute with an on-line scan speed of 0.6 s per spectrum. The spectral information can be used for rapid method optimization and verification of separation quality. Light from a 20 W Xenon flash lamp is emitted in every direction, but is monitored at right angles to the excitation beam. A parabolic reflector is inserted inside the cuvette chamber to reflect more of the light towards the photomultiplier.

(c) single- and dual-flow cells are available commercially to take account of the fluorescence of the mobile phase. The flow cells vary in size in the range 1 – 40 μl, for example, Bio-Rad provide two flow cells for their model 1700 of volumes 6.5 μl and 19 μl. The stated maximum sensitivity for the Hewlett Packard 1100 series is 10 fg anthracene with excitation at 250 nm and emission at 450 nm using a standard flow cell of 8 μl with a maximum pressure limit of 2 MPa.

Further improvements have been introduced recently, for example, the use of a laser light source which is more monochromatic than conventional light sources. Increased sensitivity can introduce problems of contaminating fluorescent material in the mobile phase. It may also be noted that, by proper choice of excitation conditions, the quantum efficiency of fluorescence of the sample can be optimized. The concentration of solute should not be so high as to cause quenching and thus reduce sensitivity or overload. It should also be mentioned that the photomultiplier or amplifier fluorescence can also be quenched by impurities and dissolved oxygen.

Compounds that do not fluoresce can be detected by carrying out pre-or post-column derivatization with fluorescamine, dansyl chloride, or other fluorophores (16).

5.3 Refractive index (RI) detectors

This type of detector measures the change in the refractive indices of the liquid in the reference and sample cells and, due to its non-specificity, is a universal detector. However, the difference in refractive indices between different solutes is very small. The sensitivity of detection is therefore much lower than the fluorescence and UV/visible detectors and is in the μg–ng range. The lower limit of detection (LLD) of a modern RI detector is in the order of 1×10^{-8} RI units. The LLD is dependent upon temperature (1–5 $\times 10^{-4}$ per degree K) and pressure fluctuations. A temperature control of $\pm$ 0.001 °C is required if a noise level of less than 1×10^{-7} RI units is to be maintained. Additionally, the detector response is affected by mobile phase composition and gradient elution is not possible with the commonly used RI detectors. The three types of RI detectors are the deflection, fresnel, and interferometric; of these, the deflection type is the most commonly used instrument. In this, a collimated light beam passes two parallel chambers in a glass prism acting as reference and sample cells. If the refractive index of solution (η_s) is equal to the refractive index of reference (η_r) the beam is slightly shifted parallel to the incident beam. If the η_s changes, then the beam is deflected and then measured by a differential photodiode. The output signal can either be positive or negative depending upon η_s being larger or smaller than η_r. An instrument of this type is supplied by Hewlett Packard (1100 series) with typical noise levels of 2.5×10^{-9} RI units and a sensitivity as high as 10 ng in carbohydrate analysis, for example, sucrose using a standard flow cell with a volume of 8 μl and a maximum pressure limit of 0.5 MPa. The optical unit and the flow cell are maintained at constant temperature by countercurrent heat exchangers.

Waters model 2410 and Bio-Rad model 1755 provide somewhat higher noise levels (2×10^{-8} RI and 4×10^{-8} RI units respectively) but similar levels of sensitivity and full-scale measurement ranges (up to 5.0×10^{-8} RI units in the case of Waters 2410).

5.4 Elecrochemical detectors

The development of electrochemical detectors has been prompted by the need for the quantitation of trace quantities of biologically important compounds, in particular the biogenic amines. These detectors offer a higher sensitivity and selectivity than the methods already discussed. Picogram and femtogram levels of electroactive compounds have been reported. The detailed theory of such instruments is outside the scope of this Chapter; the reader is referred to (17,18).

Briefly, these detectors are of two types.

(a) Bulk property detectors, of which the most commonly used are conductivity detectors, which measure the change in cell resistance. For example, the Waters 432 conductivity detector uses a multielectrode flow cell (0.6 μl volume) to minimize problems caused by resistive and capacitive noise typically characteristic of 2-electrode cells. Noise specifications of various commercially available instruments do not necessarily take account of the background conductivity. For example, the background conductivity of 18 megohm deionized water is 1-2 μS. The stated noise level with the Waters 432 is 0.005 μS with a background of 147 μS in 1 mM KCl providing a high level of sensitivity.

(b) Solute property detectors, which monitor the change in potential or current as the solute passes through the flow cell. The more popular detectors are either amperometric or coulometric. Solute is passed over an electrode held at a constant voltage that is sufficiently high to cause either reduction or oxidation; then the current produced is proportional to solute concentration. The concentration of solute does not change significantly (~5% conversion). In coulometric detectors, the solute is almost totally converted (~95%). The signal given by the latter is greater but also leads to increased background signal leading to reduced signal-to-noise radio.

Usually a multi-electrode system is used. The working electrode used in the oxidative mode is usually carbon paste (a mixture of graphite and either paraffin oil, wax, or grease) or glassy carbon. Both give low background currents and are reproducible. However, electrochemical detection is limited to phases containing <25% (v/v) methanol or 5% (v/v) acetonitrile. In the reductive mode, a mercury electrode (mercury deposited as a thin film on a gold substrate) is usually used. In the older designs, a dropping mercury electrode, usually known as polarography, was used.

It should be emphasized that the constituents of the mobile phase must be of the highest purities to minimize background currents. The voltage range within which detection of solutes can take place depends on the electrode material. For glassy carbon it is -1.0 to 1.3 V and for mercury -2 to 0.3 V with reference to a saturated calomel electrode. The thin film detector is the most commonly used, for example, Bio-Rad model 1340 uses a glassy carbon electrode with a voltage range of ±2 V and can detect 20 pg amounts/injection. Briefly, the classes of compounds that have been investigated include aromatic amines, phenols, thiols, nitro-compounds, quinones, purines ascorbate, and uric acid.

5.5 Radioactivity monitors

The technique of scintillation counting has been adapted over the last decade to on-line detection of HPLC effluents. Most radioactivity detectors use flow-cells of different types positioned between two photomultiplier tubes which usually possess reflectors to ensure high counting efficiencies. The flow cells are of three types:

(a) Homogenous: the column effluent is mixed with liquid scintillant before passing through the flow cell and is sent to waste.

(b) Heterogeneous: the eluant passes through scintillant granules inside the flow cell.

(c) High-energy: the scintillator material (liquid or solid) surrounds the flow cell through which the eluate passes.

A major limitation of these detectors is their inability to compensate for changing counting efficiency in gradient HPLC. Some detectors, however, use a standard curve to take account of this. Several detectors are now linked to microprocessors allowing detailed data collection and processing. The counting efficiencies and sensitivity of detection depend on the type of detector, type of flow cell, and the isotope, for example, sensitivity of detection for ^{14}C is 80–100 d.p.m. while for ^{3}H it is 250–400 d.p.m. in the case of the heterogeneous and homogenous detectors.

6 Practical considerations

6.1 Selection of a chromatographic mode

Generally speaking, the particular separation of a compound depends upon molecular weight, range of solubility, and its chemical structure. *Table 7* summarizes the category of compounds and the appropriate modes of separation.

6.2 Solvent selection

The criteria for selection of a suitable solvent to cause elution depend upon the physico-chemical properties of the sample and the solvent being considered. When preparing the sample for a separation, dissolve the sample in the same mobile phase as that being used to effect elution. In cases of low solubility, attempt to use a larger injection volume. If the sample is dissolved in a different solvent (for example, a stronger eluting solvent) from the mobile phase then resolution may be impaired. The choice of the eluting solvent can be facilitated by the use of the eluotropic series shown in *Table 4*. However, from a practical point of view, avoid the use of solvents of high viscosity, which can lead to high column back pressures,

Table 7 Mode selection

Sample			Mode
MW <2000	Non-ionic	High polarity	Normal phase
		Low polarity	Reversed phase
	Ionic	Acidic	Anion-exchange
			Reversed phase ion-pair
		Basic	Cation-exchange
			Reversed phase ion-pair
MW >2000	Water soluble	Ionic	Ion-exchange
		Polar	Normal phase
		Non-ionic	Reversed phase
			Affinity
			Ligand exchange
			Normal phase
			Size exclusion with aqueous mobile phase
	Water insoluble		Size exclusion with organic mobile phase
			Normal phase in organic solvents

and solvents of high volatility, since it may prove difficult to maintain a fixed composition of the mobile phase for any significant length of time. As UV is the most common method of detection, do not use solvents with a high UV cut-off point. These factors preclude the use of a number of solvents in the eluotropic series. When halide salts are included in the mobile phase, flush the whole system with plenty of water to prevent corrosion of stainless-steel tubing, column and pumps and remember that salt crystals can also form over long periods of storage.

Solvent purity can have a major effect on the column performance. The impurities in solvents can be lower or higher homologues, acids and bases, UV-absorbing compounds and water. They can also adsorb to the surface of the stationary phase and cause a change in selectivity of a column, increase in back pressures, and peak distortion. However, a large number of purified solvents, including HPLC-grade water, are now commercially available. Normal deionized water is unsuitable for use in HPLC.

6.3 De-gassing and filtration of solvents

The solvents and the sample may contain particulate matter which can impair the sealing of pump valves and clog the column frits or the surface of the column bed. Therefore, filter all solvents and samples through a 5 μm solvent-compatible filter (for example, Millipore). It is also advisable to attach additional filters (sintered) to the ends of tubing placed inside the solvent reservoir.

Dissolved gases can become problematic especially in the case of gradient systems where the gas is less soluble in a mixture than in the single solvent (for example, water/methanol or water/acetonitrile gradient) due to the exothermic mixing process. Various ways of de-gassing include boiling, vacuum degassing, agitation by sonication, and sparging with helium, argon or nitrogen. Amongst these, sonication and sparging with helium or argon are the most commonly used, although the latter are expensive but very effective. As de-gassing itself will lead to a change in composition of the mobile phase, filter and de-gass the solvents separately and then mix them. If helium or argon is used, then seal the reservoir after de-gassing. Some people continue bubbling helium at a very small flow rate throughout the HPLC separation.

6.4 Sample preparation

As crude biological samples may contain particulate matter and protein, make sure that the sample has been homogenized, centrifuged or filtered (5 μm Millipore), and deproteinized. The protein can be removed by perchloric or tricholoracetic acid. Alternatively, use methanol (1:1 ratio) and leave the sample for 5 min. In each case, remove the protein precipitate by centrifugation or filtration through a semi-permeable membrane (for example, Amicon) with the appropriate cut-off point. If the compound is in a dilute solution, then attempt concentration of the sample by evaporation, for example, using either nitrogen, vacuum or lyophilization.

6.5 Column packing

Considerable saving in cost can be made by using 'in house-packed' columns instead of commercially packed ones. Columns can be efficiently packed by using a slurry of the stationary phase. Use a high capacity, pneumatic amplifier pump (19) as HPLC pumps do not give sufficiently high impact velocities. First remove fines by suspending the support in an organic solvent, leave to stand for 15 min, and then decant the supernatant. Repeat this several times. Suspend the material in a high-density solvent and pour into a reservoir. Pump the HPLC column at about 3000 p.s.i.(206 Mpa) Fit the column with end pieces and test for proper performance. A poorly packed column can lead to column voids and compression during operation causing loss in resolution via peak splitting or peak distortion and broadening. For alternative procedures consult the review by Kaminski *et al.* (20).

6.6 Column protection

Protection of the column from particulate material by the use of on-line filters and pre-column filtration has already been mentioned. It is also important to protect the column from components of the sample which may bind to the column irreversibly and, in time, cause build up on top of the column leading to high back pressures and loss in performance. Therefore, attach a guard column between the injector and the main column. The guard column usually contains the same stationary phase as the main column. Any components with high k' value will be removed at this stage. The guard column can be replaced at a small cost by repacking it with new stationary phase. In the author's experience, adding small portions of dry matrix with a spatula and gently tapping, while rotating the column on a flat bench, proves quite satisfactory. The column can then be packed by pumping water through it at a high flow rate (a flow rate of 9 ml/min will give ~1000 p.s.i.(68.9 MPa) for a 4.6 mm internal diameter column with 5 μm particles). The guard column, however, can contribute to extra-column band-broadening effects.

The use of aqueous buffers also causes loss of column material to give a void at the top. This is especially the case with silica-based matrices above pH 7. This leads to loss in peak resolution and peak splitting. Again, place a pre-column between the pumps and the injector to pre-saturate the solvents with silica, thereby preventing the main column from dissolution.

Never allow the column to go dry as a dry stationary phase will shrink, leading to voids. Wash the column with a pure solvent such as methanol at the end of the day and cap it securely at both ends. If buffer solutions or salts have been used, wash the column thoroughly with water before storage to avoid corrosion of stainless steel. Always mark the top and bottom of the column so that the same direction of flow is always used. Only one end of the column is then contaminated by impurities. It is then easier to replace the top portion of the column stationary phase with new material.

6.7 Tubing

Use minimal lengths of stainless steel tubing interconnecting the various parts of the HPLC, as it contributes to extra-column band-broadening. Generally, 1/16 inch steel capillary tubing is used, with an internal diameter of either 0.5 or 0.3 mm.

7 Application of HPLC to enzymatic analysis

7.1 Hydrolases

7.1.1 Dihydroorotase from rat liver (EC 3.5.2.3).

In mammals, the first three reactions of pyrimidine nucleotide biosynthesis are carried out by a trifunctional enzyme, the protein Pyr 1–3. Of the three activities, the dihydroorotase catalyses the reversible cyclization of *N*-carbamyl-L-aspartate to L-dihydroorotic acid (21). At neutral pH, the equilibrium lies towards the acyclic molecule. The assay of dihydroorotase activity is based on the separation of radio-labelled substrate and product by ion-pair reverse phase HPLC.

Protocol 1

Assay of dihydroorotase

Equipment and reagents

- Centrifuge
- Reverse phase HPLC apparatus
- Hepes, pH 7.4
- L-(6-^{14}C) dihydroorotate
- Dihydroorotase
- 1% (w/v) SDS
- Elution buffer: tetrabutylammonium phosphate, acetonitrile
- Flow scintillant

1. Carry out the assay in a total volume of 100 μl containing 50 mM Hepes, pH 7.4, and L-(6-^{14}C) dihydroorotate.
2. Add 5–20 μl of enzyme to give the required percentage conversion.
3. Incubate the reaction mixture for 20–30 min at 37 °C and then terminate by adding 100 μl of 1% (w/v) SDS.
4. After a brief incubation, add 200 μl of elution buffer and centrifuge to remove precipitate.
5. Use the supernatant directly for HPLC. Carry out the separation of substrate and product on a Waters C-18 column (Nova-Pak C-18 cartridge, 0.5 × 10 cm, 5 μm particles). Use a radioactive flow detector (Flo-One/Beta system with 2.5 ml flow cell) for continuous monitoring of radioactivity. The elution buffer should be 3.5 mM tetrabutylammonium phosphate, pH 7.0, and acetonitrile (85:15).
6. Elute the column with buffer at 1.5 ml/min, while the flow scintillant (Flo-Scint 3, Radiomatic Instruments) should be pumped at 5 ml/min. The elution profile shown in *Figure 8* consists of c.p.m. versus time. The observed percentage conversion can be obtained from the chromatogram as c.p.m. carbamyl aspartate/total c.p.m. For preparation of rate liver dihydroorotase, consult the procedures described elsewhere (22).

7.1.2 Angiotensin-converting enzyme (EC 3.4.15.1 dipeptidyldipeptidase)

Angiotensin-converting enzyme converts angiotensin (I) to angiotensin (II) and the dipeptide, histidyl-leucine (His-Leu). It also degrades the vaso-depressor peptide, bradykinin (23). The enzyme is found in lung, kidney, serum, brain, and testicles (24). The assay of its activity relies upon determination of hippuric acid liberated from Hip-His-Leu (25).

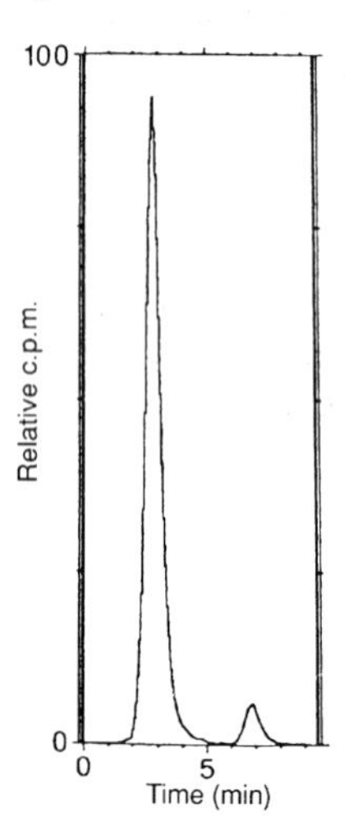

Figure 8 HPLC chromatogram of a mixture of [^{14}C]carbamyl aspartate and [^{14}C]dihydroorotate in the dihydroorotase reaction. Conditions were as described in the text (full scale = 3200 c.p.m.; total c.p.m. = 23 728; observed % conversion = 6.8%). Reproduced from (21) with permission.

Protocol 2

Assay of angiotensin-converting enzyme

Equipment and reagents

- Centrifuge
- HPLC column
- Spectrophotometer
- Angiotensin-converting enzyme
- Hip-His-Leu
- TrisfjHCl, pH 7.8, NaCl, magnesium acetate, sucrose, Nonidet-P40
- 3% (w/v) metaphosphoric acid
- Mobile phase: methanol, potassium dihydrogen orthophosphate

1. Incubate the enzyme for 30 min at 37 °C with 5 mM Hip-His-Leu in Tris–HCl, pH 7.8, containing 300 mM NaCl, 5 mM magnesium acetate, 0.25 M sucrose, and Nonidet-P40.
2. Terminate the reaction by adding 3% (w/v) metaphosphoric acid.
3. Centrifuge for 5 min before injecting 20 μl of supernatant onto the HPLC column.
4. For the purpose of separation of hippuric acid from Hip-His-Leu, use a 25 × 0.4 cm ID Nucleosil 7 C-18 column with 7.5 μm particle (Macherey-Nugel and Company).
5. The mobile phase should be methanol:10 mM potassium dihydrogen orthophosphate (1:1) and adjusted to pH 3.0 with phosphoric acid.
6. Use a flow rate of 1.0 ml/min for eluting the column.
7. Detection can be carried out by spectrophotometric means at 228 nm. Peak height can be used for the purpose of quantitation.

Figure 9a shows the chromatographic separation of a standard mixture containing hippuric acid, Hip-His-Leu and His-Leu. The upper diagrams *(Figure 9b–e)* show the formation of hippuric acid after incubation of various biological samples with Hip-His-Leu. No endogenous interfering substances are detected even when using Nonidet-P40, a detergent used for solubilizing tissue. The angiotensin-converting enzyme can be obtained from blood collected from the abdominal aorta of rat. Lung and kidney are removed immediately after sacrifice, gently rinsed with saline, chopped into small pieces, and then homogenized in Tris–HCl, pH 7.8, containing 30 mM KCl, 5 mM magnesium acetate, 0.25 M sucrose, and Nonidet-P40. The supernatant, after centrifugation at 20 000 **g**, acts as the enzyme preparation.

7.2 Isomerases

7.2.1 Diaminopimelate epimerase (EC 5.1.1.7)

LL-2,6-diaminopimelate-2-epimerase (DAP epimerase) catalyses the interconversion of the LL- and *meso*-isomers of DAP. The *meso*-DAP is then decarboxylated to yield L-lysine as the final step in the lysine biosynthetic pathway in bacteria (26, 27). Both enzymes are of interest with regard to regulation of the lysine pathway.

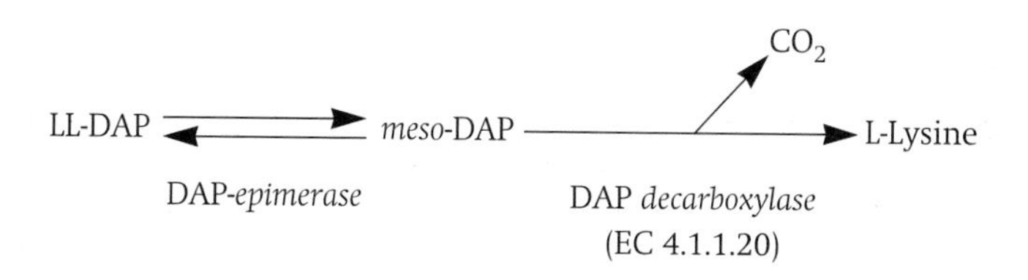

Protocol 3

Assay of diaminopimelate epimerase

Equipment and reagents

- Centrifuge
- HPLC column
- Spectrophotometer
- Pyrex tubes
- De-salted cell-free extract
- *Meso*-DAP, pyridoxal 5-phosphate, norvaline
- Potassium phosphate, pH 7.0
- 1.2 M sodium acetate, pH 5.2

1. Prepare a 1.5 ml assay mixture containing 15 μmol recrystallized *meso*-DAP, 0.1 μmol pyridoxal 5-phosphate, 3.75 μmol norvaline (internal standard), and about 0.5 mg protein as de-salted cell-free extract in 0.1 M potassium phosphate, pH 7.0.
2. Carry out the reaction at 37 °C for approximately 15 min.
3. Take 0.5 ml before and after incubation for HPLC analysis.
4. Terminate the reaction by quickly transferring samples to 10 ml Pyrex tubes each containing 1.0 ml of 1.2 M sodium acetate buffer, pH 5.2
5. Boil the tubes for 5 min and centrifuge to remove denatured protein.
6. Dilute the resulting supernatant 20-fold in water for subsequent HPLC analysis (28).

The assay relies on the separation of O-phthaldehyde (OPA) derivatives of LL-DAP and *meso*-DAP diastereomers.

Protocol 4

HPLC analysis of LL-DAP and *meso*-DAP

Equipment and reagents

- Reverse phase HPLC column
- Spectrophotometer
- Derivatizing solution: O-phthaldehyde/2-mercaptoethanol
- 2% (w/v) SDS in 400 mM sodium borate, pH 9.5
- Solvent A: 30% methanol, 70% 50 mM sodium acetate, pH 5.9
- Methanol

1. Prepare the OPA/2-mercaptoethanol derivatizing solution according to Unnithan *et al.* (29).
2. Mix a 200 μl aliquot of the deproteinized sample above with 200 μl of a 2% (w/v) solution of SDS in 400 mM sodium borate, pH 9.5, and 400 μl of the derivatizing reagent at room temperature. SDS improves the stability of the OPA-lysine derivative.
3. Inject 20 μl of this mixture after exactly 1 min.
4. Separate the derivatives on a Spherisorb C-18 ODS reverse phase column (250 × 4.5mm ID, 5 μm particles).
5. Elute with a linear gradient from 100% solvent A (30% methanol, 70% 50 mM sodium acetate, pH 5.9) to 30% solvent A and 70% methanol over a period of 35 min. The OAP derivatives can be detected by fluorescence. The excitation and emission wavelengths are 340 and 455 nm, respectively.

Figure 10 is a typical separation of OAP derivatives of LL-DAP, *meso*-DAP, norvaline, and lysine. As can be seen, by using this HPLC method both the decarboxylase and the epimerase activities can be simultaneously assayed.

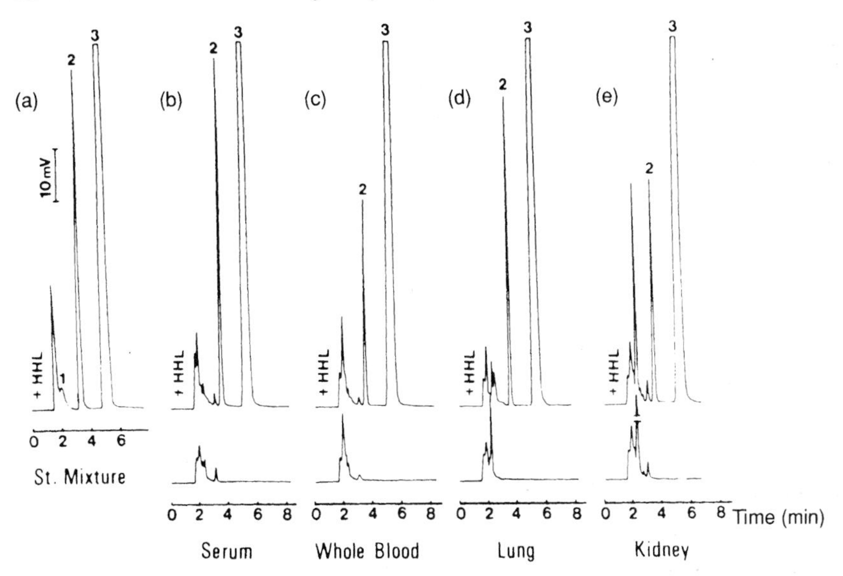

Figure 9 Chromatograms obtained from various samples incubated with (upper diagram) or without (lower diagram) Hip-His-Leu (HHL). (a) Standard mixture of 2.7 nmol His-Leu, 2.7 nmol hippuric acid and 100 nmol Hip-His-Leu. (b) A 50 μl aliquot of serum or (c) whole blood was incubated with (upper diagram) or without (lower diagram) 5 mM Hip-His-Leu according to the assay method described in the text. After 30 min, 0.75 ml of 3% (w/v) metaphosphoric acid was added and centrifuged. (d) Lung or (e) kidney was homogenized in 5 volumes of chilled Tris–HCl buffer containing 0.5% Nonidet-P40, and centrifuged. The supernatant was incubated with (upper diagram) or without (lower diagram) 5mM Hip-His-Leu. In the case of lung the supernatant was diluted 20 times with the buffer prior to incubation with Hip-His-Leu. Analytical conditions were as described in the text. Peaks: 1, His-Leu; 2, hippuric acid; 3, Hip-His-Leu. Reproduced from (25) with permission.

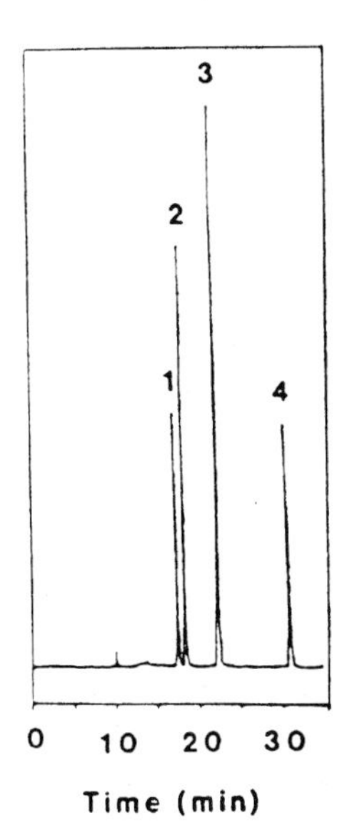

Figure 10 HPLC analysis of an enzyme reaction after 30 min incubation (see text for experimental details). The peaks are O-phthaldehyde derivatives of LL-DAP (1); *meso*-DAP (2); norvaline (3); lysine (4). Reproduced from (28) with permission.

7.3 Lyases

7.3.1 C_{17-20} Lyase (Cytochrome P-450$_{21SCC}$)

Human cytochrome P-450 $_{21SCC}$ has two activities, as shown below:

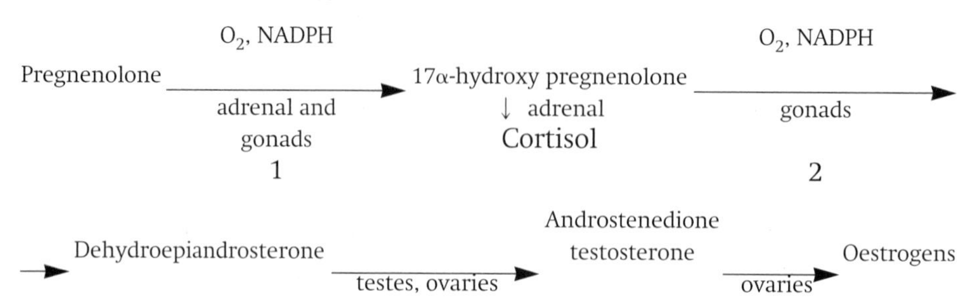

where activity (1) is that of 17α-hydroxylase and activity (2) of C_{17-20} lyase. The lyase reaction is predominant in gonads (30). The assay described here uses normal phase chromatography for the separation of steroids and on-line measurements of radioactivity (31).

Protocol 5

Assay for lyase

Equipment and reagents

- HPLC column with on-line detector
- Centrifuge
- Water bath
- Nitrogen
- (7-^{3}H)-17 α-hydroxypregnenolone
- NADPH, glucose-6-phosphate, glucose-6-phosphate dehydrogenase
- Microsomal protein
- Phosphate buffer, pH 7.4
- $CHCl_3$, methanol, water
- THF in hexane

1. Prepare an assay mixture containing 0.8 μM (7-^{3}H)-17 α-hydroxypregnenolone as the substrate, 1 mM NADPH, 5 mM gluocse-6-phosphate, 1 IU/ml of glucose-6-phosphate dehydrogenase and 0.02 mg microsomal protein. The total volume of the assay should be 100 μl in 50 mM phosphate buffer, pH 7.4, while the total ^{3}H per assay is 0.2 μCi.
2. Carry out the incubation at 34 °C for 6 min.
3. Terminate the reaction by adding 5ml of a 2:1 mixture of $CHCl_3$: methanol, and 0.9 ml water.
4. Shake the tube for 5 min and centrifuge to separate phases.
5. Discard the upper aqueous phase and wash the interface with $CHCl_3$-MeOH-H_2O (4:48:47).
6. After discarding the wash, use a stream of nitrogen to evaporate the lower phase to dryness in a water bath at 40 °C.
7. Rinse the sides of the tube with a little chloroform and repeat evaporation.
8. Dissolve the residue in THF in hexane prior to HPLC.
9. Perform the separation of steroids on a silica gel column (LiChrosorb Si-60, 5 μm, 250 × 4 mm) by eluting with a THF-hexane gradient at a flow rate of 1ml/min. Use a silica pre-column to saturate buffers with silica. The gradient conditions are 18–22% THF in hexane over 30 min and isocratic at 22% THF for 8 min. A Flo-One model HS radioactivity detector, for example, can be used for on-line detection.

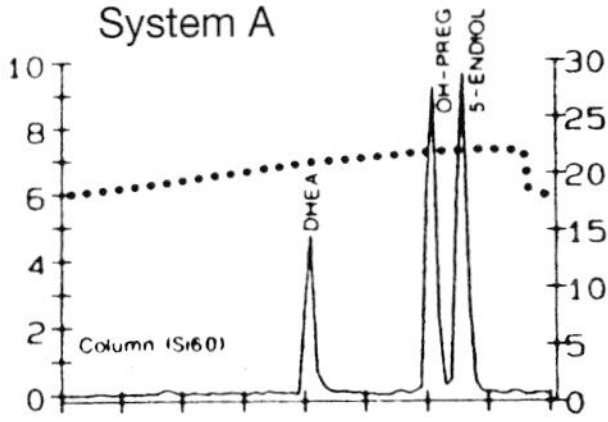

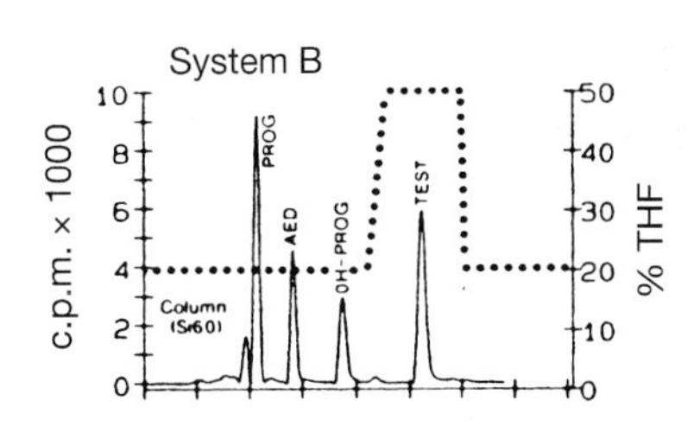

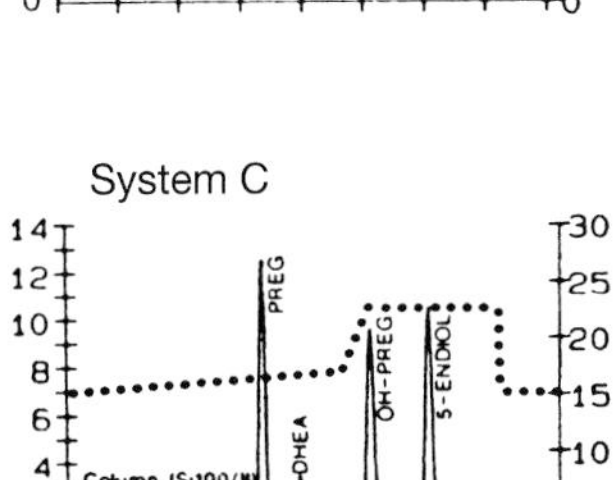

Figure 11 Chromatograms of steroid mixtures. The flow detector was in the scaler mode and reported a count every 6 s. The dotted line represents the mobile phase gradient, as percentage THF in hexane-THF. The mixture contained the following steroids: pregnenolone (PREG); dehydroepiandrosterone (DHEA); androst-5-ene-3β-diol (5-ENDIOL); progesterone (PROG); androst-4-ene-3,17-dione (AED); testosterone (TEST). Reproduced from (31) with permission.

Figure 11 shows the separation of 17 α-hydroxy pregnenolone and its products, dihydroepiandrosterone (i.e. lyase) and androstenedione [see (31) for the preparation of lyase from human testes].

7.3.2 Uroporphyrinogen decarboxylase (EC 4.1.1.37) from mouse liver

Uroporphyrinogen decarboxylase is an enzyme of heme biosynthesis and converts uroporphyrinogen, which contains eight carboxyl groups, to coproporphyrinogen, which is a tetracarboxy product (32). The reaction proceeds via the hepta-, hexa-, and pentacarboxy intermediates. The enzyme from mouse liver is assayed for its activity using reverse phase HPLC (33). A similar system has been described in the case of the enzyme from chicken erythrocytes (34). Uroporphyrinogens (I) and (III) as well as intermediates can be used as substrates for the enzyme.

Protocol 6

Assay of uroporphyrinogen decarboxylase

Equipment and reagents

- Reverse phase HPLC column
- Centrifuge
- Spectrophotometer
- UV light source
- Hamilton syringe
- Nitrogen
- Porphyrin solution
- 5% sodium amalgam in 5 mM NaOH
- 4 M H_3PO_4
- Enzyme extract
- Sodium phosphate buffer, pH 6.8
- EDTA, sodium mercaptoacetate
- HCl
- Methanol, lithium citrate, pH 3

Preparation

1 Reduce solutions of porphyrins (0.5–0.7 mmol/μl in a total volume of 100 μl) to porphyrinogens with 5% sodium amalgam in 5 mM NaOH. Carry out the reduction under N_2 until no fluorescence is observed under UV.

Protocol 6 continued

2 Remove the prophyrinogen solution from the mercury with a Hamilton syringe and neutralize with 0.5–2 μl of 4M H_3PO_4.
3 Dilute the solution with an equal volume of incubation buffer containing 0.1 mM EDTA/0.1 M sodium mercaptoacetate.

Assay

1 Mix 0.2 ml of the enzyme extract with 0.78ml of 0.1 M sodium phosphate buffer, pH 6.8, containing 0.1 mM EDTA/3 mM mercaptoacetate and keep at 4 °C.
2 After saturating the mixture with a stream of N_2 for 20 s, add 20–50 μl of the prophyrinogen and then stopper the incubation under N_2.
3 Rapidly shake at 37 °C in the dark.
4 Terminate the incubations with uroporphyrinogens after 30 min and those with pentacarboxyporphyrinogen after 10 min by rapid cooling in ice and then mixing with 50 μl of 6 M HCl.
5 Leave the mixtures in the light for at least 30 min to oxidize porphyrinogens to prophyrins and then add a further 50 μl of 6 M HCl.
6 Centrifuge the tubes and analyse the supernatants by HPLC.

HPLC analysis

1 Carry out the HPLC on a Spherisorb ODS column (25 cm × 4.6 mm ID, 5 μm particles). Use a CO:PELL ODS as a pre-column.
2 Elute the porphyrin standards as well as those formed from porphyrinogens with a linear gradient of 65–95% methanol over 20 min in 50 mM lithium citrate, pH 3.
3 Continue washing with 95% methanol for a further 5–10 min. Use a flow rate of 1 ml/min
4 The detection can be carried out by absorption at 400 nm.

Figure 12 a and *b* shows separation of a mixture of standard porphyrins (A) and the products for the decarboxylase reaction (B). The identity of HPLC peaks is given in the legend. Peak 8 corresponds to the substrate while peaks 4, 5, 6 and 7 are the products.

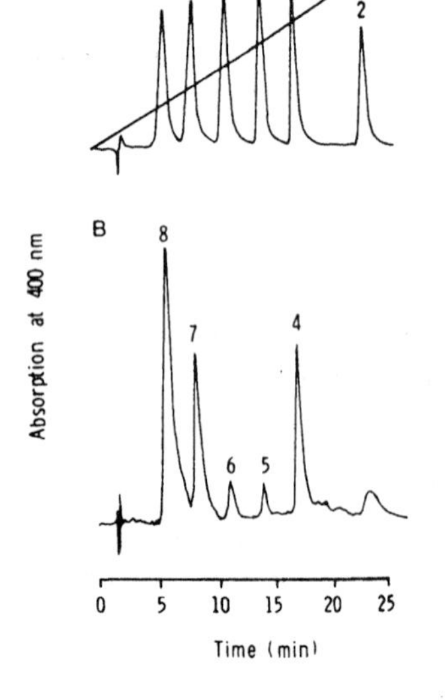

Figure 12 Separation of free porphyrins by HPLC using a Spherisorb ODS 15 μm column (25 cm × 4.6 mm ID) with a linear gradient system of methanol/50 mM lithium citrate (pH 3) from 65% to 95% methanol at 1 ml/min, as described in the text. (A) A scintillant mixture of the I isomers of uroporphyrin (8), heptacarboxyporphyrin (7), hexacarboxyporphyrin (6), pentacarboxyporphyrin (5), and coproporphyrin (4), together with mesoporphyrin IX (2). The slope of the solvent gradient system is also shown. (B) Products of an incubation of mouse liver supernatant with uroporphyrinogen (10 nmol) for 60 min as described in the text. Reproduced from (33) with permission.

7.4 Ligases

7.4.1 Glutaminyl cyclase from bovine pituitary homogenate

Glutaminyl cyclase catalyses the conversion of the peptide Gln-His-Pro-NH_2 to thyrotropin-releasing hormone (TRH) (35, 36). The method, involving dansylation of the N-terminus (37) followed by detection of the fluorescent derivative, is precluded in the present case due to the N-terminal location of gluatamine cyclization. The model peptide Gln-Leu-Tyr-Glu-Asn-Lys-OH can be used instead since it possesses a C-terminal lysine which could provide the ε-NH_2 group for dansylation (38). The dansylated derivative acts as the substrate.

The substrate and the cyclized product can be synthesized by conventional solid-phase peptide synthesis.

Protocol 7

Assay for glutaminyl cyclase

Equipment and reagents

- HPLC column
- Fluorescence detector
- Bovine extract
- Dansylated peptide
- Tris–HCl, pH 8.0 or Mops, pH 7.0
- 10 mM phenanthroline
- 24% (v/v) acetonitrile in 0.1 M sodium acetate, pH 6.5

1. Carry out an incubation of glutaminyl cyclase at 37 °C by adding up to 0.5 mg/ml of bovine extract (prepared according to reference 39) to the dansylated peptide substrate present at concentrations of 1–98 μM in 50 mM Tris–HCl, pH 8.0, or Mops, pH 7.0.
2. Remove 10–20 μl aliquots at various times and stop the reaction by the addition of 10 μl of 10 mM phenanthroline.
3. Inject 10 μl for HPLC analysis.
4. Carry out the separation at 50 °C on a thermostated Hypersil ODS column (4.6 mm ID × 100 mm, 5 μm particles). Use a Brownlee RP-18 guard column (3.2 mm ID × 15 mm) as added protection for the main column.
5. Elute the substrate and cyclized peptide product by isocratically washing the column with 24% (v/v) acetonitrile in 0.1 M sodium acetate, pH 6.5, at a flow rate of 1.2 ml/min. Detection can be accomplished by the use of a fluorescence detector with an excitation filter of 352–360 nm and an emission cut-off filter of 482 nm.

Figure 13a clearly shows the separation of substrate and product peptides, while *Figure 13b* is a time-course of enzymatic conversion of the dansylated peptide substrate.

7.4.2 δ-(L-α-aminoadipyl)-L-cysteinyl-D-valine (ACV) synthetase

ACV synthetase acts on L-α-aminoadipic acid, L-cysteine, and L-valine to form ACV, which is the precursor to the β-lactam ring in fungi and streptomyces. The assay of this enzyme depends on the separation and detection of derivatives of *O*-phthaldehyde (OPA) with the ACV and the unreacted amino acids by reverse-phase HPLC (40).

Protocol 8

Assay of ACV synthetase

Equipment and reagents

- Reverse-phase HPLC column
- Fluorescence detector
- Centrifuge
- Nitrogen
- *Streptomyces clavilígurus* cell extract
- Buffer: 150 mM KCl, 45 mM ATP, 45 mM $MgCl_2$, 15 mM EDTA, 3 mM chloramphenicol in 0.3 M Mops, pH 7.2
- 20% (w/v) trichloroacetic acid
- L-α-aminoadipic acid, L-cysteine hydrochloride, L-valine
- Performic acid
- Fluo-R reagent
- Homoserine
- 0.1 M sodium acetate, pH 6.25
- Solvent A: 0.1 M sodium acetate, pH 6.25, methanol, THF (90:9.5:0.5)
- Solvent B: methanol

Assay

1. Mix 0.5 ml of a crude or fractionated cell extract of *Streptomyces clavilígurus* with 0.5 ml of a solution containing 150 mM KCl, 45 mM ATP, 45 mM $MgCl_2$, 15 mM EDTA and 3 mM chloramphenicol in 0.3 M Mops buffer, pH 7.2.
2. Sparge the mixture with N_2 at 0 °C for 10 min and seal the tube with a rubber septum.
3. Initiate the reaction with injection of 0.5 ml of a de-gassed solution containing 15 mM L-α-aminoadipic acid, 15 mM L-cysteine hydrochloride and 15 mM L-valine adjusted to pH 7.2.
4. Incubate the reaction mixture at 27 °C with gentle agitation.
5. Terminate the reaction by injecting 0.4 ml of 20% (w/v) trichloroacetic acid and clarify the suspension by centrifugation.
6. Store at −20 °C until analysis by HPLC.

HPLC analysis

1. Oxidize a 100 μl aliquot of the sample with an equal volume of performic acid for 2.5h at 0 °C, add 2 ml water and lyophilize before re-dissolving in 100 μl water.
2. Prepare fluorescent isoindole derivative (OPA-derivative) by mixing 20 μl of above sample with 40 μl Fluo-R reagent. 20 μl homoserine solution can also be added to act as an internal standard.
3. Quench the reaction with 120 μl 0.1 M sodium acetate buffer, pH 6.25, and use 20 μl aliquots for HPLC analysis.
4. Separate the isoindole derivatives on a 45 × 4.6 mm reverse-phase Ultrasphere ODS column containing 5 μm particles.
5. Develop a binary gradient between solvent A (0.1 M sodium acetate, pH 6.25–methanol–THF, 90:9.5:0.5) and solvent B (methanol).
6. Carry out the detection with a fluorescence detector with excitation at 300–395 nm and emission at 420–650 nm.

Figure 14 is a chromatogram of an oxidized sample from a 1 h incubation mixture containing extract of mycelium. This simple and rapid assay has also been used for monitoring purification of ACV synthetase (40).

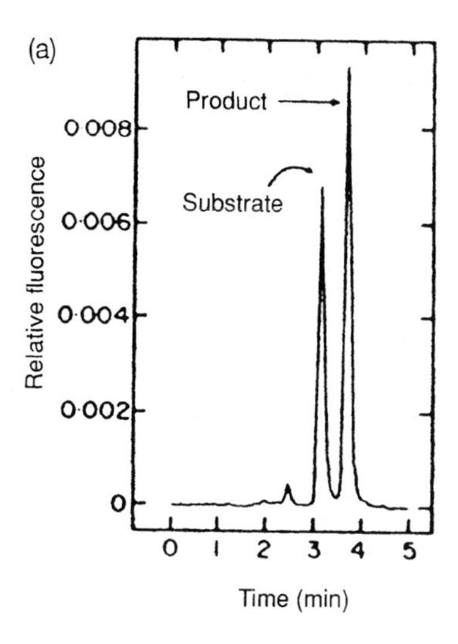

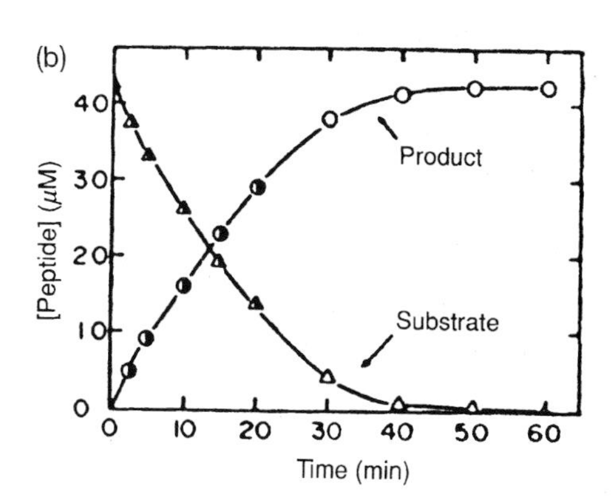

Figure 13 **(a)** HPLC separation of a reaction mixture containing the substrate Gln-Leu-Tyr-Glu-Asn-Lys-ε-(Dns)-OH and the enzymatically generated product <Glu-Leu-Tyr-Glu-Asn-Lys-ε-(Dns)-OH). Partial conversion of the substrate to product was achieved by incubation of Gln-Leu-Tyr-Glu-Asn-Lys-ε -(Dns)-OH (43 μM) with crude bovine pituitary homogenate (0.5 mg ml^{-1}) in 50 mM Tris–HCl, pH 8.0, for 10 min. HPLC conditions are as described in the text. Peak heights correspond to the injection of 80 pmol substrate and 130 pmol product. Reproduced from (39) with permission. **(b)** Time-course of enzymatic conversion of Gln-Leu-Tyr-Glu-Asn-Lys-ε-(Dns)-OH to Glu-Leu-Tyr-Glu-Asn-Lys-ε-(Dns)-OH. Reactions were initiated by the addition of enzyme preparation to achieve S-300 purified glutaminyl cyclase Peak II (160 μg ml^{-1}) and peptide substrate (43 μM), in 50 mM Tris–HCl, pH 8.0. At the indicated times, aliquots (10 μl) of reaction mixture were removed, mixed with 10 mM phenanthroline (10 μl) and subjected to HPLC analysis. The time course of product formation (o) and substrate depletion (Δ) were obtained from HPLC analysis of samples either immediately following collection (open symbols) or after a 3 h delay (filled symbols). Reproduced from (39) with permission.

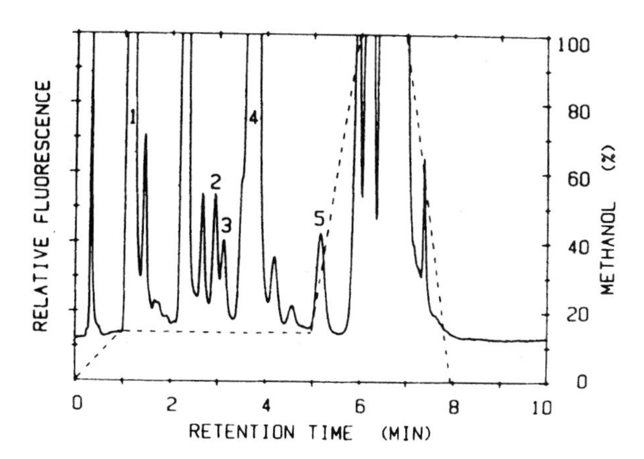

Figure 14 Chromatogram of an oxidized sample from a 1 h incubation mixture containing the crude extract from 24 h mycelium. The concentration of methanol in the gradient is shown by the dashed line. The peaks are: 1, cysteic acid; 2, ACV sulfonate; 3, unidentified but associated with cysteine; 4, α-aminoadipic acid; 5, homoserine, the internal standard. Reproduced from (40) with permission.

7.5 Oxidoreductases

7.5.1 Glutamate synthase (EC 1.4.7.1) from cyanobacteria

The assimilation of ammonia occurs mainly via the glutamine synthetase (EC 6.3.1.2) and glutamate synthase (EC 1.4.7.1) pathway and in some micro-organisms via gluatmate dehydrogenase (41). These enzymes form a key role in bridging nitrogen and carbon metabolism. Both the synthetase and synthase activities can be monitored by detecting glutamine and/or glutamate. Both metabolites can be separated by reverse phase HPLC (42).

Protocol 9

Assay of glutamate synthase

Equipment and reagents

- Reverse-phase HPLC column
- Spectrophotometer
- Centrifuge
- Potassium phosphate, pH 7.0, L-glutamine, 2-oxoglutarate, aminooxyacetate
- *Synechococcus* ferredoxin
- Glutamate synthase
- Sodium dithionite, $NaHCO_3$, HCl
- Potassium phosphate, pH 7.5
- OPA
- 20 mM sodium phosphate, pH 6.5, 22% (v/v) methanol, 2% (v/v) THF

1. Prepare an assay mixture of glutamate synthase with a total volume of 0.9 ml and containing 45 μmol of potassium phosphate, pH 7.0, 5 μmol L-glutamine, 1 μmol 2-oxoglutarate, 5 μmol aminooxyacetate, 10 nmol *Synechococcus* ferredoxin and an aliquot of enzyme.
2. Start the reaction with 0.1 mol solution containing 0.8 mg sodium dithionite in 0.12 M $NaHCO_3$ and incubate at 30 °C for 15 min. Terminate the reaction with 0.6 ml of 1 M HCl.
3. Centrifuge 0.4 ml of the sample at 12 000 **g** for 4 min and dilute 25-fold with 50 mM potassium phosphate, pH 7.5.
4. Mix 50 μl of this with 150 μl of a derivatizing solution of OPA according to (43).
5. Inject 20 μl into the injector loop after 90 s.
6. Carry out the HPLC analysis on a *μ* Bondapak C18 or a Novapak C18 (3.9 mm × 4 cm) column thermostatted at 45 °C.
7. Elute the column with 20 mM sodium phosphate, pH 6.5, containing 22% (v/v) methanol and 2% (v/v) THF at a flow rate of 1–1.5 ml/min.

Figure 15 shows separation of gluatmate and glutamine for the *in situ* assay. The amounts of glutamate and glutamine can be obtained from a calibration curve prepared with standards of these amino acids. Use amounts in the range 0–3 nmole/injection.

7.6 Transferases

7.6.1 Aryl alkylamine (serotonin) *N*-acetyltransferase (EC 2.3.1.87)

Serotonin N-acetyltransferse (NAT) catalyses the N-acetylation of serotonin to N-acetyl serotonin and is a key regulatory enzyme in the melatonin (5-methoxy-N-acetyltryptamine) pathway (44). The present assay for the NAT activity is based upon ion-pair HPLC using either fluorescence or electrochemical detection of N-acetyltryptamine (45). In the present case, only the former detection method will be described.

Protocol 10

Assay of NAT

Equipment and reagents

- Reverse phase HPLC column
- Fluorescence detector
- Sonicator
- Centrifuge
- Retinal or pineal gland tissue
- Tryptamine
- 0.25 M potassium phosphate, pH 6.5, 1.4 mM acetyl-CoA
- Perchloric acid
- Phosphoric acid, methanol, sodium octylsulphate

1. Prepare a homogenate of retinal or pineal gland tissue by sonicating in different volumes of ice-cold 0.25 M potassium phosphate buffer, pH 6.5, containing 1.4 mM acetyl-CoA to give the desired protein concentration.
2. Centrifuge at 28000 **g** for 10 min at 4°C.
3. Mix 75 μl aliquots of the resulting supernatant or whole homogenate (cultured retinal cells) with 25 μl of 8 mM tryptamine in the same buffer and incubate at 37°C for 15 min.
4. Stop the enzyme reaction by the addition of 20 μl of 6 M perchloric acid and centrifuge at 28000 **g** for 10 min at 4°C.
5. Use 10 μl aliquots for HPLC analysis.
6. Carry out the separation on a Partisphere C18 (5 μm particles, 110 × 4.7 mm) reverse phase column. Wash the column with 50 mM phosphoric acid containing 33% (v/v) methanol and 0.65 mM sodium octylsulphate, adjusted to pH 3.5, at a flow rate of 1.5 ml/min.
7. Set the excitation and emission wavelengths at 285 and 360nm respectively for detection.

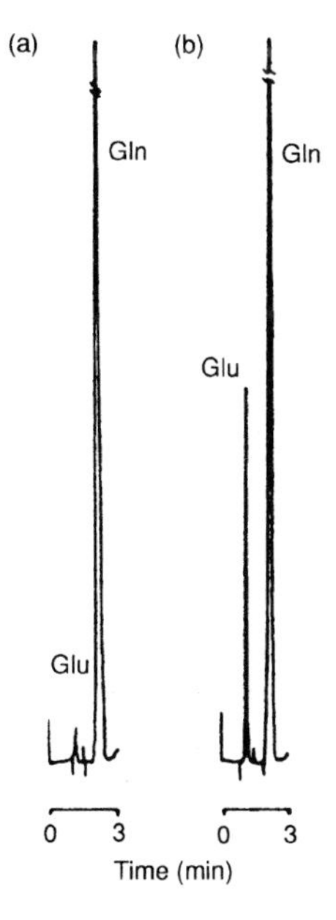

Figure 15 Chromatograms corresponding to a glutamate synthase *in situ* assay. (a) Sample taken at zero time; (b) sample taken after completion of the assay. Reproduced from (42) with permission.

The separation of *N*-acetyltryptamine from the substrate tryptamine is shown in *Figure 16*. Sodium octylsulphate acts as an ion-pairing reagent and increases the retention time of tryptamine, which elutes as a broad peak. The synthesis of *N*-acetyltryptamine standard and procedures for the preparation of tissues and cell cultures are fully described in (45).

7.6.2 Ornithine aminotransferase (OAT) (EC 2.6.1.13) from rat liver

Ornithine aminotransferase is a mitochondrial matrix enzyme catalysing the conversion of L-ornithine and 2-oxoglutarate to form glutamic-γ-semialdehyde and glutamate. The semialdehyde spontaneously cyclises to give Δ^1- pyrroline-5-carboxylic acid (P5C). The assay for the aminotransferase activity in the present case depends on the reaction of P5C with *O*-aminobenzaldehyde (OAB) and separation of the resulting dihydroquinozolinium (DHQ) by reverse phase HPLC (46)

Protocol 11

Assay of ornithine aminotransferase

Equipment and reagents

- Reverse-phase HPLC column
- Spectrophotometer
- Centrifuge
- Potassium phosphate, pH 7.4, sucrose, pyridoxal 5-phosphate
- Female rat liver
- L-ornithine, 2-oxoglutarate
- HCl, OAB
- Methanol

1. Homogenize the liver of a female rat in a 20% (w/v) solution containing 0.1 M potassium phosphate, pH 7.4, with 0.2 M sucrose and 4 μg pyridoxal 5 -phosphate/ml at 4 °C.
2. Centrifuge at 14000 **g** for 15 min to remove cell debris and particulate matter and use it for subsequent assays.
3. Prepare a fresh assay mixture containing 35 mM L-ornithine, 3.7 mM 2-oxoglutarate and 4 μg pyridoxal-phosphate/ml in 50 mM potassium phosphate, pH 7.4, in a total volume of 2 ml.
4. Incubate the enzyme at 37 °C for 0–60 min withdrawing samples periodically and terminating the reaction with 1 ml of 3 M HCl containing 7.5 mg OAB/ml.
5. Centrifuge samples to remove precipitated protein. Monitor the formation of DHQ at 440 nm using a UV/visible detector.
6. Calculate its concentration by using an absorption coefficient of 2.59 mM^{-1} cm^{-1}.
7. Elute the DHQ and OAB from a LiChrosorb C_{18} (4.6 × 250 mm, 10 μm particles) column by isocratically washing with methanol/H_2O (1:2) mixture, with detection at 254 nm. Use a flow rate of 1.5 ml/min.

Figure 17a shows the separation of DHQ and OAB standards while *Figure 17b* is a plot of OAT activity versus time. The OAT activity is represented as μmoles P5C, which can be calculated from a standard curve.

References

1. Englehardt, H. and Muller, H. (1981). *J. Chromatogr.*, **218**, 395
2. Unger, K. K., Becker, N., and Roumeliotis, P. (1976). *J. Chromatogr.*, **125**, 115.
3. Englehardt, H. and Ahr, G. (1981). *Chromatographia*, **14**, 227. *High-performance liquid chromatography* (1985). (ed. A. Hanschen, K. P. Hupe, F. Lottspeich, and W. Voelter). VCH Verlagsgesellschaft, Germany.

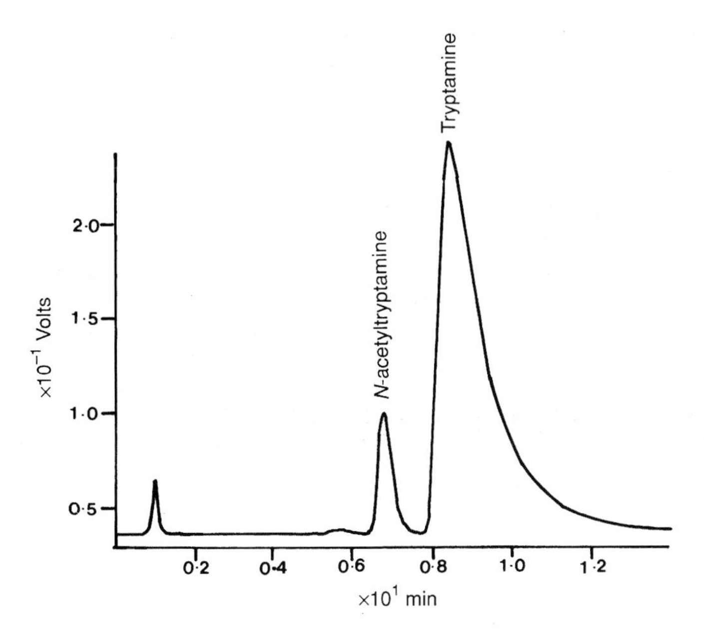

Figure 16 Representative HPLC-fluorescence chromatogram of a sample prepared from chicken pineal gland. Pineal glands isolated in the middle (6th hour) of the dark phase of the light–dark cycle were assayed for NAT activity. The enzymatic reaction was stopped by addition of 6 M perchloric acid and samples were further processed for HPLC analysis. The *N*-acetyltryptamine peak represents 448 pmol injected in a volume of 10 μl. For methodological details see text. Reproduced from (45) with permission.

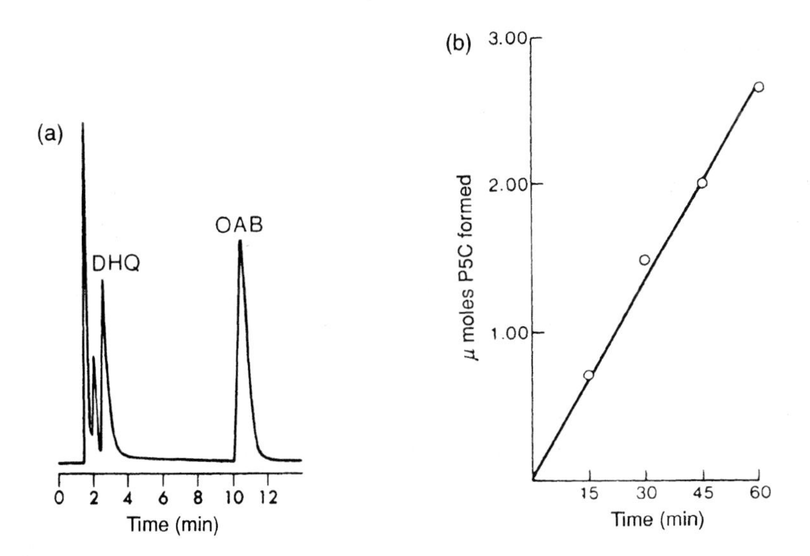

Figure 17 **(a)** Isocratic reverse-phase HPLC showing separation and detection of dihydroquinozolinium (DHQ) and *O*-aminobenzaldehyde (OAB). The column was LiChrosorb C_{18}, 10 μm and 4.6 × 250 mm, and the solvent system was 1 part methanol:2 parts H_2O, pumped at a flow rate of 1.5ml/min. Reproduced from (46) with permission. **(b)** OTA activity as determined by HPLC (see text) plotted against time. The reaction is linear for 60 min.

4. Krstulovic, A. M. and Brown, P. R. (1982). *Reserved-phase high-performance liquid chromatography – theory, practice and biomedical applications.* John Wiley and Sons, New York.
5. Horvath, C. (ed.) (1983). *High performance liquid chromatography (advances and perspectives)*, Vol. 3. Academic Press, New York.
6. Hearn, M. T. W. (ed.) (1985). *Ion-pair chromatography (theory and biological and pharmaceutical applications).* Marcel Dekker, Inc., New York.
7. Knox, J. H. and Pryde, A. (1975). *J. Chromatogr.*, **112,** 171.
8. Karch, K., Sebastian, I., and Halasz, I. (1976). *J. Chromatogr.*, **122,** 3.
9. Frank, H. S. and Evans, M. W. (1945). *J. Chem. Phys.*, **13**, 507.
10. Horvath, C., Melander, W., and Molnar, I. (1976). *J. Chromatogr.*, **125,** 129.
11. Karger, B. L., Gant, J. R., Hartknopf, A., and Weiner, P. H. (1976). *J. Chromatogr.*, **128**, 65.
12. Jandera, P., Colin, H., and Guiochon, G. (1982). *Anal. Chem.*, **54,** 435.
13. Karger, B. L., Le Page, J. N., and Tanaka, N. (1980). *High-Perf. Liq. Chromatogr.*, **1,** 113.
14. Horwath, C., Melander, W., Molnar, I., and Molnar, P. (1977). *Anal. Chem.*, **49,** 2295.
15. Snyder, L. R. and Kirkland, J. J. (1979). *Introduction to modern liquid chromatography.* Wiley-Interscience, New York.
16. Lawrence, J. L. (1987). *J. Chromatogr.* Sci., **17**, 147.
17. Kissinger, P. T. (1983). *Chromatogr.* Sci., **23**, 125.
18. Stulik, K. and Pacakova, V. (1982). *CRC Crit. Rev. Anal. Chem.*, **14**, 297.
19. Majors, R. E. (1972). *Anal. Chem.*, **44**, 1722.
20. Kaminski, M., Klawiter, J., and Kowalczyk, J. S. (1982). *J. Chromatogr.*, **243**, 225.
21. Mehdi, S. and Wiseman, S. (1989). *Anal. Biochem.*, **176**, 105.
22. Mori, M. and Tatibana, M. (1978). In *Methods in enzymology*, Vol. 51 (ed. P. Hoffee and M. E. Jones), pp. 111–21. Academic Press, San Diego, CA.
23. Sofer, R. L. (1976). *Ann. Rev. Biochem.*, **45**, 73.
24. Cushman, D. W. and Cheung, H. S. (1971). *Biochim. Biophys. Acta*, **250**, 261.
25. Horiuchi, M., Fujimara, K., Tarashima. T., and Iso, T. (1982). *J. Chromatogr.*, **233**, 123.
26. White, P. J. and Kelly, B. (1965). *Biochem. J.*, **96**, 75.
27. Asada, Y., Tanizawa, K., Kawabata, Y., Misono, H., and Soda, K. (1981). *Agric. Biol. Chem.*, **45**, 1513.
28. Weir, A. N.C., Bucke, C., Holt, G., Lilly, M. D., and Bull, A. T. (1989). *Anal Biochem.*, **180**, 298.
29. Unnithan, S., Moraga, D. A., and Schuster, S. M. (1984). *Anal. Biochem.*, **136**, 195.
30. Hall, P. F. (1986). *Steroids*, **48**, 131.
31. Schatzman, G. L., Laughlin, M. E., and Blohm, T. R. (1988). *Anal. Biochem.*, **175**, 219.
32. Jackson, A. H., Sancovich, H. A., Ferramola, P. M., Evans, N., Games, D. E., Matlin, S. A., *et al.*. (1976). *Phil. Trans,. R. Soc. Lond. Ser. B.*, **273**, 191.
33. Francis, J. E. and Smith, A. G. (1983). *Anal Biochem.*, **138**, 404.
34. Kawanishi, S., Seki, Y., and Sano, S. (1983). *J. Biol. Chem.*, **258**, 4285.
35. Busby, W. H., Quackenbush, G. E., Humm, J., Youngblood, W. W., and Kizer, J. S. (1987). *J. Biol. Chem.*, **262**, 8532.
36. Fischer, W. H. and Spiers, J. (1987). *Proc. Natl. Acad. Sci.* USA, **84**, 3628.
37. Bond, M. D., Auld, D. S., and Lobb, R. R. (1986). *Anal. Biochem.*, **155**, 315.
38. Merrifield, R. B. (1963). *J. Amer. Chem. Soc.*, **85**, 2149.
39. Consalvo, A. P., Young, S. D., Jones, B. N., and Tamburini, P. P. (1988). *Anal. Biochem.*, **175**, 131.
40. White, R. L., De Marco, A. C. Shapiro, S., Vining, L. C., and Wolfe, S. (1989). *Anal. Biochem.*, **178**, 399.
41. Stewart. G. R., Mann, A. F., and Fentem, P. A. (1980). In: *The Biochemistry of plants*, Vol. 5 (ed. P. K. Stumpf, and E. E. Conn), pp. 2271–327. Academic Press, New York.
42. Marques, S., Florencio, F. J., and Candau, P. (1989). *Anal. Biochem.*, **180,** 152.
43. Lindroth, P. and Mopper, K. (1979). *Anal Chem.*, **51**, 1667.
44. Klein. D. C., Berg, G. R., and Weller, J. L. (1970). *Science*, **168,** 979.
45. Thomas, K. B., Zawilska, J., and Iuvone, P. M. (1990). *Anal. Biochem.*, **184,** 228.
46. O'Donnell, J. J., Sandman, R. P., and Martin, S. R. (1978). *Anal Biochem.*, **90,**

Chapter 5
Electrochemical assays: the oxygen electrode

J. B. Clark
Department of Neurochemistry, Institute of Neurology, UCL Queen Square, London WC1N 3BG, UK.

1 Introduction

Until about 45 years ago the methods of choice for the measurement of oxygen consumption or evolution in biological systems were Warburg manometry or chemical techniques. However, with the development of the membrane-covered, complete polarographic oxygen electrode by L. C. Clark (1) and its subsequent modification, commercially available, robust and sensitive oxygen electrodes became available for general use. Prior to this, the use of oxygen polarography had been limited to the experts who could construct their own electrodes and the specialized recording apparatus to go with them. The advent of a means of measuring oxygen instantaneously and continuously dramatically changed approaches to the study of biological respiratory systems and more recently to those enzyme systems which utilize or produce molecular oxygen. Oxygen electrode systems are now available commercially for a range of applications, from macrotechniques for assessing oxygen content in rivers, sewage, or industrial reactors under high pressure and temperature, to micro systems for measuring oxygen content of body fluids, or oxygen utilized by subcellular organelles, e.g. mitochondria.

2 Theory and principles

The principles underlying the theory of the oxygen electrode (more correctly an oxygen sensor) have been carefully and extensively reviewed elsewhere (2, 3). Hence only a brief resumé will be presented here, relating specifically to the Clark-type electrode, which is the one most routinely in use for biochemical assays. This electrode, first described by Clark in 1956 (1), has a platinum cathode which is maintained at a negative voltage of 0.6–0.9 V with respect to a Ag/AgCl reference electrode, the whole electrode being covered by a thin polyethylene or Teflon membrane (0.3–0.13 mm). This covers a thin layer of saturated KCl which acts as the bridge between the two electrodes. This membrane, which prevents premature poisoning of the electrodes by biological material and is impermeable to water and solutes, acts as a non-conducting barrier between the electrodes and solution/gas in which the oxygen is to be measured, but is nevertheless permeable to oxygen.

The oxygen in the solution to be measured undergoes electrolytic reduction at the cathode with the production of a current which may be measured by means of an appropriate amplifier system and recorder. The reaction at the cathode follows the stoichiometry:

$$O_2 + 2H_2O + 2e^- \rightarrow 2OH^- + H_2O_2$$
$$H_2O_2 + 2e^- \rightarrow 2OH^-$$

Thus, there is a stoichiometry of 4e/mol of oxygen consumed and, perhaps more importantly for biochemical assays, the reaction is insensitive to pH.

3 Current/voltage relationships

The relationship between the applied negative voltage to the cathode and the response of the electrode in terms of current produced is such that there is a plateau response (current) between 0.6 V and 0.9 V, when the plateau (or diffusion) current is proportional to the chemical activity (or tension, partial pressure) of the reactant (O_2) at the electrode surface. As most media in which biochemical assays are conducted have a chemical activity of unity, this means that the current is proportional to the oxygen concentration at the electrode surface. It is, however, important that the solution in which the oxygen is being measured is kept stirred so that the oxygen concentration at the electrode surface is truly representative of the oxygen concentration of the whole solution. It is also important that stirring should be reproducibly controlled since there may be a 2-fold drop in current on the cessation of stirring.

4 Sensitivity

The sensitivity of an oxygen electrode to the rate of change of oxygen tension is limited by a number of factors, some of which have already been mentioned: the polarizing voltage and the rate of oxygen diffusion to the cathode surface. In general terms, for a particular set of conditions the response time will be inversely proportional to the diameter of the cathode. Given that most electrodes are of fairly standard size, perhaps more important is the thickness of the Teflon membrane stretched across the electrode surfaces. Not only will a thicker membrane produce a decreased response time, but any variation in the thickness of the film of saturated KCl solution between the membrane and cathode caused by the membrane may be a source of instability. It is therefore important to stretch the membrane taut and evenly across the electrode surface. Response times for several available electrodes are in the order of seconds, although modifications have succeeded in achieving response times as low as 1 ms (4).

Dependent on their cathode size and design, all oxygen electrodes have a small residual consumption of oxygen which gives rise to a residual current below which oxygen tensions cannot be measured. For a Clark-type electrode this will be in the order of 1×10^{-10} amp which represents an oxygen concentration of 0.0005% or a PO_2 of 0.04 mmHg (3). Electrodes will also drift over a time period and this may be of the order of 0.1 µl O_2/hour (2). Both these factors will determine the lower limit of sensitivity at which Clark-type electrodes will operate and consideration must be given to the rate of oxygen utilization by the biological system before embarking on development of a polarographic assay. It is worth noting that oxygen electrodes have a relatively high temperature coefficient and therefore must be operated under conditions of careful temperature control.

Ageing or lack of sensitivity of electrodes is caused by poisoning of the platinum electrode, particularly by biological samples containing phosphates, –SH reagents, and proteins. Much of this is prevented by the thin Teflon membrane covering the electrode, but nevertheless electrodes do age. However, they may be rejuvenated by soaking in 3% (w/v) ammonium hydroxide solution for a few minutes and subsequently rubbing the electrode surfaces with a slight abrasive paper (e.g. fine Emery paper).

5 Calibration

Clearly calibration must be done routinely and under the precise conditions of temperature and media conditions that pertain to the experimental system under investigation. This has been reviewed extensively (4). However, most biochemical assay systems are carried out in dilute aqueous salt solutions in which standard calibration conditions pertain. Either of the following procedures may be used:

(a) Set up the electrode in its chamber and introduce 1 ml of distilled water (equilibrated with air = 100%). Add a few crystals of sodium dithionite ($Na_2S_2O_4$). The oxygen concentration rapidly falls to zero with a consequent response by the recorder. This position should be adjusted on the recorder chart to zero.

(b) Carefully wash out the electrode chamber several times with distilled water to remove any remaining dithionite. Introduce a further ml of distilled water (air equilibrated), allow to equilibrate to the temperature of the electrode chamber (25 °C, electrode response will rise slightly and then plateau). Adjust recorder response to 90% by suitable sensitivity controls. This level now represents 240 μM O_2 or 480 ng atoms oxygen/ml at 25 °C. This may be calculated from the dissolved gas constants available in the literature (*Table 1*) (5, 6).

Calibration of the intervening recorder chart may then be carried out assuming linearity or alternatively by carrying out the following procedures:

(a) Pipette 0.95 ml of the experimental buffer solution into the electrode chamber and add 50 μl of a freshly-prepared solution of phenazine methosulphate (2 mg/ml water). Close electrode, allow solution to temperature equilibrate.

Table 1. (a) Volume of oxygen dissolved in aqueous medium (microliters of oxygen per millilitre at 1 atmosphere).

	Equilibrated with 100% O_2		Equilibrated with air (21% O_2)	
Temp. °C	**H_2O***	**Ringer Soln.†**	**H_2O***	**Ringer Soln.†**
15	34.2	34.0	7.18	7.14
20	31.0	31.0	6.51	6.51
25	28.5	28.2	5.98	5.92
28	26.9	26.5	5.65	5.56
30	26.1	26.0	5.48	5.46
35	24.5	24.5	5.14	5.14
37	23.9	23.9	5.02	5.02
40	23.1	23.0	4.85	4.83

* from *Handbook of Chemistry and Physics* 40th Ed., Chemical Rubber Pub. Co., Cleveland. 1958–1959.
† recalculated from Umbriet *et al.* (1964). *Manometric Methods*. 4th Ed. Burgess Pub. Co.

(b) Solubility of O_2 in buffered mitochondrial medium equilibrated with air (21% O_2).

Temp. °C	μg atoms O_2/ml*	μmoles/ml (mM)
15	0.575	0.288
20	0.510	0.255
25	0.474	0.237
30	0.445	0.223
35	0.410	0.205
37	0.398	0.199
40	0.380	0.190

* Solubility of O_2 experimentally determined by Chappell (1964). *Biochem J.*, **90**, 225, in a buffered mitochondrial medium containing NADH, inorganic phosphate, and isolated mitochondria.

These tables are taken with permission from the YSI publication on YSI model 5300 Biological O_2 monitor manual.

(b) Add to electrode chamber 10 μl of 10 mM NADH (i.e. 0.1 μmole NADH). Note that the actual NADH concentration should be determined spectrophotometrically by monitoring the fall in absorbance at 340 nm in the presence of malate dehydrogenase and oxaloacetate.

(c) The PMS will be reduced stoichiometrically by the NADH and the reduced PMS reoxidized by the oxygen in solution. From the amount of NADH injected it should be possible to calculate the theoretical oxygen uptake and hence equate that with the distance fallen on the recorder.

NADH → NAD$^+$; PMS → PMS.2H; PMS.2H → PMS; O_2 → H_2O_2

It remains to be stressed, however, that the oxygen electrode measures chemical activity and not the concentration of oxygen present. For this reason the electrode must be calibrated for the reaction medium to be used experimentally and it may not be assumed that one particular calibration holds for a different medium, since the activity coefficient is dependent, amongst other factors, on the ionic strength of the buffer.

6 Electrode systems

Several manufacturers supply oxygen electrode systems which consist of the following:

- an oxygen probe (electrode)
- a polarizing, back-off box
- a temperature-compensated incubation chamber with stirrer
- a waterbath
- a potentiometric recorder

The electrode most commonly used is the probe variety available from YSI Inc., Yellow Springs, Ohio, USA (through Clandon Scientific Ltd., Lynchford House, Lynchford Lane, Farnborough, Hants GU14 6LT) or a 1 ml probe from Instech Labs (5209, Militia Hill Rd, Plymouth Meeting, PA 19462-1216, USA). Gilson (through Anachem, Luton, Beds. LU2 0EB) and Rank Bros. Ltd. (Bottisham, Cambs, CB5 9DA) also supply electrode systems. The incubation chambers, made out of Perspex or glass, must be water-jacketed and the chamber itself must accommodate a stirrer bar. It must be possible to shut off the atmosphere by means of a sleeve. Suitable injection ports, however, must be available for the introduction of small volumes of samples or solutions.

The polarizing box is a simple circuit carrying out two functions:

- maintaining the electrode at a constant negative potential
- providing suitable sensitivity and back-off for the electrode output to a recorder.

Suitable circuit designs are provided by the manufacturers and, although they may be purchased commercially, any competent electrical workshop will be able to construct one. Most conventional laboratory recorders will be suitable to record the electrode output; ideally they should have a sensitivity better than 20 mV full scale, acceptance of a source impedance of 2K/mV, and a response time of less than 1 s. Most commercially available electrodes function with a final volume of 1–3 ml but may be adapted down to 0.25 ml with suitable modifications and ministurization of the incubation vessel

7 Polarographic assays

7.1 Tissue/organelle respiration studies

A major use to which the oxygen electrode has been put is to study oxygen consumption of various biological preparations ranging from whole cells and slices to subcellular fractions such as mitochondria, synaptosomes, and chloroplasts. In view of the wide variety of conditions, incubation media, and so on, which relate to such studies, no attempt will be made to deal with them comprehensively. A typical example of a polarographic study of mitochondrial function will be used to illustrate the use of the electrode in this context. Human skeletal muscle mitochondria were prepared (7) and stored on ice in cold isolation medium (225 mM mannitol, 75 mM sucrose, 10 mM Tris-HCl, 100 μM K^+-EDTA, pH 7.2) at a concentration of 10–15 mg mitochondrial protein/ml.

The studies were carried out in a final volume of 1 ml in a thermostatted incubation chamber (25°C) and the assay solution was well stirred. The respiration medium consisted of 100 mM KCl, 75 mM mannitol, 25 mM sucrose, 10 mM phosphate-Tris, 10 mM Tris-HCl, and 50 μM EDTA, pH 7.4, plus 0.5 mg bovine serum albumin (BSA) and approximately 0.5 mg of mitochondrial protein (see *Protocol 1*).

Protocol 1

Respiration studies using oxygen electrode

Equipment and reagents

- Oxygen electrode
- Recorder
- Dithionite
- Respiration medium
- Mitochondrial sample
- Substrates
- ADP
- FCCP

1. Adjust sensitivity of recorder to give full scale deflection equal to the oxygen content of 1 ml water at 25°C; adjust back-off sensitivity to give zero reading in the presence of dithionite (see calibration).
2. Wash electrode chamber carefully to remove all traces of dithionite, add 1 ml of respiration medium (including BSA).
3. Allow electrode to equilibrate, inject mitochondrial sample in a minimal volume (50–100 μl) and allow to re-equilibrate.
4. Add substrates in small aliquots (5 or 10 μl) to give final concentrations; 5 mM pyruvate + 2.5 mM malate or 10 mM glutamate + 2.5 mM malate or 10 mM succinate + 5 μM rotenone, and so on.
5. Allow equilibration and then add 250 nmoles ADP (in 5–10 μl) and measure stimulated rate of oxygen consumption [state 3 (8)].
6. When all the added ADP has been phosphorylated as shown by the slowing down of the respiration rate (state 4) repeat **5**) to measure further state 3 respiration.
7. The 'uncoupled' rate of respiration may be measured at the end of the run by adding 1 μM FCCP (5 μl). This will stimulate respiration until all oxygen has been used and will give an additional check on the zero calibration of the recorder.

The respiration rates can be calculated from the calibration, P/O ratios from the oxygen consumed after the addition of a defined amount of ADP, and the respiratory control ratio (measure of mitochondrial integrity) from the ratio of state 3/4 respiration.

Uncouplers (FCCP) and some inhibitors (rotenone, antimycin) are dissolved in ethanol and to remove all traces from the incubation chamber, the chamber and electrode must be rinsed in ethanol. Particular care should be taken that the ethanol does not stay in contact with the electrode membrane from more than a few seconds.

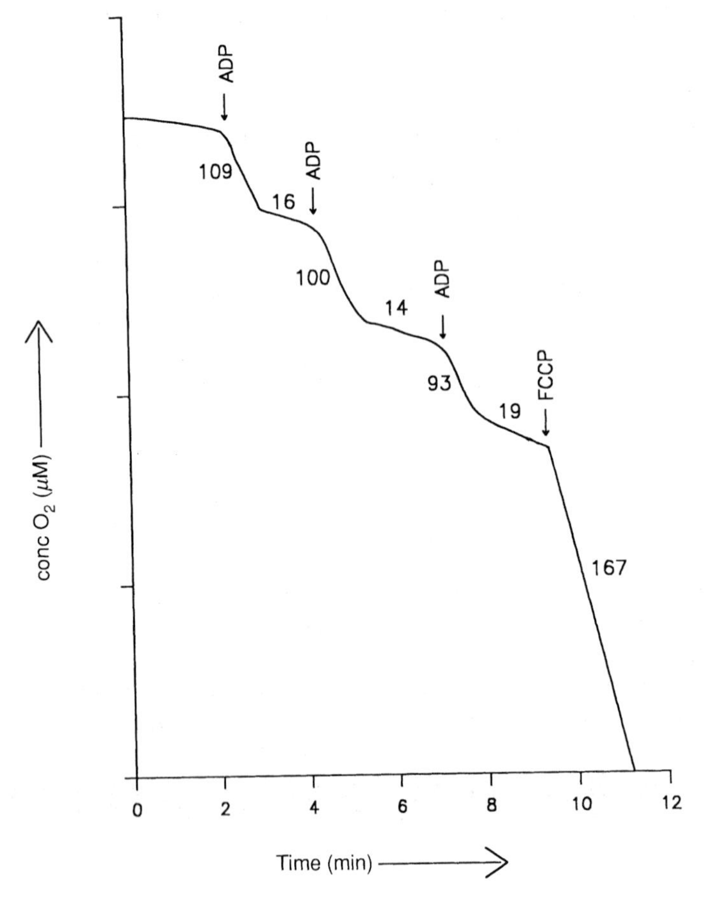

Figure 1 Polarographic studies using an oxygen electrode. The experiment was carried out at 25°C in a KCl, mannitol, sucrose-containing medium (see text) using human skeletal muscle mitochondria (0.75 mg protein). Additions were as follows (final concentrations): malate (2.5 mM), glutamate (10 mM), ADP (250 μmol where indicated), FCCP (1 μmol). The reaction volume was 1 ml and the time axis units are in minutes. The increased rates are expressed as ng atoms oxygen consumed per minute per mg mitochondrial protein.

Figure 1 shows a typical oxygen electrode trace involving the study of human skeletal muscle mitochondria. A number of modifications, particularly for small samples, have been reported in the literature and for further details the reader is referred to (9–11). Additionally, in specifically designed spectrophotometer cells, oxygen electrodes may be used in conjunction with other ion-sensitive electrodes so that simultaneous measurements of parameters such as pH, K^+ concentration, and redox state can be made as well as oxygen uptake (4).

7.2 Specific enzyme studies

Those enzymes which involve the use or production of molecular oxygen as an obligatory part of their mechanism can be assayed directly by the oxygen electrode. Such enzymes are, however, rather limited; e.g. glucose oxidase, catalase. However, modifications of oxygen electrodes, whereby an oxygen utilizing enzyme (e.g. glucose oxidase) has been impregnated into the membrane associated with the electrode, has allowed an effective glucose-sensing electrode-system to be constructed (4). Such adaptations may well be of considerable use in continuous industrial processes.

7.2.1 Glucose oxidase assay

Glucose oxidase (EC 1.1.3.4) catalyses the oxidation of D-glucose to gluconolactone as follows:

$$\text{glucose} + O_2 + H_2O \rightarrow \text{gluconolactone} + H_2O_2$$

The oxygen utilized during the reaction may be used as a measure of the enzyme's activity or alternatively, in the presence of non-saturating glucose concentrations, as a measure of glucose concentration (12). The assay is very simple and may be set up optimally in 50 mM Na acetate buffer, pH 5.6, although other buffers and a wider range of pH may be used without too much loss of activity (pH 3.5–9). Such a system was found to show a linear relationship between either glucose concentration (70 μM–30 mM) or enzyme concentration (0.1–2 units) (12).

7.2.2 Catalase assay

Catalase (EC 1.11.1.6) catalyses the breakdown of hydrogen peroxide as follows:

$$2H_2O_2 \rightarrow 2H_2O + O_2$$

It is a widely distributed enzyme across the animal and plant world and successful analysis of its kinetics was difficult before the establishment of the oxygen electrode system. All assay procedures involve the setting-up of an electrode chamber at constant temperature and with constant stirring of the reaction medium, together with the ability to close-off the reaction system to the atmosphere save for a small injection port for the addition of substrate or enzyme. Such a system has been described (13), the essentials of which are outlined below:

Protocol 2

Assay of catalase

Equipment and reagents

- Oxygen electrode
- Recorder
- Nitrogen
- 50 mM Na_2HPO_4-KH_2PO_4, pH 7.0
- H_2O_2
- Catalase

1. Set up the electrode and add 1 ml of 50 mM Na_2HPO_4-KH_2PO_4, pH 7, buffer.
2. Allow to equilibrate to 25 °C and establish recorder setting near 100%.
3. Bubble buffer system with N_2 gas until zero setting is established and close-off reaction chamber to atmosphere.
4. Add 10 μl H_2O_2 (33.5 mM final concentration) through injection port and allow to equilibrate.
5. Add 50 μl of suitably diluted catalase sample and the initial velocity of reaction may be established from the kinetics of the O_2 production rate. (Note that any non-enzymatic production of oxygen should be deducted.)

It was found (13) that an activity range of 0.01–8.4 μmol O_2/min could be measured and further increases in sensitivity could be achieved by improving the electrode/recorder amplification (down to 0.002 μmol O_2/min). This permitted the measurement of catalase activities over almost a 1000-fold concentration range and was 20 times more sensitive than other assay systems.

7.2.3 Other systems

Considerable use has been made of the oxygen electrode to set up continuous assay procedures for metabolic intermediates along similar lines to those described for the glucose electrode. These are particularly useful in clinical laboratories and include measurement of free and esterified cholesterol by the use of impregnated cholesterol oxidase and hydrolase (14), tyrosine (15), prostaglandin synthesis by the measurement of arachidonic acid (16), and uric acid using urate oxidase and monitoring the oxygen consumption (17). These systems are, however, outside the scope of this article and are only mentioned for completeness.

References

1. Clark, L. C., Jr. (1956). *Trans. Am. Soc. Artificial Internal Organs,* **2,** 41.
2. Fatt, I. (1976). *The polarographic oxygen sensor: its theory of operation and its application in biology, medicine and technology CRC Press, Cleveland, USA.*
3. Davis, P. W. (1962). In *Biological research*, Vol. IV (ed W. L. Nastuk), pp. 137–79. Academic Press, New York.
4. Lessler, M. A. (1982). *Method. Biochem. Anal.*, **28,** 173.
5. *Handbook of chemistry and physics*, (40th edn), 1958–9. Chemical Rubber Co., Cleveland.
6. Chappell, J. B. (1964). *Biochem, J.*, **90,** 225.
7. Holt, I. J., Harding, A. E., Cooper, J. M., Schapira, A. H. V., Toscano, A., Clark, J. B., and Morgan-Hughes, J. A. (1989). *Ann. Neurol.*, **26,** 699.
8. Chance, B. and Williams, G. R. (1956). *Adv. Enzymol,* **17,** 65.
9. Nakamura, M., Nakamura, M. A., and Kobayashi, Y. (1978). *Clin. Chim. Acta,* **86,** 291.
10. Lessler, M. A. and Scoles, P. V. (1980). *Ohio J. Sci.*, **80,** 262.
11. Pappas, T. N., Lessler, M. A., Ellison, E. C., and Carey, L. C. (1982). *Proc. Soc. Exp. Biol. Med.*, **169,** 438.
12. Hertz, R. and Barenholz, Y. (1973). *Biochim. Biophys. Acta,* **330,** 1.
13. Del Rio, L. A., Gomez Ortega, M., Lopez, A. L., and Lopez Gorge, J. (1977). *Anal. Biochem.*, **80,** 409.
14. Dietschy, J. M., Delente, J. J., and Weeks, L. E. (1976). *Clin. Chim. Acta,* **73,** 407.
15. Kumar, A. and Christian, G. D. (1975). *Clin. Chem.*, **21,** 325.
16. Lord, J. T., Ziboh, V. A., Blick, G., Poitier, J., Kursonoglu, I., and Penneys, N. S. (1978). *Brit. J. Dermatol.*, **98,** 31.
17. Nanjo, M. and Guilbault, G. G. (1974). *Anal. Chem.*, **46,** 1769.

Chapter 6
Electrochemical assays: the nitric oxide electrode

R. D. Hurst
Centre for Research in Biomedicine, Faculty of Applied Sciences, University of West of England, Frenchay Campus, Coldharbour Lane, Bristol BS16 1QY

J. B. Clark
Department of Neurochemistry, Institute of Neurology, UCL Queen Square, London WC1N 3BG

1 Introduction

Nitric oxide (NO) is an important, short-lived bio-regulatory messenger molecule that plays a critical part in a variety of biological functions. The selective and reliable detection of NO in biological systems is crucial in order to research and understand the role of NO. Non-electrochemical methods for the detection of NO have been reported but are not well suited for real-time measurement and are very time consuming. A few electrochemical NO sensors have also been described but have significant drawbacks in terms of both selectivity and sensitivity for NO. Moreover, these systems have been limited to researchers with a good understanding of electro-analytical theory and often require an external reference electrode, which can complicate and limit the experimental design (1, 2). Only recently has the technology become available to reliably and specifically measure NO in biological systems. World Precision Instruments (WPI) released recently the first commercially available NO electrode for general laboratory use. The system is described here and offers distinct advantages over those previously reported and enables quick, easy, and reliable NO measurement.

WPI's NO electrode comes with a range of probes of different sizes and flexibility to meet the needs of a variety of experimental applications. Additionally, various accessories for data acquisition and electrode calibration are also available. Using this system NO has been measured in a range of research applications including release from cells in culture (3–6), from beating hearts during ischaemia and reperfusion (7, 8), and in peripheral blood (9).

2 Principles of detection

The detection of NO by the probe is based on a principle extensively reviewed elsewhere and is similar to the Clark oxygen electrode (1). In brief, NO diffuses across a gas-permeable membrane and a thin film of electrolyte covering the probe and is oxidized on the electrode surface. A potential is applied to a measuring electrode relative to a reference electrode and the resulting current caused by oxidation of NO, according to the following reaction, is measured by an amplifier system and recorder:

$$NO + 4OH^- \rightarrow NO_3^- + 2H_2O + 3e^-$$

3 Principles of selectivity and sensitivity

The original probe of the electrode has a 2.0 mm shielded-probe tip and is principally discussed here. The electrode has an inherently high selectivity because the internal electrode is separated from the sample by a gas-permeable membrane. This prevents interference from larger molecules or dissolved ionic species. The selectivity of the probe for NO, over other gases which permeate the membrane, is determined by the potential applied to the electrode. The electrode is electrochemically unreactive to O_2, CO, and N_2. Although CO_2 and NO_2 interfere with the original electrode probe, this normally does not pose a significant problem. Large changes in CO_2 over a relatively short period of time can change the pH of the electrolyte contained between the membrane and the internal electrode and produce a signal. In physiological systems, however, the concentration of dissolved CO_2 generally remains fairly constant and so any signal generated by CO_2 is included in the baseline measurement. NO_2 may interfere, if it reaches the electrode surface, and only poses a real problem when gas-phase measurements are to be made (a gas-phase electrode is now available from WPI). In solution NO_2 is highly unstable and at physiological pH degrades to form nitrate and nitrite – these species do not influence the electrode.

WPI now offer even smaller 'micro' probes, which incorporate a multilayered selective coating which excludes most of the species related to NO research. These include arginine, ascorbic acid, CO, CO_2, cysteine, dopamine, ethanol, glucose and other carbohydrates, H_2O_2, methanol, *N*-acetyl cysteine, nitrate, nitrite, N_2, O_2 and proteins.

If the electrode is correctly calibrated it is extremely sensitive to NO and can measure concentrations of 1 nM–20 μM in solution. In gas mixtures the electrode can measure NO concentrations as low as 1 part/million. The instrument has a limited response time of 5–10 seconds.

4 Environmental influences

The NO electrode is sensitive to certain environmental influences, which must be critically controlled, or adjusted for, in order to make the correct interpretation of the readings obtained.

4.1 Temperature

The electrode is sensitive to temperature. An increase in the background current of the electrode is observed as temperatures increase. In order to compensate, the instrument is equipped with zero adjust controls. At physiological temperatures (37°C) however, it may not always be possible to zero the baseline directly on the meter. To eliminate this problem it is advisable to link the electrode to a chart recorder or acquisition system that imposes no limit on the magnitude of the offset that may be applied to zero the baseline. Because temperature affects the partial pressure of NO in liquid or gas samples, the permeability of the probe membrane, and the conductivities of the meter's circuit components, it should be noted that the sensitivity of the electrode to NO also changes with temperature.

4.2 Electrical interference

Because the meter detects extremely small currents generated at the electrode probe, various external electrical sources can influence the system and produce large extraneous signals. Of course, the external noise level depends upon the environment of the laboratory where the electrode is housed. If the electrical interference is excessive it may be necessary to ground and shield the instrument and sample.

There is a grounding connection on the NO meter itself and the most advisable procedure is to route all electrical equipment in the system, and the sample, to this common ground. This set-up ensures all instruments and equipment are associated with one, and only one, connection to ground – thereby avoiding ground loops. Proper grounding should eliminate most sources of extraneous noise, but in some instances protection against stray electrical fields may be necessary. The best solution then is to place all, if not most, of the instruments and the sample into an iron shield Faraday cage. This should be properly grounded and connected at one point to the common instrument ground.

The movement of people in the immediate vicinity of the electrode and meter can also cause current fluctuations from variations in the resulting stray capacitance. Although it is difficult to eliminate these effects, it is advisable to plan the placement of the system with this in mind. It may be helpful to eliminate the generation of large static charges generated by the operator's body, by using grounded wrist straps. In these circumstances, it is also wise to avoid wearing synthetic (e.g. nylon) fabrics.

5 Membrane integrity and maintenance

The electrode is a delicate and sensitive piece of equipment. Caution must be exercised to avoid damage to the gas permeable membrane covering the electrode probe. If the membrane becomes damaged sample contents will be free to react with the internal electrode surface and will cause erroneous current readings. Furthermore, organic matter can accumulate on the membrane over time and may lead to sluggish responses and/or unusually low sensitivity. For this reason the electrode probe should always be immersed in distilled water after each use.

The membrane integrity can be checked by determining that the current remains low and stable when the electrode is placed in a strong (1 M) saline solution. A damaged membrane should be replaced, but a dirty one can sometimes be cleaned by briefly immersing it in an acid or base solution. A mild protease solution can also be used to remove protein buildup.

6 Calibration

As with the oxygen electrode, calibration must be done routinely under conditions that match, as closely as possible, the experimental system under investigation. Two techniques will be discussed, one suitable for determination of NO in liquids, and another appropriate for gas phase measurements.

6.1 Calibration for liquid measurements

This procedure generates known concentrations of NO within a container supplied by WPI for calibration. A calibration curve demonstrating changes in current or peak height (if a chart recorder is used) as a function of NO concentration produced (a typical chart recorder output is depicted in *Figure 1*).

The generation of NO is based on the following equation where a known amount of KNO_2 ($NaNO_2$ can be substituted for KNO_2) is added to produce a known amount of NO.

$$2KNO_2 + 2KI + 2H_2SO_4 \rightarrow 2NO + I_2 + 2H_2O + 2K_2SO_4$$

Because stoichiometry exists between added KNO_2 and the NO generated, and because KI and H_2SO4 are added in excess, the final concentration of NO generated is equal to the concentration of KNO_2 in the solution. The NO generated in this reaction persists long enough to calibrate the instrument with ease.

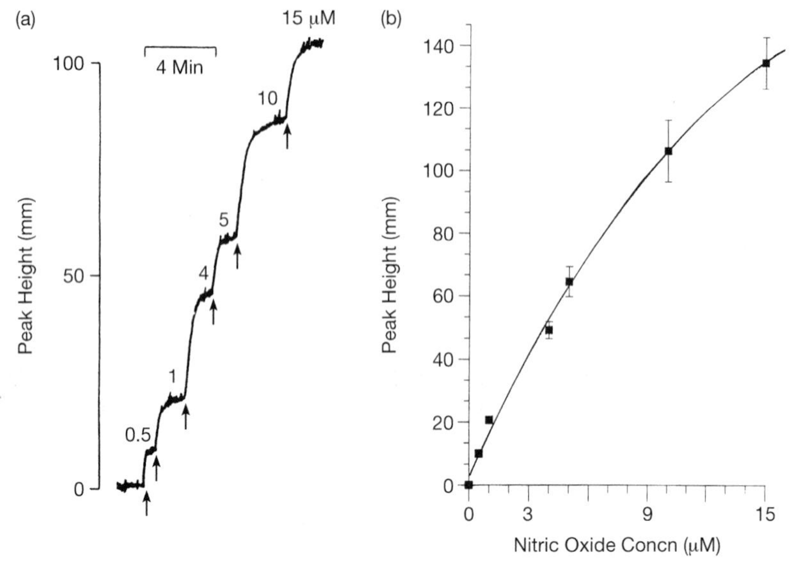

Figure 1 Calibration of the NO electrode by chemical generation of known amounts of NO. A: A representative calibration trace showing the responses of the electrode after successive injections (arrows) of $NaNO_2$ substrate. B: NO calibration curve generated from five separate calibrations. Data are mean ± SEM values.

Protocol 1

Calibration for liquid measurement

Equipment and reagents

- NO electrode
- Chart recorder/acquisition system
- Argon gas
- 0.1 M H_2SO_4, 0.1 M KCl
- KNO_2

1. First check the integrity of the gas permeable membrane as described in Section 5. Place a small magnetic stirrer and an appropriate volume of 0.1 M H2SO4 / 0.1 M KI into the electrode sample container. Gently insert the electrode probe so that it is immersed about 2–3 mm in the solution. It should not be in contact with the stir-bar. Purge the chamber with argon gas for 15 min making sure excess gas is allowed to escape. Once purging is complete and the gas is turned off, seal the chamber so that the argon is not replaced by air.
2. Zero the baseline on the meter or via the chart recorder/acquisition system, if it is being used. The level of baseline noise is variable and dependent upon the experimental arrangement, and how well the system is grounded (see Section 4.2). Generate a known concentration of nitric oxide in the solution by adding a known concentration of KNO2. Within seconds the concentration of NO throughout the solution will be uniform and be reflected by an increase in output current from the electrode. The rise in current will usually level off to a relatively stable value (Figure 1). The currents generated by the addition of successive additions of KNO2 will give the data necessary to generate a standard curve whereby chart recorder responses can be calculated back to give NO concentration. A typical calibration curve is shown in Figure 1. If the electrode is not linked to a chart recorder/acquisition system it is possible, with additional meter adjustments, to read the concentration of NO directly from the meter display.

6.2 Calibration for gas-phase measurements

This procedure requires the preparation of a stock NO gas mixture. NO gas poses a potential safety hazard and therefore it is important that the user takes appropriate precautions, follow the recommendations of the gas supplier and material data safety sheet. The procedure should be carried out in a fume hood.

The procedure requires a gas-tight glass vial (stock vial) equipped with a rubber cap in which two syringe needles (25-gauge are recommended) are inserted. One needle serves as an inlet for purging and should extend well into the container while the other, smaller needle, serves as a gas outlet. Once the NO gas stock is made the procedure also requires a calibration vial (supplied by WPI). This glass vial has a rubber cap with a radial slit through which the electrode can be inserted.

Protocol 2

Calibration for gas-phase measurements

Equipment and reagents

- NO electrode
- Chart recorder/acquisition system
- Gas-tight glass vial (stock vial)
- Two syringe needles (25-gauge)
- Gas-tight syringe
- Argon gas
- NO gas

1 Purge the stock vial with argon gas for 10–15 min. It is important to monitor the flow rate carefully so that the pressure within the vial does not get dangerously high. Then purge the vial with NO gas and, after 10–15 min, quickly remove both needles from the cap leaving the vial sealed.

2 With the electrode placed in the calibration vial insert two syringe needles through the cap as was done for the preparation of stock NO. Purge the vial gently for 10–15 minutes with argon and remove both needles.

3 Zero the baseline of the electrode on the meter or via the chart recorder/acquisition system. Once this value is stable, inject into the calibration vial an aliquot of NO stock using a gas-tight syringe. As with calibration for liquid measurements, a calibration curve can be produced by making successive NO injections. The output from the electrode should be similar to that obtained for liquid calibration (Figure 1).

Assuming a 100% purity of NO gas the concentration of NO in the calibration vial can be calculated from its volume and the amount injected. For example, the volume of the calibration vial supplied by WPI is 22.85 ml ± 0.03 ml. Injection of 0.001 ml (1 μl) of NO stock would give a final concentration of 0.001 ml/22.85 ml = 43.8 parts/million.

It should be stressed that a correction is required if the NO gas supply is not 100% pure.

7 NO and cellular respiration studies

In response to cytokine stimulation the production of NO from a variety of cells may be cytotoxic to other cells in the neighbouring vicinity. This may be particularly important in the brain, because neurons may be particularly sensitive to oxidative stress-induced damage and activated astrocytes may be a major source of the NO (10, 11). Although the mechanism of toxicity is not known, NO can inhibit certain components of the mitochondrial respiratory chain, inhibit cellular respiration and hence limit ATP production. In the studies of Brown *et al.*, the NO electrode was used in combination with the oxygen electrode to monitor cytokine-activated astrocytic NO production and respiration simultaneously (3). Subsequently

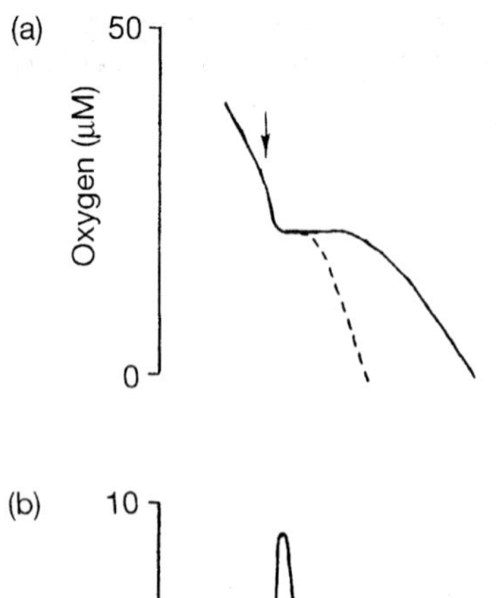

Figure 2 Chart recorder traces showing the effects of NO gas on endothelial cell oxygen consumption. Endothelial cell oxygen consumption (A) and NO concentration (B) were simultaneously measured at 37 °C. Arrows indicate addition of NO at 10 μM (dotted line) and 20 μM (solid line).

we established a similar protocol and have investigated: (a), the properties of a novel NO donor compound on cellular respiration (12), and (b), whether NO-induced BBB (blood–brain barrier) dysfunction is mediated by inhibition of endothelial cell respiration (13). As an example to illustrate a typical use of the NO electrode, the simultaneous determination of NO concentration and endothelial cell respiration using a Clark type oxygen electrode will be described here. *Figure 2* shows typical NO and oxygen electrode traces.

Human vascular endothelial cells (HUVEC-304, European Collection of Animal Cell Cultures, Wiltshire, UK) were grown to confluence in Hepes buffered M199 medium supplemented with fetal bovine serum (10%). Endothelial cells were trypsinized and kept on ice in M199 culture medium until required (<4 hours). The final monitoring volume was 250 μl in a Perspex chamber designed to accommodate both the NO and oxygen electrodes. The vessel was thermostatically regulated at 37 °C and magnetically stirred. The respiration medium consisted of oxygen-saturated Hank's balanced salt solution supplemented with 2 mM $CaCl_2$ and 20 mM Hepes (pH 7.4, 37 °C). Each electrode was individually connected to separate chart recorders. A saturated NO solution was prepared by first bubbling with oxygen-free nitrogen and then authentic NO gas. The concentration of NO was taken at saturation (room temperature) to be 2.0 mM (14).

Protocol 3

NO and O_2 electrodes for cellular respiration studies

Equipment and reagents

- See Protocol 1
- Oxygen electrode
- Respiration medium
- Cells
- NO gas

1. Calibrate the NO electrode for liquid measurements (Section 6.1).
2. Calibrate the oxygen electrode (see Chapter 5).
3. Wash electrode chamber and add 250 μl of respiration medium. Allow both electrodes to equilibrate.
4. Place an aliquot of cells (1.0×10^6) in 250 μl of respiration medium into the chamber and allow to re-equilibrate (2–4 minutes).
5. Once the baseline NO levels and the oxygen consumption rate are stable add an aliquot of NO gas.
6. At the end of each run sample the cells for protein determination.
7. Peak NO concentration can be determined from the calibration curve and cellular oxygen consumption rates calculated from the dissolved gas constant of 0.199 mM and the oxygen calibration.

References

1. Shibuki, K. (1992). *Neuroprotocols*, **1**, 151.
2. Malinski, T. and Taha, Z. (1992). *Nature*, **358**, 676.
3. Brown, G. C., Bolaños, J. P., Heales, S. J.R., and Clark, J. B. (1995). *Neurosci. Lett.*, **193**, 201.
4. Tsukahara, H., Krivenko, Y., Moore, L. C., and Goligorsky, M. S. (1994). *J. Am. Soc. Nephrol.*, **5**, 1046.
5. Tsukahara, H., Ende, H., Magazine, H. I., Bahou, W. F., and Goligorsky, M. S. (1994). *J. Biol. Chem.*, **269**, 21778.
6. Liu, Y., Shenouda, D., Bilfinger, T. V., Stefano, M. L., Magazine, H. I., and Stefano, G. B. (1996). *Brain Res.*, **722**, 125.
7. Engelman, D. T., Watanabe, M., Engelman, R. M., Rousou, J. A., Flack, J. E., Deaton, D. W., and Das, D. K. (1995). *J. Thorac. Cardiovasc. Surg.*, **110**, 1047.
8. Engelman, D. T., Watanabe, M., Maulik, N., Cordis, G. A., Engelman, M., Rousou, J. A., *et al.* (1995). *Annals of Thoracic Surg.*, **60**, 1275.
9. Rysz, J., Luciak, M., Kedziora, J., Blaszczyk, J., and Sibinska, E. (1997). *Kidney Int.*, **51**, 294.
10. Bolaños, J. P., Almeida, A., Stewart, V., Peuchen, S., Land, J. M., Clark, J. B., and Heales, S. J.R. (1997). *J. Neurochem.*, **68**, 2227.
11. Bolaños, J. P., Heales, S. J.R., Land, J. M., and Clark, J. B. (1995). *J. Neurochem.*, **64**, 1965.
12. Hurst, R. D., Chowdhury, R., and Clark, J. B. (1996). *J. Neurochem.*, **67**, 1200.
13. Hurst, R. D. and Clark, J. B. (1997). *Nitric Oxide: Biol. Chem.*, **1**, 121.
14. Archer, S. (1993). *FASEB J.*, **7**, 349.

Chapter 7
Electrochemical assays: the pH-stat

Keith Brocklehurst

Laboratory of Structural & Mechanistic Enzymology, School of Biological Sciences, Queen Mary, University of London, Mile End Road, London, E1 4NS

1 Introduction

pH-stat assays are used to monitor the progress of chemical reactions in which protons are liberated or taken up. This is achieved by measuring the quantity of base or acid that needs to be added at various times to keep the pH essentially constant. The technique was developed in 1923 by Knaffle-Lenz (1) in connection with his work on esterases. It has since been applied widely in many biochemical systems (see ref. 2 for early examples and refs. 3 - 5 for more recent examples). In a more general sense, 'stat' techniques are used to monitor an even greater variety of reactions in which a chemical species that is either generated or consumed can be detected by using electrodes. Reactions involving metal ions can be monitored by using ion-selective electrodes and reactions involving redox changes by using platinum electrodes.

The pH-stat method of determining reaction rates complements spectrophotometric methods. It requires a change in proton binding sites during reaction rather than a change in chromophoric character. Particular advantages are that it can be used to study kinetics in non-buffered solution and in stirred suspension. A consequence of the latter is that the technique is readily used to study turbid cellular extracts and immobilized enzymes or cells. In such systems in particular it may be necessary to establish conditions in which reaction rate is independent of the rate of stirring of the heterogenous reaction mixture or, at least for comparative studies, to keep the stirring rate constant (6, 7).

2 The basis of pH-stat methodology

2.1 Principle and general approach

A pH stat is a type of autotitrator that can be used to maintain a constant pH in a non-buffered solution during a reaction that involves the production or uptake of protons by addition of a solution of base or acid of known concentration (8). The amount of titrant added to maintain the pH is recorded as a function of time to provide a progress curve for the reaction which may be subjected to kinetic analysis. In many kinetic studies reaction volumes and concentrations of reactants and titrant are arranged such that there is negligible volume change during a kinetic run and thus standard kinetic equations are applied. The more general situation in which concentrations of substrate and titrant are comparable, and thus reaction is accompanied by significant volume change, has been treated analytically (9).

When using a simple pH-stat assembly with a 0.5 ml burette, 0.02–0.05 M titrant, and a reaction time restricted to a few minutes to attempt to obviate problems of enzyme instability in dilute solution, it is possible to record enzyme activities as low as about. 0.1–0.5 μmol of substrate transformed per minute. When this limiting sensitivity is not required, however, it is common practice to use assay mixtures such that reaction can be monitored by using 0.1 M titrant. When the titrant is sodium hydroxide, use of this concentration ensures that absorption of residual carbon dioxide does not seriously interfere with the assay and permits the inclusion of very dilute buffer (<1 mM) to assist in establishment of the required assay pH without perceptible influence on the measured value of the steady-state rate.

Variants on the standard pH-stat methodology include electrolytic titrant generation and spectrophotometric monitoring of pH (10).

2.2 pH-stat components and their functions

A pH-stat consists of:

- a thermostatted magnetically stirred reaction vessel
- a pH meter
- a glass electrode
- a KCl bridge
- a reference (calomel) electrode
- an electrode shield
- a controller
- a motor-driven burette
- a recorder

The principles underlying the function of each of these components have been discussed in detail by Jacobsen *et al.* (2). The electrochemical cell, comprising the glass electrode immersed in the reaction mixture, the KCl bridge and calomel electrode, is connected to the pH meter, which is itself connected to the controller. The electrode assembly needs to be accurate and stable and well shielded from the surroundings. Even with such precautions, clothing made from artificial fibres should be avoided to minimize electrostatic perturbations. The controller starts the motor-driven burette when the output potential of the pH meter changes from its fixed initial value and stops it when the initial potential is re-established. The rate of addition of acid or base (titrant) necessary to effectively maintain the pH and to provide a good estimate of the reaction rate depends upon a variety of factors (2), notably the buffer capacity and size of the sample as well as the titrant concentration. The control functions that determine the rate of addition of titrant, therefore, are of considerable importance. The traditional pH stat used an incremental technique in which the progress curve is produced in a stepwise manner through a series of alternate reagent additions and pauses. Regulation of the size of each increment of titrant and each pause was achieved by a proportional control known as a proportional band. The limitations of this approach are:

- overshooting of titration lag
- a steady-state pH that is offset from the required value
- inability to 'remember' how the system reacted previously to a given offset and subsequent addition of titrant

In more recent equipment considerable improvement in proportional control has been achieved, which eliminates the offset from the desired pH and provides a fast response to changes in the reaction mixture (11).

2.3 Some limitations and sources of error

To obtain satisfactory results in pH-stat experiments it is necessary to arrange for reliable constant stirring, stable electrode systems, adequate shielding, and an effective system for the prevention of absorption of atmospheric carbon dioxide. Other problems are existence of an unknown liquid junction potential between saturated KCl and the reaction mixture, and the tendency of the liquid from the burette tip (situated below the surface of the reaction solution) to leak. It is assumed (2) that the diffusion potential is small in dilute solutions between pH 3 and pH 11, being of the order of 0.1 mV (0.002 pH unit) in a solution containing 1 mM HCl and 0.1M KCl and that, in such solutions, its variation may be neglected. With more concentrated solutions, determination of whatever non-enzymic rate may be observed prior to addition of enzyme is considered to deal effectively with a variety of extraneous effects, including whatever change in diffusion potential may be occurring. Leakage from the submerged tip of the burette is minimized by making the density of the titrant lower than that of the reaction mixture. A potential systematic error that may arise in pH-stat assays of enzymes in haemolysates is described in Section 5.

3 Commercial and custom-made pH-stat assemblies

3.1 The range of equipment

The simplest pH stat, sometimes used in undergraduate laboratories, consists of a pH meter and electrode system, a reaction vessel mounted on a magnetic stirrer, a manually operated burette, and a stop-clock. Automatic burettes were developed in the 1930s (12, 13) and autotitration devices (14) led to the development of the automated pH stats built at the Carlsberg Laboratory in the 1950s (15). In the 1950s and 1960s many pH-stat experiments were conducted by using the Radiometer TTT1 autotitrator, an SBU1 burette, a reaction vessel such as the TTA31, and an SBR2 recorder, and reports of experiments conducted with this type of equipment and its successors (e.g. TT2, TTA80, TIM90, and VIT90 autotitrators) continue to appear in the literature. State-of-the-art, computer-linked pH-stat systems currently available commercially include those offered by Radiometer and by Metrohm UK Ltd. Some custom-made pH-stat systems that have been reported in the literature, constructed to meet special requirements in some cases, are described below. These could be used as a basis for updated procedures using modern computer technology.

3.2 Some pH-stat systems described in the literature

3.2.1 Inexpensive systems

Warner *et al* (16) describe an inexpensive pH-stat autotitrator which can be used for kinetic experiments with small reaction volumes (e.g. 2–3 ml). A digitally controlled burette adds titrant to a reaction mixture at a rate proportional to the differences between the solution pH and the specified stat pH. Hamilton glass syringes (250 μl–2.5 ml) deliver titrant at rates from 100 nl/min to about 1.5 ml/min. Advantages of this inexpensive pH system include:

(a) proportional control of the digital titrant dispensing system provides a large range of rates of titrant addition;

(b) the nature of the comparator and the proportionality of titrant addition permit relatively rapid reactions to be followed accurately without over-shooting;

(c) the digital nature of the system provides for relatively low noise and permits easy interfacing with a computer.

Detailed descriptions of the circuits and the theory of operation are given in (16). In addition, an equation relating reaction parameters (reaction volume, titrant concentration, buffer capacity, and kinetic constants) to tracking accuracy is derived and discussed.

An inexpensive, simple, comparator/control unit for a pH-stat containing solid-state circuitry is described by Job and Freeland (17). This can be used with a pH meter and a solenoid burette (18) to construct a general-purpose pH-stat–autotitrator system.

3.2.2 Automated systems dealing with multiple samples

In conventional pH-stat procedures for the determination of enzyme activities, the following operations need to be performed: the controls of the titrator and titrigraph are set; the titration assembly is brought to an initial position with the syringe burette filled with titrant, and the recorder pen set to a zero position; the pH of the substrate solution is brought to the required value and the reaction is started by addition of enzyme solution, often as the recorder pen crosses a marked line on the chart. After recording the progress curve, the reaction vessel and electrodes are rinsed, the burette is refilled, and the recorder pen repositioned in readiness for the next reaction. Keijer (19) describes equipment that can be used in an automatic procedure for large numbers of pH-stat assays of a given enzyme in sequence and illustrates its performance by a cholinesterase assay.

A more flexible automated pH-stat system involving an alternating stop and flow system, which permits the changeover from one type of enzyme assay (i.e. one type of substrate) to another, is described by Vandermeers *et al.* (20). The operation of the system is illustrated by assays of hydrolases in pancreatic homogenates and intestinal juice (lipase, trypsin, and chymotrypsin).

3.2.3 Multiple pH-stat systems: linked pH-stat systems for maintenance of substrate concentration and data acquisition and processing for several pH-stats simultaneously

A method preventing substrate depletion during a kinetic run at low substrate concentrations, by coupling an automatic titration burette to a second burette containing an equimolar solution of the substrate, is described by Konecny (21). Rousseau and Atkinson (22) describe the application of digital data logging equipment in the recording of the operations of multiple pH-stats and the development of software for the conversion of digital data to titrimetric data and then into kinetic parameters. The system allows experimental data to be sampled, collected, and analysed subsequently by off-line computer methods.

3.2.4 Computer-controlled systems

Innovations in reagent delivery systems and computer automation reported in the 1970s and 1980s provide for rapid, accurate, automated kinetic determinations by pH-stat methods. Of particular value are the development of a digital computer-controlled titrant delivery system capable of producing μl aliquots (23) and the description of a pH-stat under real-time computer control (24, 25). Real-time computer-controlled processing eliminates time-consuming calculations and permits real-time modification of parameters according to experimental demands. Feedback from experiments eliminates pH overshoot or lag and allows dynamic alteration of sampling frequency.

3.2.5 A pH-stat with spectrophotometric pH monitoring and electrolytic titrant generation

Conventional pH-stat systems utilize volumetric addition of titrant and potentiometric detection of deviations from the control point. Electrolytic titrant generation provides high sensitivity and accuracy and eliminates the need for standard solutions. It is particularly useful for kinetic studies because rates are followed continuously without dilution errors. Spectrophotometric monitoring of the operating pH can provide greater sensitivity to pH change in alkaline media and faster response times.

Karcher and Pardue (26) describe equipment for this type of pH-stat with the following characteristics: the operating point is monitored spectrophotometrically, a pulsed electrolysis current source is used rather than a continuous source, and the data are processed automatically to yield reaction rates or chemical concentration data directly. The use of pulsed electrolysis current makes the instrument more directly compatible with digital equipment and eliminates the problem of overshoot inherent in continuous current generation. The instrument was developed for the assay of acetylcholinesterase. Acetic acid produced in the reaction is monitored at 404 nm by using 4-nitrophenolate as indicator. Deviation of the transmittance at 404 nm by a preselected amount from an adjustable control point triggers a pulse of current which generates base and titrates the acetic acid produced. Electrolysis current and pulse duration are constant during each pulse of titrant and so a given number of pulses corresponds to a constant reaction interval for each rate measurement.

3.2.6 A totally electrochemical pH-stat

Potential problems with using spectrophotometric pH detection as described in Section 3.2.5 are that some pH-indicator dyes undergo redox changes or bind to proteins, both of which can cause spectral changes. A totally electrochemical device, therefore, is attractive. The major problem encountered when using such devices, both for the measurement of pH and for its restoration following a perturbation, is the interaction between the two processes; for example, the acid–base titrator described by Johansson (27) has a slow response and is susceptible to oscillation about the end-point of a titration. This device measures the electrode signal directly, amplifies it, and applies the signal from which a pre-set voltage is subtracted to a pair of generating electrodes. Adams *et al* (28), prompted by the development by Brand and Rechnutz (29) of a high impedance differential potentiometric circuit, reported the construction and evaluation of equipment in which an isolation amplifier is used to eliminate difficulties that derive from a common ground for pH-measuring and current-generating systems. Their totally electrochemical pH-stat has a minimum time constant of 2–3 s at which no overshoot is observed, and has a current efficiency of almost 100% over a wide range of concentrations.

3.2.7 A temperature-scanning pH-stat

Thermal denaturation of protein in non-buffered solution is accompanied by change in pH (30) and this phenomenon can be investigated at constant pH in a device consisting of a pH-stat, a programmable heating unit, and a temperature measuring and recording system (31).

3.2.8 A flow-through pH-stat for studies on enzymes (such as lipases) that are optimally active at low pH

Taylor (32) describes a continuous pH-stat method in which conditions for the lipase-catalysed reaction and the titration of the fatty acid products are separately controlled. This allows the use of optimal conditions for titration of fatty acids, despite their inhibitory effects on most lipases, and optimal conditions for the lipase-catalysed reaction. In this method, enzyme and substrate are pumped into a stirred emulsion reactor where the catalysed reaction occurs. The reaction mixture is then allowed to flow to a second stirred vessel for product titration.

4 General pH-stat procedure and specific protocols for some individual enzymes

4.1 Procedures

In essence the procedure for carrying out pH-stat assays is common to most reactions. A general procedure is described in *Protocol 1*, below, in terms of reactions such as ester hydrolysis in which protons are liberated. This is readily adapted for use with other reactions in which protons are taken up, or in which special requirements such as a supply of oxygen are required and where more specialized pH-stat systems, such as those involving automation or computer control, are used. Six specific protocols follow (*Protocols 2–7*) to illustrate some of the reaction types to which the pH-stat assay technique has been applied:

- glucose oxidase (33)
- dihydrofolate reductase (DHFR) (34)
- triacylglycerol lipase (35)
- acetylcholine esterase (36)
- cysteine proteinases and serine proteinases using ester substrates (37)
- urease (38)

Other pH-stat assays for which methods are described in the literature include those for:

- hydrogenases (39 – 41)
- lipoprotein lipase (42)
- proteinases using peptide substrates (43, 44)
- β-lactamases (45)
- guanine deaminase (46)
- adenosine deaminase (47)
- glutamate decarboxylase (48)
- lysine decarboxylase (49)
- asparagine and glutamine synthetases (50, 51)
- hexokinase (52)
- phosphofructokinase (53, 54)
- creatine and adenylate kinases (55)
- glucose-6-phosphate isomerase (56)
- microbial proteinases (57)
- rhodanase (58)
- protein denaturation (ovalbumin at low pH) (59)
- chemical modification of proteins (60)
- determination of thermodynamic characteristics of ATP–Mg complexes (61)
- alkaline phosphatase from human placenta (3, 62)
- studies on human ferritins (5)

In the protocols described the reaction conditions (notably temperature and concentrations of substrate and titrant) used in the references cited are given. These usually relate to pH values in the range associated with optimal enzyme activity, and values of substrate concentration that are sufficiently larger than relevant K_m values to provide rates that approximate to V_{max} values. The reaction may be started by addition of either substrate or enzyme depending upon the stabilities of the reactants at the pH employed. Starting a reaction by addition of enzyme has the advantage that conformationally labile enzymes are not kept in dilute solution prior to reaction, and that any (slow) non-enzymic background rate of reaction of substrate can be determined in the same kinetic run prior to addition of enzyme. It may be necessary to stabilize the pH of solutions prior to starting the reaction by using very dilute buffer (Section 2). When using immobilized enzymes it may be necessary to start the reaction by addition of substrate to permit some types of solid support (such as carboxymethylcellulose) to equilibrate at the desired pH before starting the reaction. In the study of immobilized enzymes, a convenient way of delivering aliquots of stock suspension to the reaction vessel is by means of a calibrated graduated 1 ml pipette from which the narrow tip has been removed. The protocols described here are readily adapted to suit other reaction conditions, e.g. for use in experiments over a range of substrate concentrations.

Many pH-stat assays are based on the production of weak acid–conjugate base equilibrium mixtures as, for example, in ester hydrolysis, where a carboxylic acid–carboxylate pair is produced. When the kinetic run is carried out at pH > $pK_{a(RCO_2H)}$ such that the pH is at least 2 units greater than the pK_a the reaction results in production of protons essentially stoichiometric with the substrate consumed, which are titrated by base in the pH-stat. At pH values nearer to the pK_a, experimentally determined reaction rates will need to be corrected for the incomplete ionization of the acid product. The correction factor will depend on the difference between the pH and the pK_a of the proton-producing product such that

$$\text{rate}_{\text{true}} = \text{rate}_{\text{observed}} \times [(1+10^{(\text{pH} - \text{p}K\text{a})})/10^{(\text{pH} - \text{p}K\text{a})}]$$

Protocol 1

General pH-stat assay for reactions producing protons as products

Reagents

- Sodium hydroxide solution (1 M)
- Sodium hydroxide solution (0.1 M)
- Potassium chloride solution (0.5 M)
- EDTA (di-Na salt) solution (0.1 M)
- Substrate solution
- Enzyme solution or suspension

Method

1. Set up the pH-stat comprising autotitrator, burette (e.g. 0.5 ml capacity), and recorder
2. Load the delivery syringe with NaOH solution of the appropriate concentration (e.g. 0.1 M).
3. Maintain the reaction vessel at the desired temperature (e.g. 25 °C) by means of a water-jacket and water-bath.
4. Pass a stream of oxygen-free nitrogen through a wash bottle containing water at the reaction temperature, and then over the surface of the fluid in the reaction vessel to exclude carbon dioxide and oxygen.
5. Prepare a reaction mixture (minus substrate or enzyme) at the required pH (e.g. by addition of small volumes of 1 M NaOH from an Agla syringe to the stirred mixture).
6. Switch on the pH-stat and record any non-enzymic reaction of substrate (e.g. started reactions) before addition of the other reactant (e.g. enzyme) in solution at the same pH as the reaction pH.
7. Record the progress curve.
8. From the settings of titrant delivery and chart speed calculate the initial rate in mol $l^{-1}s^{-1}$ (for zero-order reactions from linear traces or tangents at t = zero), or the rate constant for first-order reactions by regression analysis of conventional logarithmic plots or by computer analysis of the progress curve.

Protocol 2

pH-stat assay for glucose oxidase (33)

[see also (24) for a glucose oxidase assay by a computer-controlled pH-stat]

Reagents

- Aqueous glucose solution
- Aqueous enzyme solution
- Pure oxygen
- Aqueous sodium hydroxide solution (5 mM)

Method

1 Set up the pH-stat with:
   (a) the stirred reaction mixture (10 ml) containing glucose (e.g. 0.16 M to provide approximately saturating (V_{max}) conditions), and enzyme in deionized water, pH 5.6 (or other pH as required) at 30 °C;
   (b) a supply of pure oxygen to the bottom of the reaction vessel and an oxygen probe to check for appropriate agitation and oxygen flow rate (approx. 0.5 volumes of oxygen per volume of reaction mixture per min);
   (c) 5 mM sodium hydroxide solution in the burette.

2 Record the titration of released protons for 5 min as a zero-order (initial rate) linear progress curve.

Protocol 3

pH-stat assay for dihydrofolate reductase (34)

Reagents

- Aqueous dihydrofolate (DHF) solution (approx. 1×10^{-4}M)
- Aqueous NADPH (approx. 1×10^{-4}M)
- Aqueous Na_2SO_4 solution (0.1 M)
- Aqueous enzyme (DHFR) solution
- H_2SO_4 (at least 5×10^{-5} M)

Method

1 Set up the pH-stat as follows:
   (a) de-gas the stock solutions of DHF and NADPH (each approx. 1×10^{-4} M) and maintain under nitrogen;
   (b) use 0.1 M Na_2SO_4 as electrolyte support to avoid adsorption phenomena;
   (c) place 2 ml of each of these stock solutions in a thermostatted reaction vessel at 28 °C;
   (d) initiate reaction by injection of 1.50 μl of DHFR solution previously adjusted to assay pH (e.g. pH 7.4).

2 Maintain the pH as the reaction proceeds by automatic addition of H_2SO_4 ($\geqslant 5 \times 10^{-5}$ M) and record the addition as a function of time (for approx. 30 s in thoroughly stirred solution).

3. If the solutions are not degassed or if a nitrogen is not used, take account of the baseline drift (approx. 10^{-10} mol H^+ /min) to correct the initial value.

Protocol 4

pH-stat assay for triacylglycerol lipase (35)

(see also ref. 32 for a procedure using a flow-through pH-stat; Section 3.2.8)

Equipment and reagents

- A conventional pH-stat with a 25 ml reaction vessel thermostatted at 30 °C and a 250ml burette
- A high-speed blender
- Hydroxypropyl methyl cellulose solution
- Triolein or purified olive oil
- Sodium glycocholate
- Calcium chloride dihydrate
- CO_2-free water
- Colipase
- Sterile isotonic saline

Method

1. Add 171.4 ml of hydroxypropylmethyl cellulose solution, previously cooled to 4–8 °C, to 28.6 ml of triolein or purified olive oil in a high-speed blender.
2. Blend at high speed for 5 min, cool at 4 °C for 1 h, and repeat the emulsification process for 5 min.
3. Store the substrate emulsion (which should be stable for 5 days) in a tightly stoppered bottle at 4 °C.
4. Check the adequacy of the surface area of the substrate emulsion by ensuring that the activity of a lipase preparation (⩾4000 U/l) has the same value with emulsified substrate concentration 150 ml/l as with concentration 100 ml/l.
5. Dissolve 6.095 g of sodium glycocholate and 0.4463 g of calcium chloride dihydrate in approx 80 ml of CO_2-free water with constant stirring.
6. Transfer the solution to a 100 ml volumetric flask and make up to mark with CO_2-free water.
7. Store at 4–8 °C in a stoppered bottle protected from the light with Al foil (for up to 6 weeks if required).
8. Prepare a stock solution of colipase at a concentration of 1200 mg/l in sterile isotonic saline (assuming that lyophilized preparations of colipase are approx 60% colipase).
9. Store the solution in appropriate aliquots at −70 °C (for up to 1 year if necessary) or at 4 °C (for up to 2 weeks).
10. Prepare a secondary stock solution of colipase for immediate use by mixing 1 volume of stock solution with 1 volume of sterile saline.
11. Arrange for the final assay concentrations as:
    - substrate 100 ml/l
    - sodium glycocholate 35 mmol/l
    - $CaCl_2$ 8.5 mmol/l
    - colipase 6mg/l
12. Introduce 7.0 ml of the substrate solution into the reaction vessel.
13. Add 2.8 ml of the sodium glycocholate–$CaCl_2$ solution.
14. Add 0.10 ml of the colipase solution (600 mg/l).
15. Conduct a gentle, constant stream of nitrogen gas over the surface of the reaction mixture through a tube in the port of the cover of the reaction vessel to remove CO_2 gas and prevent CO_2 absorption.
16. Adjust the pH to 9.0 with NaOH solution (3 mM) and equilibrate for 5 min.

Protocol 4 continued

17 Add 0.10 ml of NaCl solution (155 mM) in the case of blank run, and 0.10 ml of enzyme sample to the assay mixture.

18 Readjust the pH with either HCl (50 mM) or NaOH (30 mM) until the pH meter reads between 8.99 and 9.00, and record the addition of the NaOH titrant as a function of time to provide a linear progress curve over, say, 5 min.

19 Determine blank rates in duplicate at the beginning of the day and at intervals during the day and carry out electrode maintenance if variation is noted.

20 Calculate lipase activity as:
$150 \times (T/t_T - B/t_B)$ U/litre

when U = micromoles of fatty acid produced per min; T and B are the volumes (μl) of a 15 mM solution of NaOH needed to titrate the test and blank mixtures over times t_T and t_B (min) for the test and blank mixtures, respectively. The factor of 150 derives from 0.015 (the molar concentration) of titrant and 1×10^4 which converts the 100 μl sample volume to 1 litre.

Protocol 5

pH-stat assays for acetylcholinesterase (36)

[see also ref. (26) for a procedure for this enzyme using a pH-stat with spectrophotometric pH monitoring and electrolytic titrant generation described in Section 3.2.5]

Reagents

- Aqueous acetylcholine chloride solution (0.11 M)
- Aqueous sodium hydroxide solution (1 mM)

Method

Use a conventional pH-stat and standard procedure (see *Protocol 1*) e.g. at pH 8.0 and 37 °C using 1 mM NaOH as titrant.

Protocol 6

pH-stat esterase assay for cysteine proteinases and serine proteinases with the enzymes either in solution or covalently immobilized (CI), e.g. on carboxymethyl-cellulose (CI-enzymes) (37)

Equipment and reagents

- A conventional pH-stat with a 0.5 ml burette and a micro-reaction vessel thermostatted at 25 °C
- O_2-free N_2
- Cysteine proteinases (papain, ficin, bromelain)
- Serine proteinases (trypsin and α-chymotrypsin)
- The corresponding CI-enzymes
- *N*-acetyl-L-tyrosine ethyl ester (ATEE) as substrate for α-chymotrypsin (0.5 M in 50% v/v aqueous methanol)
- α-*N*-benzoyl-L-arginine ethyl ester (BAEE) as substrate for the other enzymes (0.25 M in water)
- Aqueous NaOH (0.1 M)

Protocol 6 continued

Method

1 Pass a stream of O_2-free nitrogen first through a wash bottle containing water at 25 °C, and thence over the surface of the reaction vessel to exclude CO_2 and O_2.

2 Use a reaction volume of 5 ml and 0.1 M NaOH as titrant.

3 For assays with BAEE, place the following in the reaction vessel: 1 ml of 0.5 M KCl, 0.5 ml of 0.1 M EDTA, *x* ml of standard suspension of a CI-enzyme (e.g. 1 ml containing approx. 50 mg of CI-bromelain or 5–10 mg of CI-trypsin) or enzyme solution, and (1.95 - *x*) ml of deionized water, and equilibrate to pH 7 using 1 M NaOH added from an Agla micrometer syringe.

4 When equilibration is complete (approx. 5 min) add 2.0 ml of the substrate solution and rapidly readjust the pH to 7.0 using the Agla syringe.

5 Switch on the pH-stat and record for sufficient time to obtain a linear trace (with CI-enzymes, a degree of curvature may occur in the early stage of reaction due to final re-equilibration, particularly with CM-cellulose derivatives).

6 For assays with ATEE, follow the same procedure as for BAEE but use 1.0 ml of substrate solution and 2.95 × ml of water.

7 Use chart speeds of 1–4 cm/min.

Protocol 7

pH-stat assays for urease (38)

Equipment and reagents

- A conventional pH-stat with a 20 ml glass reaction vessel thermostatted at 38 ± 0.1 °C and a 0.5 ml burette
- A combination electrode sodium phosphate buffer (0.05 M, pH 7.0)
- Buffer A: sodium phosphate buffer (0.02M, pH 7.1) containing 1 mM EDTA and 1 mM-2-mercaptoethanol)
- Aqueous dithiothreitol (1 mM) containing 0.5 mM EDTA
- Aqueous urea (0.05 M) prepared from recrystallized urea and boiled-out distilled water
- Urease
- Aqueous HCl (0.01M)

Method

1 Standardize the electrode frequently at the assay temperature with 0.05 M phosphate buffer (pH 7.0).

2 For routine assays, use 10 ml aliquots of 0.05 M recrystallized urea in boiled-out distilled water, thermostatted at 38 ± 0.1 °C for at least 15 min in 20 ml glass vessels.

3 Place a combination electrode (used only for urease assays), a glass stirrer, and the burette tip into the reaction vessel.

4 Rinse the above three items in buffer between assays.

5 Add 20 μl of 1 mM dithiothreitol containing 0.5 mM EDTA just prior to addition of urease solution to the reaction mixture.

6 Start the reaction by addition of urease in buffer A and record the uptake of acid at pH 7.0.

Protocol 7 continued

7 For a convenient rate, use of 1 IU of enzyme activity (1 IU of urease activity is that amount of enzyme that causes the decomposition of 1 mmol of urea/min under the standard assay conditions of 38 °C, pH 7.0, 0.05 M urea, 2 μM dithiothreitol).

8 Correct the experimentally determined uptake of acid at pH 7.0 for the ionization of the cationic and ammonia products by using $-d[\text{urea}]/dt = (d[H^+]/dt)/(2f_n - f_c)$ where $f_c = (1 + 2K_2/[H^+])/(1 + [H^+]/K_1 + K_2/[H^+])$, $f_n = [H^+]/(K_3 + [H^+])$, and K_1, K_2 and K_3 are the acid dissociation constants of $H_2CO_3^-$ and NH_4^+ respectively.

5 A systematic error in pH-stat assays of enzymes in haemolysates

In many pH-stat procedures, a stream of O_2-free nitrogen is continuously passed over or through the reaction mixture to prevent the absorption of atmospheric carbon dioxide. Newman and Nimmo (63) pointed out that a problem arises when this procedure is applied to reaction mixtures containing oxyhaemoglobin, such as when erythrocyte acetylcholinesterase is being determined. In such circumstances, the oxyhaemoglobin slowly loses its oxygen ligand and becomes more basic. This decreases the amount of base needed to neutralize the acid liberated in the reaction being assayed. One answer for such systems is to use a stream of oxygen instead of nitrogen to prevent absorption of carbon dioxide.

6 Concluding comment

The pH-stat assay is a convenient and sensitive technique that is applicable to a wide range of enzyme-catalysed reactions in which there is a net change in hydrogen ion concentration. Equipment is available, both commercially and described in the literature, that permits automation of the technique at a number of levels, and its computer control. The technique is applicable not only to reactions in solution but also to reactions in stirred suspension such as those involving cellular extracts and immobilized enzymes and cells.

Acknowledgements

It is a pleasure to acknowledge the opportunity to become familiar with this technique during my time in the Biochemistry Department of the Medical College of St. Bartholomew's Hospital, University of London, where it was a major tool used by the late Professor Eric Crook, Garth Kay, Bob Bywater, P. V. Sundaram and Chris Wharton in the 1960s in the early development of the field of enzymes on solid supports.

References

1. Knaffle-Lenz, E. (1923). *Arch. Pathol. Pharmakol.*, **97,** 242.
2. Jackobsen, C. F., Leonis, K., Linderstrom-Lang, K., and Ottesen, M. (1957). In *Methods of biochemical analysis* (ed. D. Glick), Vol. IV, pp. 171–210. Interscience Publishers Inc., New York.
3. Roig, M. G., Serrano, M. A., Bello, J .F., Cachaza, J. M., and Kennedy, J. F. (1991). *Polymer Int.*, **25,** 45.
4. Tani, T., Fujii, S., Inoue, S., Ikeda, K., Iwama, S., Matsuda, T., *et al.* (1995). *J. Biochem.*, **117,** 176.
5. Yang, X., Chen-Barrett, Y., Arosio, P., and Chasteen, N .D. (1998). *Biochemistry,* **37,** 9743.
6. Hornby, W. E., Lilly, M. D., and Crook, E. M. (1966). *Biochem. J.*, **98,** 420.
7. Lilly, M. D., Regan, D. L., and Dunnill, P. (1974). *Enzyme Eng.*, **2**, 245.

8. Bucher, T., Hofner, H., and Romayrere, J-F. (1974). In *Methods of enzymatic analysis*, 2nd edn (ed. H. U. Bergmeyer and K. Gawehn), pp. 254–61. Academic Press, Inc., New York.
9. Tsibanov, V. V., Loginova, T. A. and Neklyudov, A. D. (1982). *Russ. J. Phys. Chem.*, **56,** 718.
10. Karcher, R. E. and Pardue, H. L. (1971). *Clin. Chem.*, **17,** 214.
11. Saunders, I. (1990). *Lab. Prod. Technol.*, Feb., 10.
12. Whitnah, C. H. (1933). *Ind. Eng. Chem. Anal. Ed.*, **5,** 352.
13. Longsworth, L. G. and MacInnes, D. A. (1935). *J. Bacteriol.*, **29,** 595.
14. Lingane, G. (1949). *Anal. Chem.*, **21,** 497.
15. Jacobsen, C. F. and Leones, J. (1951). *Compt. Rend. Lab. Carlsberg, Ser. Chim.*, **27,** 333.
16. Warner, B. D., Boehme, G., Urdea, M. S., Pool, K. H., and Legg, J. I. (1980). *Anal. Biochem.*, **106,** 175.
17. Job, R. and Freeland, S. (1977). *Anal. Biochem.*, **79,** 575.
18. Ke, B. (1975). *Bioelectrochem. Biochem. Bioenerg.*, **2,** 93.
19. Keijer, J. H. (1970). *Anal. Biochem.*, **37,** 439.
20. Vandermeers, A., Lelotte, H., and Christophe, J. (1971). *Anal. Biochem.*, **42**, 437.
21. Konecny, J. (1977). *Biochim. Biophys. Acta*, **481,** 759.
22. Rousseau, I. and Atkinson, B. (1980). *Analyst*, **105,** 432.
23. Hieftje, G. M. and Mandarano, B. M. (1972). *Anal. Chem.*, **44,** 1616.
24. Lemke, R. E. and Hieftje, G. M. (1982). *Anal. Chim. Acta*, **141,** 173.
25. Forensen, J. K. and Stockwell, P. B. (1975). *Automatic chemical analysis*, pp. 45–54. Wiley, New York.
26. Karcher, R. E. and Pardue, H. L. (1971). *Clin. Chem.*, **17,** 214.
27. Johansson, G. (1965). *Talanta*, **12,** 111.
28. Adams, R. E., Betso, S. R., and Carr, P. W. (1976). *Anal. Chem.*, **48,** 1989.
29. Brand, M. J. D. and Rechnitz, G. A. (1969). *Anal. Chem.*, **41,** 1185.
30. Bull, H. B. and Breese, K. (1973). *Arch. Biochem. Biophys.*, **158,** 681.
31. Eigtved, P. (1981). *Biochem. Biophys. Methods*, **5,** 37.
32. Taylor, F. (1985). *Anal. Biochem.*, **148,** 149.
33. Iturbe, F., Orgega, E., and Lopez-Munguia, A. (1989). *Biotechnol. Techniques*, **3,** 19.
34. Gilli, R., Sari, J. C., Sica, L., Bourdeaux, M., and Briand, C. (1986). *Anal. Biochem.*, **152,** 1.
35. Tietz, N. W., Astles, J. R., and Shuey, D. F. (1989). *Clin. Chem.*, **35**, 1688.
36. Garcia-Lopez, J. A. and Monteoliva, M. (1988). *Clin. Chem.*, **34,** 2133.
37. Crook, E. M., Brocklehurst, K., and Wharton, C. W. (1970). In *Methods in enzymology* (ed. G. E. Perlmann and L. Lorand), Vol. 19, p. 963. Academic Press, London.
38. Blakeley, R. L., Webb, E. C., and Zerner, B. (1969). *Biochemistry*, **8,** 1984.
39. Aquirre, R., Hatchikian, E. C., and Monson, P. (1983). *Anal. Biochem.*, **131,** 525.
40. Prince, R. C., Linkletter, S. J. G., and Dutton, P. L. (1981). *Biochim. Biophys. Acta*, **635,** 132.
41. Glick, B. R., Martin, W. G., and Martin, S. M. (1980). *Can. J. Microbiol.*, **26**, 1214.
42. Chung, J. and Scanu, A. M. (1974). *Anal. Biochem.*, **62,** 134.
43. Milhaly, E. (1978). In *Applications of proteolytic enzymes to protein structure studies*, (ed. R. C. East), Vol. 1, p. 129. CRC Press, Cleveland, USA.
44. Rothenbuhler, E. and Kinsella, J. E. (1985). *J. Agric. Chem.*, **33,** 433.
45. Hou, J. P. and Poole, J. W. (1972). *J. Pharmaceutical Sci.*, **61,** 1594.
46. Bieber, A. L. (1971). *Anal. Biochem.,*, **43,** 247.
47. Garth, Von H. and Zoch, E. (1984). *J. Clin. Chem. Clin. Biochem.*, **22**, 769.
48. Salvadori, C. and Fasell, P. (1970). *Ital. J. Biochem.*, **19,** 193.
49. Vienozinskiene, J., Januseviciute, R., Paulinkonis, A., and Kazlauskas, D. (1985). *Anal. Biochem.*, **146,** 180.
50. Albert, R. A. (1969). *J. Biol. Chem.*, **244,** 3290.
51. Wedler, F. C. and McClune, G. (1974). *Anal. Biochem.*, **59,** 347.
52. Hammes, G. G. and Kochavi, D. (1962). *Amer. Chem. Soc.*, **84,** 2076.
53. Dyson, J. E. and Noltmann, E. A. (1965). *Anal. Biochem.*, **11,** 362.
54. Lorenz, I. (1972). *Math-Naturwiss.*, **21,** 551.
55. Mahowald, T. A., Noltmann, E. A. and Kuby, S. A. (1962). *J. Biol. Chem.*, **237,** 1535.
56. Dyson, J. E. and Noltman, E. A. (1968). *J. Biol. Chem.*, **243,** 1401.

57. Alkanhal, H. A., Frank, J. F., and Christen, G. L. (1985). *J. Food Protection*, **48,** 351.
58. Cannella, C., Pensa, B., and Pecci, L. (1975). *Anal. Biochem.*, **68,** 458.
59. Ottensen, M. and Wallevik, K. (1968). *Biochim. Biophys. Acta.*, **160,** 262.
60. Morgan, P. H. and Hass, G. M. (1976). *Anal. Biochem.*, **72,** 447.
61. Sari, J.-C., Ragot, M., and Belaich, J.-P. (1973). *Biochim. Biophys. Acta*, **305,** 1.
62. Roig, M. G., Serrano, M. A., Bello, J. F., Cachaza, J. M., and Kennedy, J. F. (1991). *Polymer Int.*, **25,** 75.
63. Newman, P. F. J. and Nimmo, I. A. (1980). *J. Clin. Pathol.*, **33,** 1009.

Chapter 8

Enzyme assays after gel electrophoresis

Gunter M. Rothe
Department of General Botany, Faculty of Biology, Johannes Gutenberg-University, Saarstraße 21, 55099 Mainz, Germany

1 Introduction

Enzyme visualization after electrophoresis involves five stages: (a) preparation of crude enzyme extracts, (b) electrophoresis, (c) enzyme visualization, (d) documentation and (e) interpretation of data. While different electrophoretic methods may lead to similar separation patterns, extraction of enzymes from a biological source affords special attention. Therefore, a brief summary concerning enzyme extraction is given in the following chapter. However, detailed methods to isolate subcellular particles from mammalian or plant tissues are not dealt with, and the reader is referred to reference (1), for example. Also, the preparation of gels for starch-, polyacrylamide- and Cellogel-electrophoresis are not considered. Protocols to set up such gels are to be found in several text books (1–5).

2 Preparation of enzyme extracts

2.1 Extraction of microorganisms

Microorganisms such as bacteria, algae, moulds and others, may be ruptured by sonication, by passage through a French press (6, 7), by blending with glass beads (8), or by digesting the cell walls enzymically and later rupturing the cells osmotically (9).

Protocol 1

Rupture of microorganisms

Equipment and reagents

- Blender
- Proteinase inhibitors
- Microorganisms
- Extraction buffer: 100 mM phosphate, pH 7.0 containing a reducing agent

About 100–500 mg of microorganisms are suspended in 1–2 ml of buffer and then ruptured. An extraction buffer that may be used is 100 mM phosphate, pH 7.0, containing a reducing agent (0.1–1 mM 2-mercaptoethanol, or 0.05–0.1 mM dithiothreitol (or dithioerythritol) or 0.1 mM ascorbic acid) and, if necessary, one or several proteinase inhibitors (1–10 mM EDTA, 1 mM *p*-hydroxymercuribenzoate, 1 mM phenyl-methylsulphonylfluoride (use with caution!)). To rupture yeast cells blending of frozen material is recommended.

Modified from a method given in ref. (1), with permission.

Protocol 2

Rupture of yeast cells

Equipment and reagents

- Waring blender
- Yeast cells
- Liquid nitrogen
- Tris–HCl, pH 8.1
- $MgCl_2$
- DTT
- Proteinase inhibitors

One part of crumbled yeast cells are frozen in 1H parts (w/v) of liquid nitrogen. Then, the liquid nitrogen and frozen yeast are poured into a stainless steel Waring blender and homogenized for 4 min at 1 min intervals. After each minute, the frozen yeast powder is scraped off the inner surface of the container. Then the fine frozen powder is suspended in 5 mM Tris–HCl buffer, pH 8.1, containing 10 mM MgCl2, 1 mM dithiothreitol and a proteinase inhibitor, allowed to thaw, stirred for 1 h, and centrifuged.

Modified from a method given in refs (1, 10), with permission.

Note: The problem with yeast cells is that they are rich in proteinase activities (1). To decrease these activities one or several proteinase inhibitors (e.g. diazoacyl-norleucine methyl ester (1 mM) plus 1 mM Cu21, dimethyl-dichlorovinyl-phosphate (1 mM), EDTA (1–10 mM), p-hydroxy-mercuribenzoate (1 mM), pepstatin (10 mg/ml), o-phenanthroline (1 mM), or phenylmethylsulphonylfluoride (1 mM)) are added to the yeast suspension before homogenization (1).

2.2 Animal soft tissues

Except for very small animals a definite organ should be used for enzyme extraction (2). Preferably fresh material should be used, otherwise tissue blocks should be stored at −80 °C (or in liquid nitrogen). Before storage the vascular system or organs should be freed from blood with 1.8 % saline (1, 2).

Protocol 3

Rupture of animal soft tissues

Equipment and reagents

- Potter-Eveljhem homogenizer
- Sonicator
- Fresh tissue
- Extraction medium: triethanolamine-HCl, $MgCl_2$, EDTA, DTT, pH 7.5
- Triton X-100, deoxycholate

About 1–2 g of fresh tissue is cut into small pieces and suspended in 8–20 ml of distilled water or extraction medium (1 in 10 dilution). Homogenization is achieved with a Potter-Eveljhem glass homogenizer. The ground-glass tube is cooled and homogenization is performed for a few minutes, rotating the pestle at about 1000 r.p.m. If, for example, enzymes from mitochondria have not been set free, ultrasonication for about 1 min may be performed afterwards. Non-ionic detergents such as Triton X-100 (0.1 w/v %) or deoxycholate (10 μl of a 10% (w/v) solution per ml of homogenate) may be added to the extraction medium to solubilize membrane-bound enzymes, avoiding ultrasonic vibrations. The following extraction medium has been suggested for the extraction of enzymes from muscle, liver, mammary gland and brain: 50 mM triethanolamine–HCl, 2 mM MgCl2, 1 mM EDTA, 2 mM dithiothreitol, adjusted to pH 7.5 with KOH.

Modified from a method given in ref. (1), with permission.

Methods to isolate subcellular particles from mammalian tissues can be found in ref. (1).

2.3 Mammalian blood

Protocol 4

Preparation of mammalian blood serum

Equipment and reagents

- Glass tube
- Centrifuge
- Fresh whole blood
- Cellulose acetate or agar electrophoresis equipment

Fresh whole blood is drawn without anticoagulant and placed in a glass tube. As soon as the clot forms the sample is centrifuged and the supernatant serum is used for electrophoresis. Upon prolonged storage of serum several enzymes lose activity. Cellulose acetate or agar electrophoresis are preferred to starch gel electrophoresis (1).

Modified from a method given in ref. (1), with permission.

Protocols to isolate blood cells such as leucocytes, platelets, lymphocytes, polymorphs, natural killer cells or monocytes can be found, for example, in ref. (1).

2.4 Insects

Insects may be immobilized with ether or CO_2. The gas is allowed to dissipate before they are used. They may be homogenized as follows:

Protocol 5

Extraction of insects

Equipment and reagents

- Plastic centrifuge tube
- Glass rod
- Insect
- Buffer solution: saline solution, or heparin, or Tris/EDTA/NADP

One insect is put in a (5 ml) plastic centrifuge tube and a drop of buffer (e.g. the gel buffer used in electrophoresis) is added and homogenization performed with a glass stirring rod. Afterwards an additional 0.1 ml of buffer is added and then the homogenate is left to stand for a little while on ice to sediment larger particles. As buffer media, the following are recommended: (a) saline solution: 8.5 g NaCl in 1 litre of dH_2O (distilled water), or (b) 1.0 g heparin in 100 ml dH_2O or (c) 1.2 g Tris, 0.37 g EDTA in 1 litre of dH_2O, pH adjusted to pH 6.8 with concentrated HCl, plus 4 ml of 1% NADP in dH_2O.

Modified from a method given in ref. (2), with permission.

2.5 Plant tissues

Seeds, vegetative buds and cambium of plants are mostly free of large amounts of protein-interfering substances and are therefore preferably used as enzyme source. These tissues may be extracted with a simple acid (acetate-phosphate 100 mM, pH ~5) or neutral (phosphate 100 mM, pH ~7) buffer (1). Such buffers can also be used for herbs which are substantially free of phenols, such as spinach or pea leaves (1). Phosphate buffers often retain the catalytic ability of enzymes better than buffers of Tris (hydroxymethyl)-aminomethane do. However, phosphate buffers are to be avoided if metal-ion-dependent enzymes are to be studied.

Protocol 6

Extraction of plant seeds

Equipment and reagents

- Mortar and pestle
- Seed embryos
- Sodium phosphate, pH 7.3, sucrose, 2-mercaptoethanol

To extract seeds of *Camelia japonica* L., for example, embryos were removed and ground in a chilled mortar with a pestle in 50 mM Na-phosphate, pH 7.3, containing 5 w/v % sucrose and 0.1% 2-mercaptoethanol.

Modified from a method given in refs (1, 10), with permission.

Most plants store considerable amounts of phenolics in the vacuoles of their leaves and roots (11) while their protein contents are low (11, 12). Many also store terpenoids and resin acids (13, 14). When these compounds are set free upon homogenization, they denature the intrinsic proteins, unless protective agents are added to the extraction medium. It cannot be expected that a certain extraction procedure will serve to extract all enzymes in an active state. Different pH-levels may result in differential extraction of isozyme groups (14–16) and may affect lability of enzymes after extraction (17).

The formation of complexes between proteins and phenolic compounds can be suppressed by a number of phenolic-adsorbents, such as polyvinylpyrrolidone (18), in a soluble form (PVP) or in an insoluble form (PVPP, Polyclar AT) (1), casein (19), bovine serum albumin (19) or resins (11). The phenol-scavenger PVP is frequently used but is not equally effective with all types of phenols (1). The use of resins to bind phenols has been reported (11), but isozymes with low isoelectric points may be selectively removed from extracts under conditions of low ionic strength (20). Thiol reagents such as 2-mercaptoethanol, cysteine or dithiothreitol are used in extraction media to inhibit the enzymatic turnover of phenol oxidases, which oxidize phenols to quinones in the presence of O_2. The quinones tend to form covalent bonds with proteins and thus irreversibly inactivate them (21, 22). Since the various phenol oxidases are active only in the presence of O_2 it is good practice to prepare enzyme extracts in the cold and under an atmosphere of N_2 (1). Besides PVP (or PVPP), a non-ionic detergent is often added to a medium to extract enzymes from woody plants because these surfactants overcome phenol inhibition of enzymes (20) and liberate proteins from cell membranes (23).

Protocol 7

Extraction of enzymes from woody plants

Equipment and reagents

- Eppendorf tubes
- Cooling device
- Motor-driven grinding cone
- Centrifuge
- Fresh tissue
- Liquid nitrogen
- Extraction medium

The following protocol has been used to extract enzymes from buds and mycorrhizal roots of European beech (*Fagus sylvatica* L.):

Vegetative buds or mycorrhizal roots (diam. < 1 mm) having a fresh weight of at least 50 mg are put in a 1.5 ml Eppendorf Safe-Lock tube, frozen for 10 s in liquid N_2 and used immediately or stored

Protocol 7 continued

at −80 °C. Then the tubes are put in a cooling device and cooled from underneath with liquid nitrogen. A motor-driven grinding cone adapted in shape to the tube and rotating at 700 r.p.m. is used to homogenize the material to a fine powder. The cone is later cleaned in boiling water and a fresh cone used for every extraction. The Eppendorf tube containing the homogenized sample is put on ice for 1 min and then provided with 100 μl extraction medium per 50 mg of sample material together with 7.5 mg of wet PVPP (polyclar AT). Afterwards the slurry is thoroughly mixed with a spatula and stored on ice for a maximum of 10 min. The extraction medium consists of (mg/100 ml of 100 mM Na-phosphate buffer, pH 7.0): cysteine (30), 2-mercaptobenzothiazole (3.3), Na-metabisulfite (95), EDTA-Na_2 (186), $MgCl_2$-$6H_2O$ (102), NADP (39.2), NAD (35.8), sucrose (14000), bovine serum albumin (500) and Tween[R] 80 (500). The homogenized sample is centrifuged for 25 min at 4 °C and 5000 r.p.m. Aliquots of 20 μl of the supernatant are transferred into 0.5 ml Eppendorf tubes and stored at −30 °C.

Method developed by O. Fiedler, C. von Meltzer and G. M. Rothe.

3 Principles of enzyme visualization

The basic principle of enzyme visualization *in situ* is to present an enzyme with a solution containing an enzyme-specific substrate. Demonstration of an enzyme is achieved if the catalytic action of the enzyme on this substrate produces a coloured reaction product. Often, however, the primary reaction products are colourless and require coupling with a visualizing agent to generate a coloured product. To demonstrate oxidative enzymes, mostly tetrazolium salts are used. These salts are colourless and water-soluble in an oxidized state, but turn into a deeply coloured water-insoluble deposit, known as formazan, when reduced. Transfer of electrons from NAD(P) to a tetrazolium salt requires a redox element such as phenazine methosulfate.

Hydrolytic enzymes are mostly visualized by use of diazonium salts. These salts will react with enzymatically released naphthol from corresponding esters or glycosides to form an intensely coloured insoluble azo dye. The coupling rate of diazonium salts to naphthol derivatives depends on their chemical nature and the pH of the incubating medium. Therefore, if the pH value of the enzymic reaction fits that of the diazonium salt, it may be added to the substrate solution, but if not, the coupling reaction is performed after the enzymic reaction has taken place for a certain while.

3.1 Methods to visualize oxidative enzymes

Dehydrogenases may be assayed by two different methods: (a) observation of the change in fluorescence of the pyridine nucleotides NAD(P) or NAD(P)H participating in the enzymic reaction, or (b) visualization of the oxidation reaction by use of a tetrazolium salt which turns coloured when reduced.

3.1.1 The tetrazolium salt method to assay dehydrogenases

Oxidoreductases catalyze redox reactions according to the general equation:

$$\text{Substrate-H}_2 + \text{acceptor for hydrogen} \rightarrow \text{product} + \text{acceptor-H}_2.$$

The acceptor for hydrogen can be a co-substrate, O_2, or an artificial indicator molecule. Oxidoreductases that transfer hydrogen to oxygen are called 'oxidases', while those which reduce a pyridine nucleotide coenzyme are named 'dehydrogenases' (1). If dehydrogenases are to be visualized *in vitro*, a hydrogen-transferring substance must also be included in the staining system (1). As the hydrogen transferring substance, the molecule phenazine methosulfate

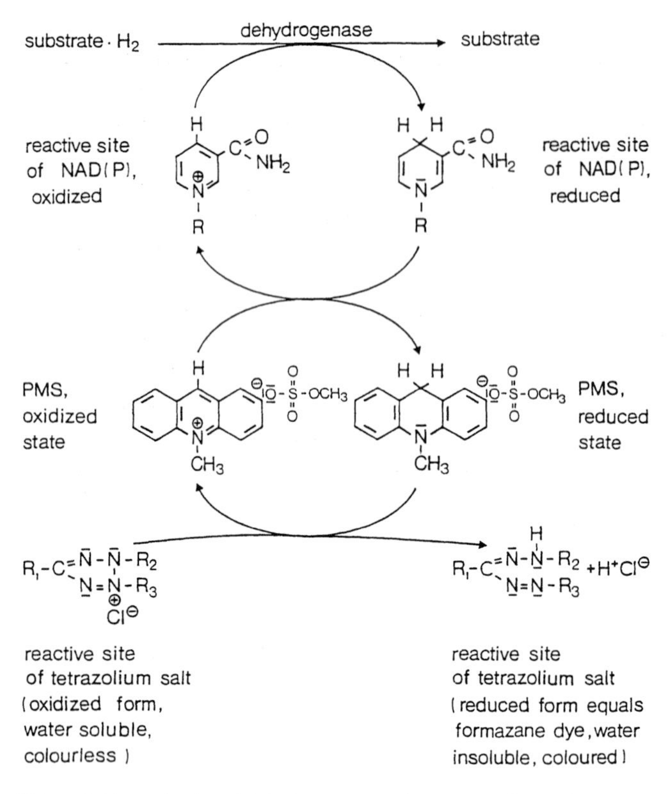

Figure 1 Determination of dehydrogenases using a tetrazolium salt as chromogenic substance and PMS as electron carrier. Figure taken from ref. (1) with permission.

(*N*-methyldibenzopyrazine methyl sulphate) (PMS) is mostly used. PMS is capable of accepting the hydrogen from NAD(P) and passing it over to a tetrazolium salt (*Figure 1*). Mg^{2+} ion concentrations exceeding 10 mmol/l inhibit the redox capability of PMS (25). In case these concentrations are needed in an enzyme reaction PMS must be replaced by the enzyme diaphorase (1). Tetrazolium salts can also be reduced non-enzymatically at alkaline pH values or by substances with free SH-groups such as cysteine, reduced glutathione or 2-mercaptoethanol (1). Dithiothreitol and 2-mercaptoethanol are uncharged at pH <8 so that they do not migrate into the electrophoretic support medium at these pH values. But in starch gel electrophoresis they migrate to the cathode by electroendosmosis where they may cause an unspecific colour reaction when they meet a tetrazolium salt (1).

Especially when assaying small amounts of dehydrogenases, the more sensitive monotetrazolium salts such as MTT (3-(4,5-dimethyl-thiazolyl-2)-2,5-diphenyltetrazolium bromide) should be used (*Figure 2*). Reduced tetrazolium salts can be re-oxidized by strong oxidation reagents such as HNO_2 or the enzyme superoxide dismutase (1). Sometimes it may happen that a positive reaction is observed in a reaction with a tetrazolium salt even if no substrate is included into the incubation solution ('nothing dehydrogenase reaction'). False positive dehydrogenase reactions may be observed with unpurified lactate dehydrogenase, alcohol dehydrogenase, glutamate dehydrogenase or malate dehydrogenase (1). Sometimes the false positive reaction can be traced back to the presence of traces of alcohol and alcohol dehydrogenase (1). Drying the starch used in starch gel electrophoresis (and other material) can eliminate the 'nothing dehydrogenase reaction' (1).

3.1.2 The assay of oxidases

The determination of oxidases and peroxidases are often performed with 3-amino-9-ethylcarbazole as electron donator (*Figure 3*) but the reagent should be used with care (1). 3,3′,5,5′-Tetramethylbenzidine has been recommended as substrate for peroxidases because it has been shown to be non-carcinogenic in animal tests (1).

Enzyme reactions in which H_2O_2 is liberated may be visualized by the addition of peroxidase and a chromogenic hydrogen donator such as 3-amino-9-ethyl-carbazol. Coupled enzyme tests of this kind were set up, for example, for D-(L)-amino acid oxidase, β-D-fructo-furanosidase, glucose oxidase, β-glucosidase, glycollate oxidase, peptidases and xanthine oxidase (1).

Phenolases oxidize phenolic compounds to quinones, which may be coloured. Phenolases comprise enzymes such as catechol oxidase, laccase, ascorbate oxidase, *o*-aminophenol oxidase and monophenol monooxygenase (tyrosinase). The plant enzyme catechol oxidase accepts as substrate 3,4-dihydroxyphenyl alanine (DOPA) but this compound may also be used by peroxidase when H_2O_2 is present (H_2O_2 is produced when DOPA is oxidized by catechol oxidase) (24). Consequently, peroxidases may be visualized upon staining for phenolases. Catechol oxidases are copper-containing enzymes and can be completely inhibited by 10 mM diethyldithiocarbamate, whereas the iron-containing peroxidases remain unaffected (1).

3.2 Methods to visualize transferases

Specific methods to localize the catalytic activities of transferases are not in existence. Transferases are mostly visualized by use of a coupled assay, i.e., one of the reaction products of a

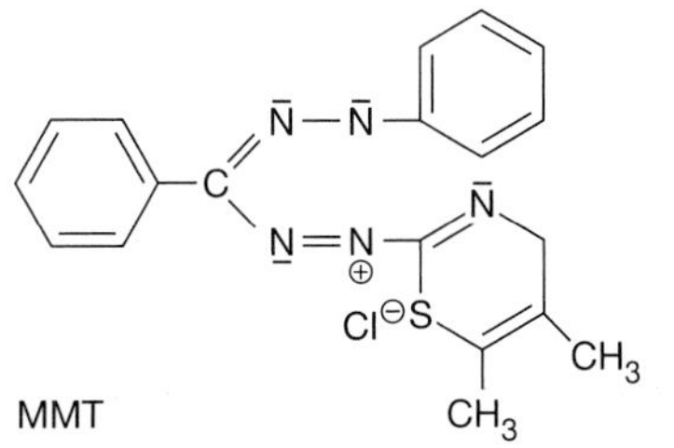

MMT

Figure 2 Formula of the tetrazolium salt MTT. Figure taken from ref. (1) with permission.

Carbazol method

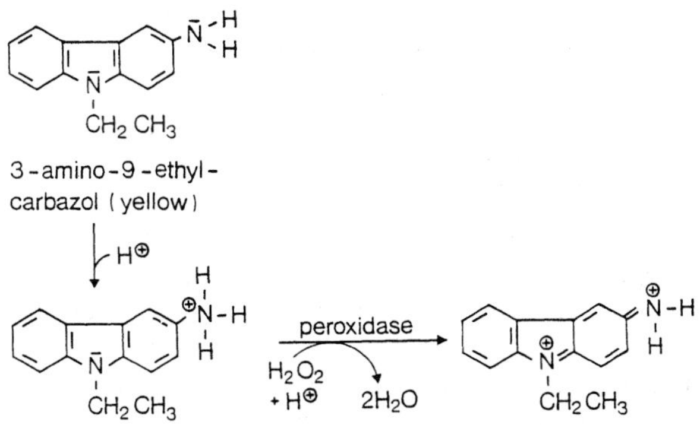

3-amino-9-ethyl-carbazol (yellow)

peroxidase

3-amino-9-ethyl-carbazol, oxidized (brown)

Figure 3 Scheme of reaction when using 3-amino-9-ethyl-carbazole to visualize peroxidases. Figure taken from ref. (1) with permission.

transferase reaction is used as substrate for a coupled dehydrogenase reaction which is made visible (1).

3.3 Methods to visualize hydrolases

Hydrolases catalyze the hydrolytic cleavage of C–O, C–N, C–C, and some other bonds, including those of phosphoric anhydrides. According to the substrates used, hydrolases are also named esterases, glycosidases, amylases, etc. Of the various methods to detect hydrolytic enzymes the most famous are: (a) the use of a substrate which results in a fluorescent (e. g. methylumbelliferone) compound when enzymatically hydrolyzed (*Figure 4*) and (b) the use of naphthol derivatives as substrate and coupling the enzymatically hydrolyzed naphthol to a diazonium-compound to produce a coloured (diazo-)dye (*Figure 5*) (1).

3.3.1 The umbelliferone method to detect hydrolases

A very sensitive method to visualize hydrolases is the use of esters or glycosides of 4-methylumbelliferone (*Figure 4*). The enzymatically released 4-methylumbelliferone moiety fluoresces under long-wave UV-light (around 350 nm). A problem in visualizing hydrolases by this method, however, may result from the fact that 4-methyl-umbelliferone fluoresces exclusively at alkaline pH, while many hydrolases exhibit their pH optimum at an acidic pH (1). To circumvent this problem, the enzyme reaction is carried out at an acidic pH while the detection of 4-methylumbelliferone is performed after having alkalified the electrophoretic support medium, e. g. by exposing it to ammonia vapours, or by incubating it into an alkaline buffer solution (1, 25). Methylumbelliferone has two disadvantages with respect to enzyme visualization following electrophoretic enzyme separation: (a) it is highly soluble so that it diffuses rapidly out of large pore gels and (b) it is unstable in alkaline solutions, which results in background colouring. Therefore, processed gels should be evaluated as soon as possible after electrophoresis (1).

3.3.2 The azo-coupling method to detect hydrolases

The method of azo-coupling uses derivatives of α- or β-naphthol as substrates. During enzymatic hydrolysis the α- or β-naphthol compound of the corresponding acid or amide is liberated. By the process of azo-coupling the non-coloured naphthol compound is bound non-enzymatically to a diazonium salt. The basic structure of diazonium salt is $R-N^{+}\equiv N|$ (*Figure 5*), which on formation to azo groups (−N=N−), confer colour. The colour of diazo dyes depends both on the diazonium salt used and the compound to be coupled to it (*Figure 6*). Coupling of naphthol derivatives to diazonium salts can be performed by two different methods: (a) the diazonium salt is directly included in the test system, or (b) it is added after sufficient α- or β-naphthol has formed. Post-coupling must be applied when the coupling salt contains a heavy metal that inhibits the enzyme reaction, or if optimum enzyme reaction occurs above pH 6, since diazonium salts are rapidly hydrolyzed in aqueous solutions of this pH range (1). If derivatives of 4-methoxy-β-naphthylamine are used as substrate, enzymatically liberated methoxy naphthylamine can be examined under UV light before gels are reacted with a suitable diazonium salt (1).

Several hydrolases can also be visualized by coupling one of the reaction products to a dehydrogenase and its coenzyme, such as adenosylhomocysteinase, alkaline phosphatase, arginase, cystidine deaminase, dipeptidase and tripeptide aminopeptidase (1). Hydrolases that can be detected by including a peroxidase and a chromogenic hydrogen donor into the incubation medium include aminoacylase, β-D-fructofuranosidase, α-D-glucosidase and peptidases (1).

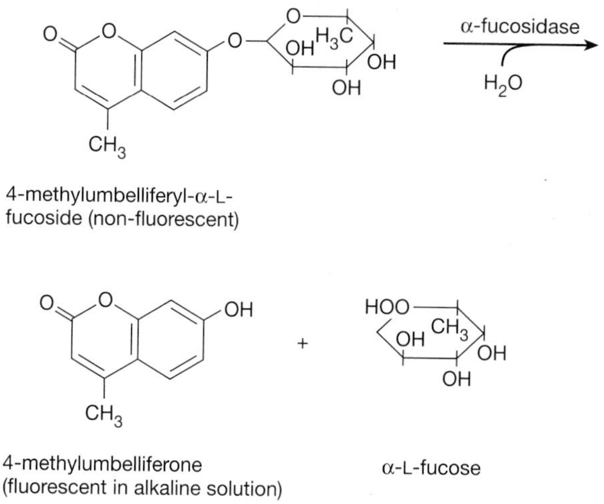

α-fucosidase

H_2O

4-methylumbelliferyl-α-L-fucoside (non-fluorescent)

4-methylumbelliferone (fluorescent in alkaline solution)

α-L-fucose

Figure 4 Course of the chemical reaction to detect the enzyme α-fucosidase by use of the substrate 4-methylumbelliferyl-α-L-fucoside. Figure taken from ref. (1) with permission.

(a)

esterase

α–naphthyl butyrate

α-naphthol + butyric acid

non-enzymatic reaction

α–naphthol

diazo dye (e.g. Fast Blue RR)

diazo chromophore

(b)

leucine aminopeptidase

L-leucine-β-naphthylamide

β-naphthol + leucinamide

non-enzymatic reaction

β-naphthol

diazonium salt (e.g. FAST RED TR)

diazo chromophore

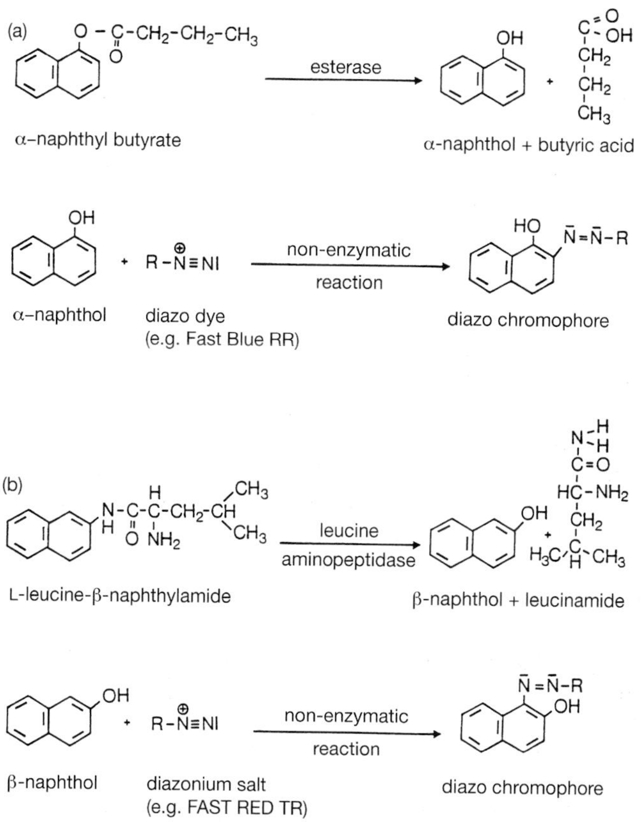

Figure 5 Course of the chemical reactions to detect hydrolases (esterase, leucine aminopeptidase) by the method of azocoupling. The enzymatically liberated naphthol is coupled to a diazonium salt (Fast Blue RR, Fast Red TR). Figure taken from ref. (1) with permission.

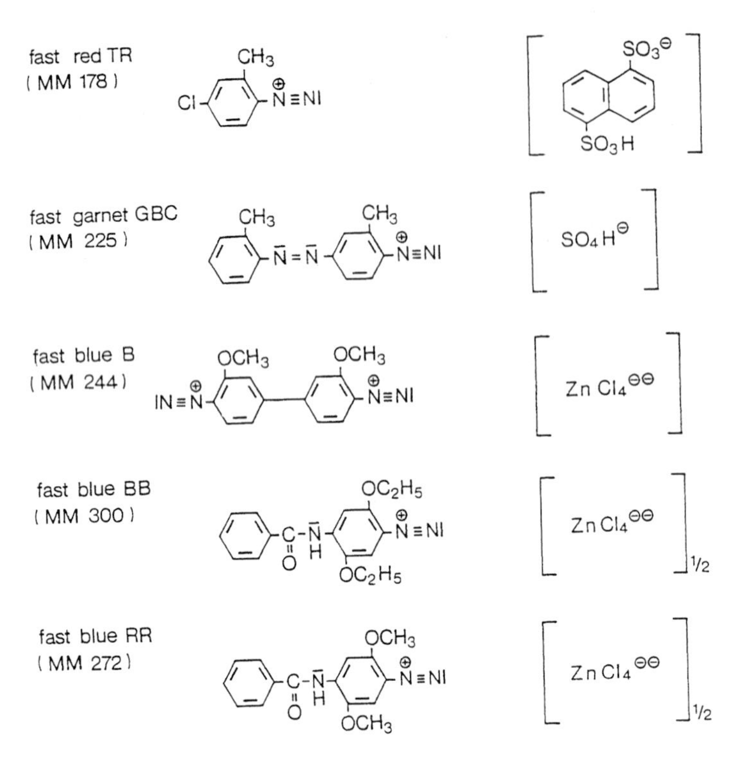

Figure 6 Diazonium salts used for the detection of naphthols. Figure taken from ref. (1) with permission.

3.4 Methods to visualize lyases, isomerases and ligases

No specific methods are available to visualize lyases, isomerases and ligases. Methods which were originally used to detect oxido-reductases or hydrolases can be modified to detect them. Lyases can be visualized (a) in a coupled test system using a dehydrogenase as indicator reaction (aconitate hydratase, anthranilate-5-phosphoribosylpyrophosphate phosphoribosyltransferse (multi enzyme complex), anthranilate synthase, argininosuccinate lyase, enolase, fructose-bisphosphate aldolase, fumarate hydratase), (b) by following the colour change of 2,6-dichlorophenolindophenol (citrate synthase, glyoxalase I), (c) staining of liberated phosphate (chorismate synthase, cystathionine-β-synthase, pyruvate decarboxylase, phospho-2-keto-3-deoxyheptonate aldolase), or (d) by other methods (carbonate dehydratase, dTDP-glucose-4,6-dehydratase, threonine dehydratase) (1). Only a few isomerase enzymes have been visualized following electrophoretic separation (glucose-phosphate isomerase, mannose-phosphate isomerase, triosephosphate isomerase). Location of these enzymes was made visible by using several auxiliary enzymes and a dehydrogenase as reaction indicator. Very few staining systems have been developed to visualize ligases; these are, for example, tryptophanyl synthetase and glutamine synthetase (1).

4 A compilation of protocols to visualize enzymes following electrophoretic separation

Table 1 indicates in alphabetical order enzymes which have been visualized following electrophoretic separation. Also indicated are the methods by which the listed enzymes were electrophoretically separated: (C) Cellogel^{R}, (D) polyacrylamide gel or (S) starch gel electro-

Table 1 Enzymes which have been visualized after electrophoresis

Name of enzyme	EC number	System[a]			Protocol
		C	D	S	
Acetylcholinesterase	3.1.1.7	+		+	
α-*N*-Acetyl-D-glucosaminidase	3.2.1.50		+		
β-*N*-Acetyl-D-glucosaminidase	3.2.1.30	+	+	+	
β-*N*-Acetyl-D-hexosaminidase	3.2.1.52	+	+	+	
Acid phosphatase	3.1.3.2		1	1	*
Aconitate hydratase	4.2.1.3			+	
Acylphosphatase	3.6.1.7	+			
Adenosine deaminase	3.5.4.4			+	
Adenosinetriphosphatase	3.6.1.3		+		
Adenosylhomocysteinase	3.3.1.1		+		
Adenylate kinase	2.7.4.3	1		1	*
Alanine amino-transferase	2.6.1.2	+		+	
Alanine dehydrogenase	1.4.1.1			+	
Alcohol dehydrogenase	1.1.1.1			2	*
Aldehyde dehydrogenase	1.2.1.3			+	
Aldolase	4.1.2.13			3	*
Alkaline phosphatase	3.1.3.1	2	1	4	*
Amine dehydrogenase	1.4.99.3		+		
Amine oxidase (copper-containing)	1.4.3.6		+	+	
D-Amino-acid oxidase	1.4.3.3			+	
L-Amino-acid oxidase	1.4.3.2		+		
Aminoacylase	3.5.1.14	+			
Aminopeptidase (cytosol)	3.4.11.1		1	5	*
Aminotransferases	2.6.1.(1–6)	+			
AMP deaminase	3.5.4.6			+	
α-Amylase	3.2.1.1	2	1		*
Anthranilate phosphoribosyltransferase	2.4.2.18		+		
Anthranilate synthase	4.1.3.27		+		
Arginase	3.5.3.1			+	
Argininosuccinate lyase	4.3.2.1			+	
Arylsulphatase	3.1.6.1	+	+		
Aspartate aminotransferase	2.6.1.1	3		2	*
D-Aspartate oxidase	1.4.3.1			+	
Carbonate dehydratase. ($NADP^+$)	4.2.1.1		1	6	*
Carboxylesterase	3.1.1.1		1	2,7	*
Catalase	1.11.1.6		1	8	*
Catechol oxidase	1.10.3.1		+		
Cathepsin B	3.4.22.1			+	
Cellulase	3.2.1.4		+		
Cholinesterase	3.1.1.8			+	
Chymotrypsin	3.4.21.1	+	+		
Citrate synthase	4.1.3.7	+			
Creatine kinase	2.7.3.2			+	
3′,5′-Cyclic-nucleotide phosphodiesterase	3.1.4.17			+	
Cystathionine β-synthase	4.2.1.22		+		
Cystyl aminopeptidase	3.4.11.3		+		

Table 1 *(Continued)*

Name of enzyme	EC number	System[a] C	D	S	Protocol
Cytidine deaminase	3.5.4.5			+	
Deoxyribonuclease I	3.1.21.1			+	
Diaphorase	1.6.4.3	4			*
Dihydrouracil dehydrogenase ($NADP^+$)	1.3.1.2		+		
Dipeptidase	3.4.13.11		1	9	*
Endo-β-*N*-acetylglucosaminidase	3.2.1.96	+			
Enolase	4.2.1.11	+		+	
β-D-Fructofuranosidase	3.2.1.26		+		
Fructokinase	2.7.1.4			+	
Fructose-bisphosphatase	3.1.3.11		+		
Fructose-bisphosphate aldolase	4.1.2.13			+	
L-Fucose dehydrogenase	1.1.1.122		+		
α- L -Fucosidase	3.2.1.51			+	
Fumarate hydratase	4.2.1.2	+			
Galactokinase	2.7.1.6			+	
α-D-Galactosidase	3.2.1.22	+	+	+	
Glucose dehydrogenase	1.1.1.47			+	
Glucose oxidase	1.1.3.4		+		
Glucose 1-phosphate uridylyltransferase	2.7.7.9	+	+		
Glucose 6-phosphate dehydrogenase	1.1.1.49	5		10	*
Glucose-phosphate isomerase	5.3.1.9	5		11	*
α- D-Glucosidase	3.2.1.20			+	
β- D-Glucuronidase	3.2.1.31			+	
L-Glutamate dehydrogenase (NADP)	1.4.1.4		2	12	*
Glutamine synthetase	6.3.1.2		+		
Glutaminyl-peptide γ-glutamyl transferase	2.3.2.13			+	
Glutathione peroxidase	1.11.1.9			+	
Glutathione reductase	1.6.4.2	6		13	*
Glyceraldehyde-phosphate dehydrogenase	1.2.1.12		1	14	*
Glycerol-3-phosphate dehydrogenase (NAD)	1.1.1.8			+	
Glycollate oxidase	1.1.3.1		+		
Guanine deaminase	3.5.4.3			+	
Guanylate kinase	2.7.4.8	+		+	
Hexokinase	2.7.1.1			3	*
Homoserine dehydrogenase	1.1.1.3		+		
3-Hydroxybutyrate dehydrogenase	1.1.1.30			+	
β-Hydroxysteroid dehydrogenase	1.1.1.51		+		
3-Hydroxyacyl-CoA dehydrogenase	1.1.1.35	+			
3α-Hydroxysteroid dehydrogenase	1.1.1.50	+			
Hypoxanthine phosphoribosyltransferase	2.4.2.8	+		+	
L-Iditol dehydrogenase	1.1.1.14			+	
Inorganic pyrophosphatase	3.6.1.1			+	
Isocitrate dehydrogenase (NADP)	1.1.1.42	7		15	*
Lactate dehydrogenase	1.1.1.27	7		16	*
Lactose synthase	2.4.1.22		+		
Lactoyl-glutathione lyase	4.4.1.5			+	

Table 1 *(Continued)*

Name of enzyme	EC number	System[a] C	D	S	Protocol
Leucine dehydrogenase	1.4.1.9			+	
Lysine 2-mono oxygenase	2.13.12.2		+		
Malate dehydrogenase	1.1.1.37	8		2	*
Malate dehydrogenase (oxalo-acetate-decarboxylating, NADP)	1.1.1.40	9		17	*
Mannitol dehydrogenase	1.1.1.67		+		
Mannosephosphate isomerase	5.3.1.8	+		+	
α-D-Mannosidase	3.2.1.24		+	+	
Melilotate 3-monooxygenase	1.14.13.4		+		
Monophenol monooxygenase	1.14.18.1		+		
NADH dehydrogenase	1.6.99.3			18	*
NADPH dehydrogenase	1.6.99.1		3		*
NAD(P) nucleosidase	3.2.2.6	+			
Nitrate reductase (NADH)	1.6.6.1		+		
Nitrogenase	1.18.2.1		+		
Nucleoside triphosphatase	3.6.1.15		+		
Nucleosidetriphosphate-adenylate kinase	2.7.4.10			+	
Nucleosidetriphosphate pyrophosphatase	3.6.1.19			+	
5′-Nucleotidase	3.1.3.5		+		
Oestradiol-17β-dehydrogenase	1.1.1.62		+		
Penicillinase	3.5.2.6			+	
Peptidases	3.4.11(13)	10		19	*
Peroxidase	1.11.1.7		4		*
Phosphodiesterase I	3.1.4.1	+	+		
6-Phosphofructokinase	2.7.1.11			20	*
Phosphoglucomutase	2.7.5.1	11		2	*
Phosphogluconate dehydrogenase (decarboxylating)	1.1.1.44	5		21	*
Phosphoglycerate kinase	2.7.2.3	+		+	
Phosphoglyceromutase	2.7.5.3			2	*
Phosphorylase	2.4.1.1		+		
Polyribonucleotide nucleotidyl transferase	2.7.7.8		+		
Purine nucleoside phosphorylase	2.4.2.1	+		+	
Pyridoxal kinase	2.7.1.35			+	
Pyruvate kinase	2.7.1.40	+		+	
Retinol dehydrogenase	1.1.1.105			+	
Ribonuclease (pancreatic)	3.1.27.5		+		
Ribosephosphate pyrophosphokinase	2.7.6.1	+			
RNA nucleotidyl transferase	2.7.7.6		+		
Sucrose phosphorylase	2.4.1.7		+		
Superoxide dismutase	1.15.1.1	5		22	*
Tetrahydrofolate dehydrogenase	1.5.1.4	+			
Threonine dehydratase	4.2.1.16		+		
Transaldolase	2.2.1.2			+	
Transketolase	2.2.1.1			+	
Triacyl glycerol lipase	3.1.1.3	+			

Table 1 *(Continued)*

Name of enzyme	EC number	System[a]			Protocol
		C	D	S	
Triosephosphate isomerase	5.3.1.1	12		23	*
Tripeptide aminopeptidase	3.4.11.4	+			
Trypsin	3.4.21.4		+		
UDPglucose-hexose-1-phosphate uridylyl-transferase	2.7.7.12			+	
Urease	3.5.1.5		+		
Xanthine oxidase	1.2.3.2		+		

[a] C= Cellogel[R] electrophoresis, D = disc electrophoresis, S = starch gel electrophoresis, * = staining protocols for these enzymes are given below.

phoresis. Staining protocols for some 30 often-recorded enzymes follow. The corresponding electrophoretic separation systems are listed after the protocols, in Boxes 1, 2 and 3. The protocols are taken from *Electrophoresis of Enzymes, Laboratory Methods* (1) and staining protocols for the remaining enzymes can be found in this publication.

4.1 Staining protocols

Abbreviations used in the following protocols are: AOL, agar overlay; FM, flow method; MOL, membrane overlay; POL, paper overlay; UTL, ultrathin layer. BIS is *N,N'*-methylenebisacrylamide; %T is (g acrylamide + g BIS)/100 ml, for PAGE (polyacrylamide gel electrophoresis).

Acid phosphatase (3.1.3.2)

Equipment and reagents

- Starch gel
- Fast Garnet GBC or Fast Blue BB
- Citrate buffer, pH 4.5
- *o*-Naphthyl phosphate
- Acetone

(Other names: acid phosphomonoesterase, phosphomonoesterase, glycerophosphatase, orthophosphoric monoester phosphohydrolase (acid optimum))

Reaction scheme

α-Naphthyl phosphate + $H_2O \xrightarrow{ACP}$ α-naphthol + P_i

α-Naphthol + Fast Blue BB → *diazo-dye* (coloured)

Protocol

Dissolve 10 mg Fast Garnet GBC (or Fast Blue BB)-salt in 10 ml of 0.05 M citrate buffer, pH 4.5, and add 0.4 ml *o*-naphthyl phosphate-Na_2 (1% in 50% acetone).

Filter, and drop on the cut surface of a processed starch gel. The appearance of blue (or red) bands indicates the presence of active enzyme(s) (3, 26, 27).

Note: If the pH of the electrophoresis buffer is greater than 7, incubate gel in the citrate buffer for 30 min at 5°C, before staining. For alternative staining techniques see ref. (1).

Electrophoresis

Starch gel, pH 7, 13 V/cm, 5 h, 4°C; system S_1 (FM)

Adenylate kinase (2.7.4.3)

Equipment and reagents

- Starch gel
- ADP, NADP, Nitro BT, PMS
- $MgCl_2$, glucose
- Tris–HCl, pH 8
- Hexokinase
- Glucose-6-phosphate dehydrogenase

(Other names: myokinase, ATP:AMP phosphotransferase)

Reaction scheme

$$ADP + ADP \xrightarrow{AK} ATP + AMP,$$

Glucose + ATP (+ hexokinase) → glucose-6-phosphate + ADP

Glucose-6-phosphate + NADP (+ glucose-6-phosphate dehydrogenase) → 6-phosphogluconate + NADPH

NADPH + PMS → NADP + $PMS_{red.}$

$PMS_{red.}$ + Nitro BT → PMS + *formazane* (blue coloured)

Protocol

Dissolve 10 mg ADP, 7.5 mg NADP, 7.5 mg Nitro BT, 0.5 mg PMS, 10 mg $MgCl_2$-$6H_2O$ and 22.5 mg glucose in 10 ml of 200 mM Tris–HCl buffer, pH 8, and add 85 Units of hexokinase and 40 Units of glucose-6-phosphate dehydrogenase. Sites of enzyme activity are indicated by the appearance of blue bands (3, 4, 26).

Electrophoresis

Starch gel, pH 7, 13 V/cm, 5 h; system S_1 (AOL)

Alcohol dehydrogenase (1.1.1.1)

Equipment and reagents

- Starch gel
- Ethanol, butanol, or octanol
- NAD, Nitro BT, PMS
- Tris–HCl, pH 8

Reaction scheme

$$\text{An alcohol} + NAD \xrightarrow{ADH} \text{an aldehyde or ketone} + NADH$$

NADH + PMS → NAD + $PMS_{red.}$

$PMS_{red.}$ + Nitro BT → PMS + *formazane* (blue-coloured)

Protocol

Mix 0.4 ml 95% ethanol (butanol or octanol), 5 mg NAD, 3 mg Nitro BT and 0.2 mg PMS with 9.6 ml of 28 mM Tris–HCl buffer, pH 8. Put the mixture on the cut surface of a processed starch gel (3, 26).

Electrophoresis

Starch gel, pH 8, 13 V/cm, 5 h; system S_2 (AOL or FM)

Aldolase (4.1.2.13)

Equipment and reagents

- Starch gel
- Tris–HCl, pH 8
- NAD, PMS, MTT
- Fructose-1,6-diphosphate
- Sodium arsenate
- Glyceraldehyde-3-phosphate dehydrogenase
- Agar solution

(Other name: fructose-bisphosphate aldolase)

Reaction scheme

D-Fructose-1,6-diphosphate $\xrightarrow{\text{ALD}}$ D-glyceraldehyde-3-phosphate + dihydoxyacetone phosphate

D-Glyceraldehyde-3-phosphate + NAD (+ glyceraldehyde-3-phosphate dehydrogenase) → 3-phospho-D-glycerol-phosphate + NADH

NADH + PMS → NAD + PMS_{red}

PMS_{red} + MTT → PMS + MTT_{red} (blue)

Protocol

Add to 25 ml of 100 mM Tris–HCl, pH 8.0, 100 mg fructose-1,6-diphosphate-Na_3-8 H_2O, 20 mg NAD, 60 mg arsenate-Na, 50 μl glyceraldehyde-3-phosphate dehydrogenase (800 Units/ ml), 0.5 ml PMS (5 mg/ml in H_2O) and 1.5 ml MTT (5 mg/ml in H_2O). Mix with 25 ml of Agar solution (2%; 50°C) and pour on cut starch gel (2, 3).

Electrophoresis

Starch gel, pH 7.4, 4 V/cm for 17 h; system S_3 (AOL)

Alkaline phosphatase (3.1.3.1)

Equipment and reagents

- Starch gel
- Tris–HCl, pH 8
- Borate buffer, pH 9.7
- β-Naphthyl-phosphate
- $MgSO_4$
- Fast Blue B or Fast Blue RR
- Methanol, water, acetic acid

(Other names: alkaline phosphomonoesterase, phospho-monoesterase, glycerophosphatase, orthophosphoric monoesterphosphohydrolase (alkaline optimum))

Reaction scheme

β-Naphthyl-phosphate-Na_2 + H_2O $\xrightarrow{\text{AKP}}$ orthophosphate + β-naphothol

β-Naphthol + Fast Blue RR → *diazo-dye* (coloured)

Protocol

Dissolve in 10 ml of 60 mM borate buffer, pH 9.7, 5 mg ß-naphthyl-phosphate-Na, 12 mg $MgSO_4$-$7H_2O$ and 5 mg Fast Blue B- (or Fast Blue RR)-salt. After development of coloured bands, gels are washed with methanol/water/ acetic acid (5:5:1) (3, 4, 26). Further methods are described in refs (29) and (30). If pH of the electrophoresis buffer is less than 7, pre-incubate gel for 30 min at 5°C in Tris buffer before staining.

Electrophoresis

Starch gel, pH 8.6, 5 V/cm, 5 h; system S_4 (AOL); systems C_2 and D_1 (31–33)

Aminopeptidase (cytosol) (3.4.11.1)

Equipment and reagents

- Starch gel
- Acetate buffer, pH 6.5
- L-leucyl-4-methoxy-2-naphthylamide-HCl
- NaCl, NaCN
- Fast Blue B
- Cupric sulfate

Reaction scheme

L-Leucyl-4-methoxy-2-naphthylamide + H_2O $\xrightarrow{APD}$ L-leucinamide + 4-methoxy-2-naphthol

4-Methoxy-2-naphthol + Fast Blue B → a *diazo-dye* (coloured)

Protocol

Add to 5 ml of 100 mM acetate buffer, pH 6.5, 1 ml of L-leucyl-4-methoxy-2-naphthylamide-HCl (4 mg/ml), 3.5 ml NaCl (850 mg/100 ml), 0.5 ml NaCN (100 mg/100 ml) and 5 mg Fast blue B. After staining, the gel may be rinsed in saline for several minutes and then transferred for a few minutes to a solution of 100 mM cupric sulfate. Cu^{2+} chelates with the dye formed on coupling 2-naphthyl amine with Fast Blue B producing a shift in colour from red to purple (22, 34, 35).

Electrophoresis

Starch gel, pH 7.4, 4 V/cm, 18 h, 4°C; system S_5 (FM); system D_1 (11.7% T, 5% BIS; 3 A/gel (diameter 6 mm, length 65 mm) for 150 min (31–33)

α-Amylase (3.2.1.1)

Equipment and reagents

- Mortar and pestle
- Glass plate
- Moist chamber with transparent cover
- Blue starch granules
- Cellulose acetate membrane or PAA-gel plate
- Special Agar-Noble

(Other names: diastase, glycogenase)

Reaction scheme

The enzyme hydrolyzes 1,4-α-glycosidic linkages in polysaccharides containing 3 or more 1,4-α-linked D-glucose units.

Protocol

Blue starch tablets are grounded to a fine powder with a pestle and mortar and mixed with 5 ml of boiled 2% special Agar-Noble per tablet. The mixture is boiled for a further 30 min and then poured onto a glass plate (10 × 30 cm) to make a smooth surface. Then the plate is set up in a moist chamber with a transparent cover so that it can be observed during the incubation period. The cellulose acetate membrane or the PAA-gel plate is now placed on the blue starch gel plate. Active amylase enzymes will be seen as white bands on a blue background (36).

Electrophoresis

Cellulose acetate, pH 7.5, 300 V, 3 h; system C_2 (AOL)

Aspartate aminotransferase (2.6.1.1)

Equipment and reagents

- Starch gel
- Tris–HCl, pH 8.5
- L-aspartic acid
- 2-oxoglutaric acid
- Fast Blue B

(Other names: glutamic-oxaloacetic transaminase, glutamate aspartate transaminase, L-aspartate: 2-oxoglutarate amino-transferase)

Reaction scheme

2-Oxoglutarate + aspartate $\xrightarrow{\text{AAT}}$ glutamate + oxaloacetate

Oxaloacetate + Fast blue B → *diazo-dye* (coloured)

Protocol

Add to 35 ml of 100 mM Tris–HCl buffer, pH 8.5, 7.5 ml L-aspartic acid (200 mg dissolved in 7.5 ml of a 100 mM Tris–HCl buffer adjusted to pH 8 with KOH), 7.5 ml 2-oxoglutaric acid (110 mg dissolved in 7.5 ml 100 mM Tris–HCl buffer adjusted to pH 8 with KOH) and 250 mg Fast Blue B. Filter and drop on the cut surface of a processed starch gel. Enzymatically liberated oxaloacetate binds to the diazonium salt Fast Blue B resulting in a blue-coloured dye (3).

Electrophoresis

Starch gel, pH 8, 6 V/cm, 18 h; system S_2 (FM); system C_3 (37, 38)

Carbonate dehydratase (4.2.1.1)

Equipment and reagents

- Starch gel
- Porous membrane
- Long-wavelength UV light source
- 4-Methylumbelliferyl acetate
- Acetone
- Phosphate buffer, pH 6.5

(Other names: carbonic anhydrase, carbonate hydrolase)

Reaction scheme

4-Methylumbelliferyl acetate + H_2O $\xrightarrow{\text{CA}}$ acetate + *4-methylumbelliferone* (fluorescent)

Protocol

The staining method is based on the fact that carbonic anhydrase can act as an esterase (EC 3.1.1.1): 10 mg 4-methylumbelliferyl acetate are dissolved in a few drops of acetone and then mixed with 100 ml of 100 mM phosphate buffer, pH 6.5. A porous membrane is impregnated with the staining solution and overlayed on the cut surface of a processed starch gel. Inspection under long-wavelength UV-light indicates active enzyme zones as white fluorescent bands (3, 26).

Electrophoresis

Starch gel, pH 8.6, 12 V/cm, 5 h; system S_6 (MOL); system D_1 (31, 32)

Carbonate dehydratase (4.2.1.1) (alternative stain)

Equipment and reagents

- Starch gel
- Long-wavelength UV light source
- Fluorescein diacetate
- Acetone
- Phosphate buffer, pH 6.5

Reaction scheme

Fluorescein diacetate + H_2O $\xrightarrow{CA}$ acetate + *fluorescein* (fluorescent)

Protocol

The staining is based on the fact that carbonic anhydrase can also act as an esterase (3.1.1.1): 10 mg of fluorescein diacetate are dissolved in a few drops of acetone and then mixed with 100 ml of 0.1 M phosphate buffer, pH 6.5. The solution is dropped on the cut surface of a processed starch gel, which is incubated at 37°C. Yellow fluorescent zones are inspected under long-wavelength UV light. 4-Methylumbelliferyl acetate is the preferred substrate for human CA_1-isozymes whereas fluorescein diacetate is preferentially hydrolyzed by human CA_2-isozymes. CA_1 and CA_2-isozymes are inhibited by 1 mM acetazolamide ('diamox') and this specific inhibitor is useful to distinguish the carbonic anhydrase isozymes from other esterases of similar electrophoretic mobilities. The enzyme from parsley fails to catalyze the hydrolysis of *p*- and *o*-nitrophenyl acetates and is reversibly inhibited by *p*-chloromercuribenzoate or imidazole-buffers (26, 39).

Electrophoresis

Starch gel, pH 8.6, 12 V/cm, 5 h; system S_6 (MOL)

Carboxylesterase (3.1.1.1)

Equipment and reagents

- Starch gel
- UV light source
- *N,N*-dimethylformamide
- 4-Methylumbelliferyl acetate
- Phosphate buffer, pH 6.5

Reaction scheme

4-Methylumbelliferylacetate (-butyrate) + H_2O $\xrightarrow{EST}$ acetate (butyrate) + *4-methylumbelliferone* (fluorescent)

Protocol

In 0.5 ml of *N,N*-dimethylformamide 1 mg of 4-methylumbelliferyl acetate (-butyrate) is dissolved. The solution is slowly mixed with 10 ml of 100 mM phosphate buffer, pH 6.5, and poured on the surface of a processed starch gel. Fluorescent zones, inspected under UV light, indicate the location of active enzyme molecules (3, 26).

Electrophoresis

Starch gel, pH 7.2, 5 V/cm, 17 h; system S_2, S_7 (FM); system D_1 (31, 32)

Catalase (1.11.1.6)

Equipment and reagents

- Starch gel
- Phosphate buffer, pH 7.4
- H_2O_2
- $FeCl_3$, $K_3(Fe^{III}(CN)_6)$

Reaction scheme

$2H^+ + H_2O_2 + 2\ Fe^{3+} \rightarrow 2\ Fe^{2+} + 2H_2O$ (no colour change in the presence of $K_3\ [Fe^{III}(CN)_6]$ and $FeCl_3$)

$3\ Fe^{2+} + 2\ [Fe^{III}(CN)_6]^3 \rightarrow Fe_3\ [Fe^{III}(CN)_6]_2$ (dark green dye)

Protocol

Add to 45 ml of 100 mM phosphate buffer, pH 7.4, 5 ml of a 3% H_2O_2 solution in water. Drop the solution on the cut surface of a processed starch gel and incubate for 15 min. Then rinse the gel with water and afterwards dip it in a freshly prepared mixture of equal volumes of a 2% $FeCl_3$ solution and a 2% $K_3(Fe^{III}(CN)_6)$ solution. Mix by gently agitating the container for a few minutes. Finally remove the stain. Zones of enzyme activity appear as yellow bands on a blue-green background (3, 26).

Electrophoresis

Starch gel, pH 8.6, 6 V/cm, 17 h, 4°C; system S_8 (FM); system D_1, 3 mA/gel (diameter: 6 mm, length 75 mm) for 150 min (≈1 U/gel) (31, 32)

Diaphorase (1.6.4.3)

Equipment and reagents

- Cellogel
- Tris–HCl, pH 8.5
- NADH, MTT
- 2,6-dichloro-indophenol
- Agar solution

(Other names: dihydrolipoamide reductase (NAD), lipoyl dehydrogenase, lipoamide dehydrogenase (NADH), lipoamide reductase (NADH))

Reaction scheme

$$\text{NADH} + \text{2,6-dichloroindophenol-Na} \xrightarrow{\text{DIA}} \text{DCIPH} + \text{NAD}$$

$$\text{DCIPH} + \text{MTT} \rightarrow \text{DCIP} + \text{MTT}_{red}\ \text{(blue)}$$

Protocol

Add to 1 ml of 100 mM Tris–HCl, pH 8.5, 1.5 ml NADH (3 mg/ml), 5 drops of 2,6-dichloro-indophenol-Na (1 % in H_2O, freshly filtered) and 5 drops of MTT (1 % in H_2O). Mix with 2 ml of liquid agar solution (20 mg/ml, 60 °C) and pour on the gel surface. Incubate until dark blue bands appear on a blue background (2, 3, 40).

Electrophoresis

Cellogel^R, pH 7.8, 10 V/cm; system C_4 (AOL)

Dipeptidase (3.4.13.11)

Equipment and reagents

- PAGE
- Disc gels
- L-alanyl-β-naphthylamide
- Tris–HCl, pH 7.1
- Acetone
- Fast Red B

Reaction scheme

L-Alanyl-β-naphthylamide + $H_2O \xrightarrow{DIP}$ L-alanine + ß-naphthol

β-naphtol + Fast Red B → *diazo-dye* (coloured)

Protocol

Dissolve 100 mg L-alanyl-β-naphthylamide in 10 ml of acetone and add 1 ml to 10 ml of 100 mM Tris–HCl buffer, pH 7.1. Incubate disc gels at 37 °C for 45 min. Then decant the substrate solution and replace with 10 ml of a 100 mM Tris–HCl buffer, pH 7.1, containing 100 mg Fast Red B. After 10 min the coupling solution is decanted and replaced by water (41, 42).

Electrophoresis

PAGE, pH 8.9; system D_1 (FM); system S_9 (3)

Glucose-6-phosphate dehydrogenase (1.1.1.49)

Equipment and reagents

- Starch gel
- Glucose-6-phosphate
- NADP, PMS, $MgCl_2$
- Nitro BT
- Tris–HCl, pH 8.0

Reaction scheme

D-Glucose-6-phosphate + $NADP^+ \xrightarrow{G6PDH}$ D-glucono-δ-lactone-6-phosphate + NADPH

NADPH + PMS → NADP + $PMS_{red.}$

$PMS_{red.}$ + Nitro BT → PMS + *reduced Nitro BT* (blue)

Protocol

40 mg glucose-6-phosphate-Na_2, 3 mg NADP, 3 mg Nitro BT, 0.2 mg PMS and 10 mg $MgCl_2$–$6H_2O$ are dissolved in 10 ml 0.04 M Tris–HCl buffer, pH 8.0. Blue bands indicate enzyme activity (3–5).

Electrophoresis

Starch gel, pH 8, 13 V/cm, 5 h; system S_{10} (AOL); system C_5 (5, 27, 43–46)

Glucose-phosphate isomerase (5.3.1.9)

Equipment and reagents

- Starch gel
- Nitrocellulose membrane
- O-ring with silicone lubricant
- Tris–citric acid buffer, pH 8.0
- Glucose-6-phosphate dehydrogenase
- D-fructose-6-phosphate
- NADP, PMS, MTT, $MgCl_2$
- Water, methanol, glacial acetic acid

Reaction scheme

D-Fructose-6-phosphate $\xrightarrow{\text{GPI}}$ D-glucose-6-phosphate

D-glucose-6-phosphate + NADP (+ glucose-6-phosphate dehydrogenase) → D-glucono-δ-lactone-6-phosphate + NADPH

NADPH + PMS → NADP + $PMS_{red.}$

$PMS_{red.}$ + MTT → PMS + *reduced MTT* (blue-coloured)

Protocol

3.2 ml of 332 mM Tris–citric acid buffer, pH 8.0, are added to 38.4 ml dH_2O. Then are added: 1 ml glucose-6-phosphate dehydrogenase (50 U/ml), 1 ml D-fructose-6-phosphate (grade I, 75 mg/ml), 1 ml NADP-Na_2-$2H_2O$ (10 mg/ml), 1 ml PMS (1.8 mg/ml), 1 ml MTT (5 mg/ml) and 6.4 ml $MgCl_2$-6 H_2O (50.75 mg/ml). Following electrophoresis a nitrocellulose membrane (diam. 47 mm, 0.45 or 0.20 μm pore size, Sartorius) attached to an O-ring with silicone lubricant is placed on a micro starch gel, the O-ring up-side. The assembly is left at room temperature for approximately 10 min while buffer from the gel soaks through the membrane. When the membrane is uniformly wet, the well is filled with stain. After staining, the stain is removed, the membrane is rinsed with water and the O-ring removed. The membrane is then lifted from the gel and washed for 1 h with a mixture of water, methanol and glacial acetic acid (5:5:1 (v/v)). The silicone lubricant is removed and the membrane is dried at 37°C (43–45).

Electrophoresis

Starch gel, pH 7.2, 33 V/cm, 2 h, 4°C, system S_{11} (MOL); system C_5 (5, 27, 43–46)

L-Glutamate dehydrogenase (NADP) (1.4.1.4)

Equipment and reagents

- Starch gel
- L-glutamic acid
- Tris–HCl, pH 7.6
- NADP, MTT, PMS
- Agar solution

Reaction scheme

L-Glutamate + H_2O + $NADP^+$ $\xrightarrow{\text{GLDH}}$ L-oxoglutarate + NH_3 + NADPH

NADPH + PMS → NADP + $PMS_{red.}$

$PMS_{red.}$ + MTT → PMS + *reduced MTT* (coloured)

Protocol

70 mg L-glutamic acid–Na salt are dissolved in 20 ml 0.5 M Tris–HCl, pH 7.6, then 5 mg NADP– Na_2 in 1 ml H_2O, 0.5 mg MTT in 1 ml H_2O and 5 mg PMS in 1 ml H_2O are added. 20 ml of a 2% agar solution cooled to 60 °C are mixed with the solution and poured on the cut starch gel surface (3, 4).

Electrophoresis

Starch gel, pH 8.1, 3 V/cm, 17 h; system S_{12} (AOL); system D_2 (47).

Glutathione reductase (1.6.4.2)

Equipment and reagents

- Starch gel
- Tris–HCl, pH 8.0
- Glutathione, NADPH
- 2-nitrobenzoic acid, EDTA
- Agar solution

Reaction scheme

Oxidized glutathione + NADPH $\xrightarrow{\text{GR}}$ NADP + 2 reduced glutathione

reduced glutathione + 2 nitrobenzoic acid → *coloured dye*

Protocol

Add to 6 ml of 200 mM Tris–HCl buffer, pH 8.0, 1 ml oxidized glutathione (67 mg/ml), 1 ml NADPH–Na_4 (5 mg/ml), 1 ml 2-nitrobenzoic acid (3 mg/ml) and 1 ml EDTA (193 mg/ml). EDTA and 2-nitrobenzoic acid are added to the buffer and heated only as much as necessary to bring the dye into solution. After the mixture reaches a temperature of 47°C, NADPH and glutathione are added. Then the staining solution is mixed with 10 ml of a 2% agar solution cooled to 45°C. The solution is poured on the cut surface of a processed starch gel. Yellowish bands mark the site of enzyme activity. Bands may also appear in the absence of substrate (4).

Electrophoresis

Starch gel, pH 8, 8–10 V/cm, 4 h; system S_{13} (AOL); system C_6 (26)

Glyceraldehyde-phosphate dehydrogenase (1.2.1.12)

Equipment and reagents

- Starch gel
- Tris–HCl, pH 7.5
- D-glyceraldehyde-3-phosphate
- NAD, MTT, PMS
- Sodium pyruvate
- Na_2HAsO_4
- Agar solution

Reaction scheme

D-Glyceraldehyde-3-phosphate + orthophosphate + NAD $\xrightarrow{\text{GAPDH}}$ 3-phospho-D-glyceroylphosphate + NADH

NADH + PMS → NAD + $PMS_{red.}$

$PMS_{red.}$ + MTT → PMS + *reduced MTT* (blue-coloured)

Protocol

Add to 25 ml of 50 mM Tris–HCl buffer, pH 7.5, 2.5 μmol D-glyceraldehyde-3-phosphate, 30 mg NAD, 50 mg $Na_2HAsO_4 \cdot 7H_2O$, 50 mg pyruvate-Na_2, 1 ml MTT (5 mg/ml in H_2O) and 0.5 ml PMS (2 mg/ml in H_2O). Mix and add to 25 ml of a 2% agar solution cooled to 55°C. Prepare the glyceraldehyde-3-phosphate from the diethylacetal barium salt using DOWEX 50 following the instructions of the manufacturer and assay the product by the method described in refs (3, 48).

Electrophoresis

Starch gel, pH 8.6, 4 V/cm, 17 h, 4°C, system S_{14} (AOL)

Hexokinase (2.7.1.1)

Equipment and reagents

- Starch gel
- Tris–HCl, pH 7.5
- α-D-glucose
- ATP, NADP, MTT, PMS
- Glucose-6-phosphate dehydrogenase
- $MgCl_2$

Reaction scheme

ATP + α-D-glucose $\xrightarrow{HK}$ ADP + glucose-6-phosphate

Glucose-6-phosphate + NADP (+ glucose-6-phosphate dehydrogenase) → 6-phosphogluconate + NADPH

NADPH + PMS → NADP + $PMS_{red.}$

$PMS_{red.}$ + MTT → PMS + *reduced MTT* (blue-coloured)

Protocol

Add to 40 ml of 100 mM Tris–HCl buffer, pH 7.5, 900 mg α-D-glucose, 40 mg ATP-Na_2-$3H_2O$, 1.5 ml NADP-Na_2-$2H_2O$ (5 mg/ml H_2O), 1 ml MTT (5 mg/ml H_2O), 0.5 ml PMS (5 mg/ml H_2O), 40 μl glucose-6-phosphate dehydrogenase (140 U/ml) and 10 ml $MgCl_2$-$6H_2O$ (4.06 mg/ml). Pour on the cut surface of a processed starch gel and observe the formation of blue bands. In human tissues up to four different isozymes have been observed. Isozyme III is inhibited by the glucose concentration given here while the other isozymes are not. Isozyme III is active when 9 mg glucose is used instead of 900 mg glucose. Isozymes I–III can also use fructose as a substrate, while isozyme IV cannot (3).

Electrophoresis

Starch gel, pH 7.4, 20 V/cm, 4.5 h, 4°C; system S_3 (FM)

Isocitrate dehydrogenase (NADP) (1.1.1.42)

Equipment and reagents

- Starch gel
- Tris–HCl, pH 8.0
- Sodium isocitrate, pH 7.0
- NADP, Nitro BT, PMS
- $MgCl_2$

Reaction scheme

threo-D_s-Isocitrate + $NADP^+$ $\xrightarrow{ICDH}$ 2-oxoglutarate + CO_2 + NADPH

NADPH + PMS → NADP + $PMS_{red.}$

PMS_{red} + MTT → PMS + reduced *Nitro BT* (blue-coloured)

Protocol

8 ml of substrate solution consisting of 0.1 M isocitrate-Na_3, pH 7.0, are mixed with 15 mg NADP, 15 mg Nitro BT, 1 mg PMS, 50 mg $MgCl_2$-$6H_2O$, 10 ml 0.2 M Tris–HCl buffer, pH 8.0, and 32 ml H_2O. The solution is poured on the cut surface of a processed starch gel. Incubation is performed at 37°C in the dark (3–5).

Electrophoresis

Starch gel, pH 8, 13 V/cm, 5 h; system S_{15} (FM); system C_7 (26)

Lactate dehydrogenase (1.1.1.27)

Equipment and reagents

- Starch gel
- Porous membrane
- Tris–HCl, pH 8.0
- Calcium L-lactate
- NAD, MTT, PMS
- Agar solution
- Sodium pyruvate
- NADH

Reaction scheme

Lactate + NAD $\xrightarrow{\text{LDH}}$ pyruvate + NADH

NADH + PMS → NAD + $PMS_{red.}$

PMS_{red} + MTT → PMS + reduced *Nitro BT* (blue-coloured)

Protocol

Add to 20 ml of 50 mM Tris–HCl buffer, pH 8.0, 100 mg L-lactate-Ca-$5H_2O$, 10 mg NAD, 1 ml MTT (5 mg/ml), 0.5 ml PMS (5 mg/ml) and 20 ml 2% agar solution cooled to 45 °C. Often weak bands also occur in the absence of lactate. This 'nothing-dehydrogenase' reaction is probably due to substrate bound to enzyme protein. The reverse reaction (pyruvate + NADH $\xrightarrow{\text{(LDH)}}$ lactate + NAD) can also be used to detect active enzyme bands. The following method can be used to detect this reaction: 20 ml of 50 mM Tris–HCl buffer, pH 8.0, containing 100 mg pyruvate-Na and 10 mg NADH are used to impregnate a porous membrane, which is placed on the cut surface of a processed starch gel. Active enzyme bands appear as non-fluorescent zones on a fluorescent background. LDH-X is relatively more active with the substrates α-hydroxy butyrate and α-hydroxy valerate instead of lactate (3).

Electrophoresis

Starch gel, pH 7, 4 V/cm, 17 h; system S_{16} (AOL); system C_7 (26)

Malate dehydrogenase (1.1.1.37)

Equipment and reagents

- Starch gel
- Tris–HCl, pH 8.0
- Na_2CO_3
- L-malic acid
- NAD, Nitro BT, PMS

Reaction scheme

L-Malate + NAD $\xrightarrow{\text{MDH}}$ oxaloacetate + NADH

NADH + PMS → NAD + $PMS_{red.}$

$PMS_{red.}$ + Nitro BT → PMS + reduced *Nitro BT* (blue-coloured)

Protocol

1.215 g of Na_2CO_3-H_2O are dissolved in 5 ml dH_2O and cooled in an ice bath; then 1.34 g of L-malic acid are added with stirring and the solution is then made up to 10 ml with H_2O. The staining solution consists of 1 ml of this solution and 9 ml of a 0.04 M Tris–HCl buffer, pH 8.0, containing 5 mg NAD, 3 mg Nitro BT and 0.3 mg PMS (4, 5, 50).

Electrophoresis

Starch gel, pH 8, 13 V/cm, 5 h; system S_2 (AOL); system C_8 (27)

Malate dehydrogenase (oxaloacetate-decarboxylating) (NADP) (1.1.1.40)

Equipment and reagents

- Starch gel
- Tris–HCl, pH 7.0
- L-malic acid
- $MgCl_2$
- NADP, MTT, PMS
- Agar solution

Reaction scheme

L-Malate + NADP $\xrightarrow{\text{MDH}}$ pyruvate + CO_2 + NADPH

NADPH + PMS → NADP + $PMS_{red.}$

PMS_{red} + MTT → PMS + reduced *Nitro BT* (blue-coloured)

Protocol

Dissolve in 20 ml of 100 mM Tris–HCl buffer, pH 7.0, 100 mg L-malic acid and readjust pH to 7.0 with NaOH. Then add 2.5 ml $MgCl_2$_$6H_2O$ (40.6 mg/ml), 1 ml NADP-Na_2-$2H_2O$ (5 mg/1 ml H_2O), 1 ml MTT (5 mg/ml in dH_2O), 0.1 ml PMS (5 mg/ml in dH_2O) and finally mix with 25 ml of a 2% agar solution cooled to 45°C. Pour on the cut surface of a processed starch gel and observe the formation of blue bands. To solubilize the human enzyme an emulsifier is used when preparing tissue homogenates (3, 5).

Electrophoresis

Starch gel, pH 8.6, 10 V/cm, 5 h, 4°C; system S_{17} (AOL); system C_9 (26)

NADH dehydrogenase (1.6.99.3)

Equipment and reagents

- Starch gel
- Tris–HCl, pH 8.5
- NADH
- 2,6-dichlorophenol indophenol
- Agar solution

Reaction scheme

NADH + 2,6-dichlorophenol indophenol $\xrightarrow{\text{NDH}}$ NAD + reduced 2,6-dichlorophenol indophenol

reduced 2,6-dichlorophenol indophenol + MTT → 2,6-dichlorophenol indophenol + *reduced MTT* (blue-coloured)

Protocol

To 50 ml of Tris–HCl buffer, pH 8.5, 10 mg NADH-Na_2–$3H_2O$ and 2.5 ml 2,6-dichlorophenol indophenol (2 mg/ml) are added. This solution is mixed with 50 ml of a 2% agar solution cooled down to 45°C. Alternative methods have been described (4), but the method given here is to be preferred (3).

Electrophoresis

Starch gel, pH 8.0; system S_{18} (AOL)

NADPH dehydrogenase (1.6.99.1)

Equipment and reagents

- PAGE
- Phosphate buffer, pH 7.5
- NADPH
- Neotetrazolium chloride
- Triton X-100
- Formalin
- Formate buffer, pH 3.5

Reaction scheme

NADPH + neotetrazolium chloride $\xrightarrow{\text{NDPH}}$ NADP + reduced *neotetrazolium chloride* (red-coloured)

Protocol

Add to 3 ml of 5 mM phosphate buffer, pH 7.5, 1 ml NADPH-Na_4 (1.66 mg/ml) and 1 ml neotetrazolium chloride (2.51 mg/ml). Immerse processed gels into the staining solution and incubate for 1 min at 37°C. The enzymic reaction is terminated by the addition of a solution consisting of 40 ml dH_2O, 3.6 ml 10% Triton X-100, 5 ml 40% formalin and 10 ml of 1 M formate buffer, pH 3.5 (51).

Electrophoresis

PAGE, pH 8.9 (5% T, 0.061% BIS) 3 mA/gel; system D_3 (FM)

Peptidases (A, B, C, E, F and S) (3.4.11.* or 13.*) and peptidase D (3.4.13.9)

Equipment and reagents

- Starch gel
- Na_2HPO_4, pH 7.5
- Dipeptide (tripeptide)
- Snake venom L-amino acid oxidase
- Peroxidase
- $MgCl_2$
- 3-amino-9-ethylcarbazole
- Agar solution

(Other names: dipeptidases, tripeptidases, aminopeptidases; proline dipeptidase for peptidase D)

Reaction scheme

A dipeptide + H_2O $\xrightarrow{\text{PD}}$ L-amino acids

(A tripeptide + H_2O $\xrightarrow{\text{PD}}$ L-amino acid + dipeptide)

L-amino acid + O_2 (+ L-amino acidoxidase) → keto acid(s) + NH_3 + H_2O_2

H_2O_2 + 9-aminoethyl carbazole (+ peroxidase) → H_2O + *oxidized 9-aminoethyl carbazole* (brown-coloured)

Protocol

35 ml of 20 mM Na_2HPO_4, pH 7.5, 20 mg dipeptide (tripeptide), 50 μl snake venom L-amino acid oxidase (approx. 15 U/ml), 100 μl peroxidase (2500 U/ml), 0.5 ml 100 mM $MgCl_2$–$6H_2O$ and 0.5 ml 3-amino-9-ethylcarbazole (25 mg/ml). Mix with 30 ml of a 2% Agar solution (3).

Electrophoresis

Starch gel, pH 7.4, 5 V/cm, 18 h, 4°C; system S_{19} (AOL); system C_{10} (26)

Peroxidase (1.11.1.7)

Equipment and reagents

- PAGE
- Citrate buffer, pH 4.0
- H_2O_2
- Guajacol (or *o*-dianisidine in ethanol)

Reaction scheme

$$4\ H_2O_2 + 4\ \text{guajacol} \xrightarrow{\text{POD}} 8\ H_2O + \text{tetrahydroguajacol (red-coloured)}$$

Protocol

Reaction mixture: 60 ml of 100 mM citrate buffer, pH 4.0, 80 μl H_2O_2 and 40 μl guajacol (or 3 ml *o*-dianisidine, 0.5 % (w/v) in 99.5 % ethanol). The gels are washed with citrate buffer, pH 4.0, before they are placed into the reaction mixture (1).

Electrophoresis

PAGE, pH 8.3, system D_4 (FM)

6-Phosphofructokinase (2.7.1.11)

Equipment and reagents

- Starch gel
- Porous membrane
- Tris–HCl, pH 8.0
- Fructose-6-phosphate
- ATP, NAD
- $MgCl_2$
- Na_2HAsO_4
- 2-mercaptoethanol
- Aldolase
- Triose-phosphate isomerase
- Glyceraldehyde phosphate dehydrogenase

Reaction scheme

$$\text{Fructose-6-phosphate} + \text{ATP} \xrightarrow{\text{PFK}} \text{ADP} + \text{fructose-1,6-diphosphate}$$

Fructose-1,6-diphosphate (+ FDP-aldolase) → dihydroxyacetone phosphate + glyceraldehyde-3-phosphate

Dihydroxyacetone phosphate (+ triose-phosphate isomerase) → glyceraldehyde-3-phosphate

Glyceraldehyde 3-phosphate + arsenate + NAD (+ glyceraldehyde-3-phosphate dehydrogenase) → 3-phosphoglyceroylarsenate + *NADH* (fluorescent)

Protocol

Add to 20 ml of 100 mM Tris–HCl buffer, pH 8.0, 12 mg fructose-6-phosphate-Na_2-H_2O, 12 mg ATP-Na_2-$3H_2O$, 7 mg NAD-$3H_2O$, 40 mg $MgCl_2$-$6H_2O$, 200 mg Na_2HAsO_4-$7H_2O$, 20 μl 2-mercaptoethanol, 400 μl aldolase (90 U/ml), 50 μl triose-phosphate isomerase (10000 U/ml) and 50 μl glyceraldehyde phosphate dehydrogenase (800 U/ml). The reaction mixture is used to impregnate a porous membrane which is placed on the cut surface of a processed starch gel. Active enzyme zones appear as fluorescent bands (3).

Electrophoresis

Starch gel, pH 7.75, 8 V/cm, 17 h, 4°C; system S_{20} (MOL)

Phosphoglucomutase (2.7.5.1)

Equipment and reagents

- Starch gel
- Tris–HCl, pH 7.0
- Glucose-1-phosphate
- $MgCl_2$
- Glucose-6-phosphate dehydrogenase
- NADP, MTT, PMS

Reaction scheme

α-D-Glucose-1,6-bisphosphate + α-D-glucose-1-phosphate $\xrightarrow{\text{PGM}}$ α-D-glucose-6-phosphate + α-D-glucose-1,6-bis-phosphate

Glucose-6-phosphate + NADP (+ glucose-6-phosphate dehydrogenase) → 6-phosphogluconate + NADPH

NADPH + PMS → NADP + $PMS_{red.}$

$PMS_{red.}$ + MTT → PMS + *reduced MTT* (blue-coloured)

Protocol

Dissolve 60 mg glucose-1-phosphate-Na_2-$4H_2O$ (containing at least 1% glucose-1,6-bisphosphate) in 2.0 ml of 0.2 M Tris–HCl, buffer, pH 7.0. Add 10 mg $MgCl_2$-$6H_2O$, 1 ml glucose-6-phosphate dehydrogenase (160 U/ml), 3 mg NADP, 4 mg MTT and 0.2 mg PMS to 8 ml dH_2O. Mix both solutions and pour over the cut surface of a processed starch gel. Active enzyme is indicated by the appearance of blue bands (5, 52).

Electrophoresis

Starch gel, pH 8, 13 V/cm, 5 h; system S_2 (AOL); system C_{11} (26)

Phosphogluconate dehydrogenase (decarboxylating) (1.1.1.44)

Equipment and reagents

- Starch gel
- Tris–HCl, pH 8.0
- 6-phosphogluconate
- NADP, Nitro BT, PMS
- $MgCl_2$

Reaction scheme

6-Phospho-D-gluconate + $NADP^+$ $\xrightarrow{\text{PGDH}}$ D-ribulose-5-phosphate + CO_2 + NADPH

NADPH + PMS → NADP + $PMS_{red.}$

$PMS_{red.}$ + Nitro BT → PMS + reduced *Nitro BT* (blue-coloured)

Protocol

Dissolve 100 mg 6-phosphogluconate-Na_3, 15 mg NADP, 15 mg Nitro BT, 1 mg PMS and 50 mg $MgCl_2$-$6H_2O$ in 10 ml of a 0.2 M Tris–HCl buffer of pH 8.0. Then add 40 ml dH_2O. The solution is poured on the cut surface of a processed starch gel, which is incubated at 37°C (5).

Electrophoresis

Starch gel, pH 8, 13 V/cm, 5 h; system S_{21} (FM); system D_1 (31, 32)

Phosphoglyceromutase (2.7.5.3)

Equipment and reagents

- Starch gel
- UV light source
- Tris–HCl, pH 8.0
- 2-phospho-D-glycerate
- NADH, ATP
- $MgCl_2$, EDTA
- Phosphoglycerate kinase
- Glyceraldehyde phosphate dehydrogenase

Reaction scheme

2,3-Bisphospho-D-glycerate + 2-phospho-D-glycerate $\xrightarrow{PGM}$ 3-phospho-D-glycerate-2,3-bisphospho-D-glycerate

3-Phospho-D-glycerate + ATP (+ phospho-glycerate kinase) → ADP + 3-phospho-D-glyceroyl-phosphate

3-Phospho-D-glyceroylphosphate + NADH (+ glyceraldehyde phosphate dehydrogenase) → D-glyceraldehyde 3-phosphate + *NAD* (non-fluorescent)

Protocol

Dissolve 25 mg 2-phospho-D-glycerate-Na_3-$6H_2O$, 30 mg NADH-Na_2, 20 mg ATP-Na_2-$3H_2O$, 40 mg $MgCl_2$-$6H_2O$ and 2 mg EDTA-Na_2-$2H_2O$ in 10 ml 0.1 M Tris–HCl, pH 8.0, and add 640 Units phosphoglycerate kinase and 200 Units glyceraldehyde-phosphate dehydrogenase. The solution is dropped on the cut surface of a processed starch gel. Upon illumination with UV light non-fluorescent bands appearing on a fluorescent background indicate active enzyme molecules (3).

Electrophoresis

starch gel, pH 8, 13 V/cm, 5 h; system S_2 (FM)

Superoxide dismutase (1.15.1.1)

Equipment and reagents

- Starch gel
- Tris–HCl, pH 8.0
- MTT, PMS
- Agar solution

Reaction scheme

MTT + PMS + day light → PMS + *reduced MTT* (blue coloured)

Reduced MTT + O_2^- $\xrightarrow{SOD}$ *MTT* (colourless)

Protocol

Add to 25 ml of 50 mM Tris-HCl buffer, pH 8.0, 1 ml MTT (5 mg/ml) and 1 ml PMS (5 mg/ml). Mix the staining solution with 25 ml of a 2% agar solution and apply on the cut surface of a processed starch gel. Expose the agar-overlayered gel for several minutes to daylight, then incubate at 37°C. Enzyme zones appear as white bands on a blue background (3).

The nitro-blue tetrazolium technique (53) is unsuitable for the detection of the fast CN-insensitive form of superoxide dismutase (54).

Electrophoresis

Starch gel, pH 7, 12 V/cm, 4h, 4°C; system S_{22} (AOL); system C_5 (5, 27, 43–46)

Triosephosphate isomerase (5.3.1.1)

Equipment and reagents

- Starch gel
- Triethanolamine–HCl, pH 8.0
- EDTA
- Glyceraldehyde-3-phosphate
- NADH
- α-glycerophosphate dehydrogenase
- Agar solution

Reaction scheme

D-Glyceraldehyde-3-phosphate $\xrightarrow{\text{TPI}}$ dihydroxyacetone phosphate

Dihydroxyacetone phosphate + NADH (+ α-glycerophosphate dehydrogenase) → α-glycerophosphate + *NAD* (non-fluorescent)

Protocol

Add to 20 ml of 100 mM triethanolamine–HCl buffer, pH 8.0, containing 5 mM EDTA, 2 ml of 30 mM glyceraldehyde-3-phosphate (prepared from the diethylacetal barium salt according to the suppliers method), 20 mg NADH-Na_2-$3H_2O$ and 20 μl α-glycerophosphate dehydrogenase (80 U/ml). Mix with 20 ml of a 2% agar solution cooled to 45°C and pour on the cut surface of a processed starch gel. Observe the formation of non-fluorescent bands on a fluorescent background. An alternative method has been described by (3).

Electrophoresis

Starch gel, pH 9.3, 8 V/cm, 18 h, 4°C; system S_{23} (AOL); system C_{12} (26, 27)

4.2 Buffer systems for electrophoresis

Box 1: Buffer systems used in Cellogel^R electrophoresis

[remarks in square brackets - see 'key' below] (reference number in round brackets)

C_1

Add B to A until pH equals 7.8

A: 20 mM Tris

B: saturated solution of citric acid

5 mm [1]; 2 h [2]; [3] (26)

C_2

Add B to A until pH equals 8.8

A: 500 mM Tris

B: 1 M HCl (31.43 ml HCl, 32% in 1 litre dH_2O)

5 mm [1]; 3.5 h [2]; 200 V; (33)

C_3

Add B to A until pH equals 7.8

A: 50 mM Tris

B: 1 mM H_3PO_4 (0.057 ml H_3PO_4, 85% (w/v) in 1 litre dH_2O)

5 mm [1]; 4 h [2]; (150 V); [4] (37, 38)

Box 1 continued

C_4

50 mM Tris, 1 mM EDTA-Na_2, 1 mM $MgCl_2$, 20 mM maleic acid, pH 7.8 (40)

C_5

61.4 mM Tris, 4 mM EDTA, 13.6 mM citric acid, pH 7.5

5 mm [1]; 2.5 h [2]; [3] (5, 27, 43–46)

C_6

Add B to A until pH equals 7.5

A: 40 mM Tris, 4 mM EDTA

B: 40 mM citric acid, 4 mM EDTA-Na_2

5 mm [1]; 3h [2]; [3] (26)

C_7

Add A to B until pH equals 7.0

A: 10 mM Na_2HPO_4-$2H_2O$

B: 1.54 mM citric acid.

5 mm [1]; 3 h [2]; [3] (26)

C_8

Add A to B until pH equals 7.5

A: 33.67 mM Tris, 4 mM EDTA

B: 6.3 mM citric acid,1.3 mM EDTA-Na_2

5 mm [1]; 2.5 h [2]; [3] (27)

C_9

Add A to B until pH equals 7.5

A: 100 mM boric acid

B: saturated Tris solution

5 mm [1]; 3.5 h [2]; [3] (26)

C_{10}

20 mM Tris, 20 mM Veronal, 1 mM $MgCl_2$–$6H_2O$, pH 8.0, add 0.2 ml 1 M 2-mercaptoethanol to 1 litre of buffer just before use

Middle of separation distance [1]; 3 h [2]; [3] (26)

C_{11}

Add B to A until pH equals 7.0

A: 10 mM Na_2HPO_4-$2H_2O$

B: saturated citric acid solution

5 mm [1]; 3 h [2]; [3] (26)

C_{12}

2 mM Tris, 2 mM veronal, 0.1 mM $MgCl_2$-$6H_2O$, pH 8.0, add 0.2 ml 1 M 2-mercaptoethanol per litre of buffer, just before use (26, 27).

Box 1 continued

Remarks

[1] Point of application (distance from the cathode)

[2] Running time

[3] The sample buffer consists of a 5 mM phosphate-buffer, pH 6.4, containing 1 mM EDTA-Na_2, 1 mM 2-mercaptoethanol, 0.1 mM diisopropyl fluorophosphate and 0.02 mM NADP

[4] The sample buffer consists of 5 mM Tris-phosphate, pH 7.8, containing 10 mM α-ketoglutarate, 1 mM pyridoxal-5-phosphate, and 0.1% Triton X-100

Box 2: Buffer systems used in disc electrophoresis

(Reference in round brackets), ddH_2O is bidistilled water

D_1

Electrode buffer: 0.6 g Tris, 2.88 g glycine in 1 litre of ddH_2O, pH 8.3

Large pore gel: 1 part of solution A, 2 parts of solution B, 1 part of solution C, 4 parts of solution D

A: 48 ml 1 M HCl, 5.98 g Tris, 0.46 ml TEMED to 100 ml in ddH_2O, pH 6.7

B: 10 g acrylamide, 2.5 g BIS to 100 ml in ddH_2O

C: 4 mg riboflavine in 100 ml of ddH_2O

D: 40 g sucrose to 100 ml in ddH_2O

Small pore gel: 1 part of solution E, 2 parts of solution F, 1 part ddH_2O, 4 parts of solution G

E: 48 ml 1 M HCl, 36.3 g Tris, 0.23 ml TEMED to 100 ml in ddH_2O, pH 8.9

F: 28 g acrylamide, 0.735 g BIS to 100 ml in ddH_2O

G: 140 mg ammonium peroxodisulfate to 100 ml in ddH_2O (31, 32)

D_2

Electrode buffer: 5.52 g diethylbarbituric acid, 1 g Tris in 1 litre dH_2O, pH 7.0

Large pore gel: 1 vol A, 2 vol B, 1 vol C and 4 vol D

A: 39 ml 1 M H_3PO_4, 4.95 g Tris, 0.46 ml TEMED in 100 ml dH_2O

B: 10 g acrylamide, 2.5 g BIS in 100 ml dH_2O

C: 4 mg riboflavine in 100 ml dH_2O

D: 40 g sucrose in 100 ml dH_2O.

Separation gel: 1 vol E, 2 vol F, 1 vol dH_2O, 4 vol G

E: 48 ml 1 N HCl, 6.85 g Tris, 0.46 ml TEMED in 100 ml ddH_2O, pH 7.5

F: 30 g acrylamide, 0.8 g BIS in 100 ml ddH_2O

G: 140 mg ammonium peroxidosulfate in 100 ml ddH_2O (freshly prepared) (47)

D_3

As system D_1 but including 0.1% Triton X-100 into the gels and electrode buffer (51)

D_4

Electrode buffer: 140 mM ß-alanine, 350 mM acetic acid, pH 4.5

Large pore gel: 60 mM KOH, 63 mM acetic acid, pH 6.8 (acrylamide + BIS = 3.125 g/100 ml; acrylamide: BIS = 10:2.5)

Small pore gel: 60 mM KOH, 376 mM acetic acid, pH 4.3 (acrylamide + BIS = 7.7 g/100 ml; acrylamide: BIS = 30:0.8) (55)

Box 3: Buffer systems used in starch gel electrophoresis

S_1

Electrode buffer: 130 mM Tris, 43 mM citrate, pH 7.0

Gel buffer: 9 mM Tris, 3 mM citric acid, pH 7.0

11 V/cm for 5 h (with cooling)

S_2

Electrode buffer: 500 mM Tris, 16 mM EDTA-Na_2, 650 mM boric acid, pH 8.0

Gel buffer: 1 in 10 diluted electrode buffer

11 V/cm for 5 h (with cooling)

S_3

Electrode buffer: 100 mM Tris, 100 mM maleic anhydride,10 mM EDTA, 10 mM $MgCl_2$, pH 7.4.

Gel buffer: 1 in 10 diluted electrode buffer

15 V/cm for 4 h (with cooling)

S_4

Electrode buffer: 300 mM boric acid adjusted with 1 M NaOH to pH 8.0

Gel buffer: 76 mM Tris, 7 mM citric acid, pH 8.6

8–10 V/cm for 4 h (with cooling)

S_5

Electrode buffer: 100 mM Tris, 100 mM NaH_2PO_4 adjusted with 1 M NaOH to pH 7.4

Gel buffer: 1 in 20 diluted electrode buffer

15 V/cm for 4 h (with cooling)

S_6

Electrode buffer: 900 mM Tris, 500 mM boric acid, 20 mM EDTA, pH 8.6, diluted 1 in 14 before use

Gel buffer: 1 in 40 diluted stock solution

5 V/cm for 17 h (with cooling)

S_7

Electrode buffer: 100 mM Tris and 100 mM maleic anhydride adjusted to pH 7.2 with 10 M NaOH

Gel buffer: 1 in 10 diluted electrode buffer

17 V/cm for 4 h (with cooling)

S_8

Electrode buffer: 900 mM Tris, 500 mM boric acid, 20 mM EDTA, pH 8.6, diluted 1 in 7 before use

Gel buffer: 1 in 10 diluted stock solution

5 V/cm overnight (with cooling)

S_9

Electrode buffer: 100 mM phosphate pH 6.5

Gel buffer: 1 in 10 diluted electrode buffer

11 V/cm for 5 h (with cooling)

Box 3 continued

S_{10}

Electrode buffer: 40 mM LiOH, 440 mM boric acid, pH 7.2

Gel buffer: 1 volume electrode buffer, 9 volume distilled water and 90 volume of a 15 mM Tris, 4 mM citric acid buffer of pH 7.2

10 V/cm overnight (with cooling)

S_{11}

Electrode buffer: 47 mM citric acid adjusted with Tris to pH 7.2

Gel buffer: 7 ml of electrode buffer diluted in 250 ml of water

10 V/cm overnight (with cooling)

S_{12}

Electrode buffer: 100 mM Tris, 100 mM NaH_2PO_4 adjusted to pH 8.1 with 1 M NaOH

Gel buffer: 1 in 10 diluted electrode buffer

3 V/cm overnight (with cooling)

S_{13}

Electrode buffer: 410 mM citric acid adjusted to pH 8.0 with NaOH

Gel buffer: 5 mM DL-histidine-HCl, pH 8.0 adjusted with 2 M NaOH

5 V/cm overnight (with cooling)

S_{14}

Cathodal buffer: 661 mM Tris, 83 mM citric acid, pH 8.6 containing 60 mg NAD in 100 ml of buffer

Anodal buffer: cathodal buffer without NAD

Gel buffer: 10 ml of anodal buffer diluted to a final volume of 275 ml and addition of 25 mg EDTA-Na_2. When preparing the starch gel 30 mg NAD in 2 ml dH_2O are added to 200 ml of cooked starch suspension just prior to degassing fully

10 V/cm overnight (with cooling)

S_{15}

Electrode buffer: 500 mM Tris, 16 mM EDTA-Na_2, 650 mM borate, pH 8.0

Gel buffer: 1 in 10 diluted electrode buffer including 0.035 mg NADP per ml of gel

3 V/cm for 17 h (with cooling)

S_{16}

Electrode buffer: 200 mM sodium phosphate, pH 7.0

Gel buffer: 1 in 20 diluted electrode buffer

7.5 V/cm for 14 h (with cooling)

S_{17}

Electrode buffer: 54 mM Tris, 23.5 mM citrate, pH 8.6

Gel buffer: 1 in 10 diluted electrode buffer

5 V/cm overnight

S_{18}

Electrode buffer: 410 mM sodium citrate, 410 mM citric acid, adjusted to pH 8.0

Box 3 continued

Gel buffer: 5 mM histidine, adjusted to pH 8.0 with 2 M NaOH
5 V/cm overnight

S_{19}

Electrode buffer: 100 mM Tris, 100 mM NaH_2PO_4, pH 7.4
Gel buffer: 1 in 20 diluted electrode buffer
5 V/cm for 18 h (with cooling)

S_{20}

Electrode buffer: 100 mM Tris–phosphate, pH 7.75
Gel buffer: 1 in 10 diluted electrode buffer; before degassing the starch gel, 2-mercaptoethanol to a final concentration of 10 mM and ATP to a final concentration of 0.2 mM are added
8 V/cm for 17 h (with cooling)

S_{21}

Electrode buffer: 500 mM Tris, 16 mM EDTA-Na_2, 650 mM borate, pH 8.0
Gel buffer: 1 in 10 diluted electrode buffer, containing 20 mg NADP per ml of gel
8 V/cm for 18 h (with cooling)

S_{22}

Electrode buffer: 100 mM phosphate buffer, pH 7.0
Gel buffer: 1 in 10 diluted electrode buffer
3–6 V/cm for 16 h (with cooling)

S_{23}

Electrode buffer: 110 mM Tris, 4 mM EDTA, adjusted to pH 9.3 with HCl
Gel buffer: 1 in 10 diluted
5 V/cm for 16 h (with cooling)

References

1. Rothe, G. M. (1994). *Electrophoresis of enzymes, laboratory methods*, p. 19. Springer Verlag, Berlin.
2. Pasteur, N., Pasteur, G., Bonhomme, F., Catalan, J., and Britton-Davidian, J. (1988). *Practical isozyme genetics*, p. 61. Ellis Horwood Limited, Chichester.
3. Harris, H. and Hopkinson, D. A. (1976). *Handbook of enzyme electrophoresis in human genetics*, p. 1–1. American Elsevier Publ. Comp., New York.
4. Brewer, G. J. and Sing, C. F. (1970). *An introduction to isozyme techniques*, p. 53. Academic Press, New York.
5. Siciliano, M. J. and Shaw, C. R. (1976). In *Chromatographic and electrophoretic techniques* (ed. I. Smith), Vol. 2, p. 185. Heinemann, London.
6. Cooper, T. G. (1977). *Tools in biochemistry*, p. 355. Wiley, New York.
7. Evans, W. H. (1979). In *Laboratory techniques in biochemistry and molecular biology* (ed. T. S. Work and E. Work), Vol. 7, p. 11. Elsevier-North Holland, Amsterdam.
8. Kemmerer, V., Griffin, C. C., and Brand, L. (1975). In *Methods in enzymology* (ed. W. A. Wood), Vol. 42, p. 91. Academic Press, New York.
9. South, D. J. and Reeves, R. E. (1975). In *Methods in enzymology* (ed. W. A. Wood), Vol. 42, p. 187. Academic Press, New York.

10. Suelter, C. H. (1985). *A practical guide to enzymology. Biochemistry: a series of monographs*, p. 64. John Wiley & Sons, New York.
11. Esterbauer, H., Grill, D., and Beck, G. (1975). *Phyton*, **17**, 87.
12. Pitel, J. A. and Cheliak, W. M. (1985). *Physiol. Plantarum*, **65**, 129.
13. Martin, B. and Bassham, J. A. (1980). *Physiol. Plantarum*, **48**, 213.
14. Weimar, M. and Rothe, G. M. (1986). *Physiol. Plantarum*, **69**, 692.
15. Bucher-Wallin, I. K., Bernhard, L., and Bucher, J. B. (1979). *Eur. J. For. Pathol.*, **9**, 6.
16. Imbert, M. P. and Wilson, L. A. (1972). *Phytochemistry*, **11**, 29.
17. Hoyle, M. C. and Routley, D. G. (1974). In *Mechanisms of regulation of plant growth* (ed. R. L. Bieleski, A. Ferguson, and R. Cresswell), Vol. 12, p. 743. Royal Society of New Zealand, Wellington.
18. Schneider, V. and Hallier, U. W. (1970). *Planta*, **94**, 134.
19. Haard, N. F. and Tobin, C. L. (1971). *J. Food Sci.*, **36**, 854.
20. Hoyle, M. C. (1978). *Physiol. Plantarum*, **42**, 315.
21. Wendel, J. F. and Parks, C. R. (1982). *Heredity*, **73**, 197.
22. Feret, P. P. (1971). *Silvae Genetica*, **20**, 46.
23. Craker, L. E., Gusta, L. V., and Weiser, C. J. (1969). *Can. J. Plant Sci.*, **49**, 279.
24. Van Loon, L. C. (1971). *Phytochemistry*, **10**, 503.
25. Shaw, C. R. and Prased, R. (1970). *Biochem. Genet.*, **4**, 297.
26. Van Someren, H., van Henegouwen, H. B., Los, W., Wurzer-Figurelli, E., Doppert, B., Veroloet, M., and Meera Khan, P. (1974). *Humangenetik*, **25**, 189.
27. Meera Khan, P. (1971). *Arch. Biochem. Biophys.*, **145**, 470.
28. Weinbaum, G. and Markman, R. (1966). *Biochim. Biophys. Acta*, **124**, 207.
29. Fisher, Z. A., Turner, B. M., Dorkin, H. L., and Harris, H. (1974). *Ann. Hum. Genet. Lond.*, **3**, 341.
30. Klebe, R. J., Schloss, J., Mock, L., and Link, C. R. (1981). *Biochem. Genet.*, **19**, 921.
31. Gabriel, O. and Wang, S. F. (1969). *Anal. Biochem.*, **27**, 545.
32. Davis, B. J. (1964). *Ann. N. Y. Acad. Sci.*, **121**, 404.
33. Posen, S., Neale, F. C., Path, M. C., Birkett, D. J., and Brudenellwoods, S. (1967). *Am. J. Clin. Pathol.*, **48**, 81.
34. Nachlas, M. M., Moris, B., Rosenblatt, D., and Seligman, A. M. (1960). *J. Biophys. Biochem. Cytol.*, **7**, 261.
35. Baker, I. P. (1974). *Biochem. Genet.*, **12**, 199.
36. Takeuchi, T., Matsushima, T., Sugimura, T., Kozu, T., and Takemoto, T. (1974). *Clin. Chim. Acta.*, **54**, 137.
37. Dikov, A. L. and Lolova, I. S. (1974). *Acta. Histochem.*, **51**, 102.
38. Lolova, I. and Dikov, A. (1975). *Acta. Histochem.*, **53**, 12.
39. Tobin, A. J. (1970). *J. Biol. Chem.*, **245**, 2656.
40. Richardson, B. J., Baverstock, P. R., and Adams, M. (1986). *Allozyme electrophoresis. A handbook for animal systematics and population studies*, p. 198. Academic Press, Sidney.
41. Beck, C. S., Hasinoff, C. W., and Smith, M. E. (1968). *J. Neurochem.*, **15**, 1297.
42. Adams, C. W. M. and Glenner, G. G. (1962). *J. Neurochem.*, **9**, 233.
43. Tsuyuki, H., Roberts, E., Kerr, R. H., and Ronald, A. P. (1966). *J. Fish. Res. Bd. Can.*, **23**, 929.
44. Peterson, A. C., Frair, P. M., and Wong, G. G. (1978). *Biochem. Genet.*, **16**, 681.
45. Melrose, T. R., Brown, C. G. D., and Sharma, R. D. (1980). *Res. Vet. Sci.*, **29**, 298.
46. Lowenstein, A., Spielman, L., and Mowshowitz, D. B. (1982). *Anal. Biochem.*, **120**, 66.
47. Kimura, K., Miyakawa, A., Imai, T., and Sasakawa, T. (1977). *J. Biochem.*, **81**, 467.
48. Schneider, A. S. (1969). In *Biochemical methods in red. cell genetics* (ed. Y. Y. Yunis), Vol. XIII, p. 189. Academic Press, New York.
49. Nimmo, H. G. and Nimmo, G. A. (1982). *Anal. Biochem.*, **121**, 17.
50. Brewbaker, J. L., Upadhya, M. D., Mäkinen, Y., and Mc Donald, T. (1968). *Physiol. Plant.*, **21**, 930.
51. Ichihara, K., Kusunose, E., and Kusunose, M. (1973). *Eur. J. Biochem.*, **38**, 463.
52. Spencer, N., Hopkins, D. A., and Harris, H. (1968). *Ann. Hum. Genet.*, **32**, 9.
53. Bauchamp, C. and Fridovitch, I. (1971). *Biochim. Biophys. Acta*, **44**, 276
54. De Rosa, G., Duncan, D. J., Keen, C. L., and Hurley, L. S. (1979). *Biochim. Biophys. Acta.*, **566**, 32.

Chapter 9
Techniques for enzyme extraction

Nicholas C. Price

IBLS Division of Biochemistry and Molecular Biology, Joseph Black Building, University of Glasgow, Glasgow G12 8QQ, UK

Lewis Stevens

Department of Biological Sciences, University of Stirling, Stirling FK9 4JR, UK

1 Introduction: scope of chapter

This chapter discusses the techniques used to extract enzymes from cells. The principal aim of such procedures is to obtain the enzyme in as high a yield as possible consistent with the retention of maximal catalytic activity. The various procedures involved in subsequent purification of enzymes to homogeneity as a prelude to detailed structural and functional analysis are not discussed here; full details of the methods involved are given elsewhere (1, 2).

Section 2 deals with the choice of the tissue and methods for disruption of that tissue and the rupture of the cells. It is by no means always necessary to break open cells in order to obtain enzymes; many enzymes, particularly hydrolases, are secreted from cells or tissues and can be purified directly from a culture filtrate or supernatant. In addition to general methods for the rupture of cells, mention is made of specific procedures for plant cells and microorganisms (fungi, bacteria etc.) which pose special problems. Plants and bacteria possess tough cell walls which must be broken in order to liberate the cell contents. Fungi possess vacuoles which contain large quantities of proteolytic enzymes, which might damage the enzymes being extracted. Plant cells often contain large quantities of phenolic compounds, which can interfere with extraction of enzymes. In recent years it has become more common to exploit the power of recombinant DNA technology to overexpress a protein of interest in a host organism that can be grown conveniently on a large scale. Purification of the enzyme of interest is then made considerably easier because of its high abundance in the initial extract (often up to 20–30% of the total cell protein). Several factors are involved in the choice of suitable host systems (3–5).

Section 3 is concerned with the protection of enzyme activity during and after disruption of cells. Damage to enzymes can result from a number of causes, such as proteolysis, oxidation of thiol groups, thermal denaturation, etc. Although a number of protocols are available to minimize such damage, it must be emphasized that each extraction should be investigated in preliminary experiments in order to establish optimum conditions.

Section 4 deals with the assays of enzyme activities in crude (unfractionated) cell extracts. The general principles of enzyme assays have already been discussed in Chapter 1 of this volume; this section will outline a number of special considerations which apply in the case of crude extracts.

Finally, two short Appendices deal with (a) buffers and the control of pH, and (b) the determination of protein concentration. Control of pH is crucial for preserving enzyme activity, and protein estimation is important for assessing the efficiency of cell breakage and establishing characteristics of the enzyme such as its specific activity as the purification proceeds.

2 Disruption of tissues and cells

2.1 Choice of tissue

The choice of tissue for extraction of enzymes depends on a number of factors. Classically the choice would have been made on the grounds of availability, cost, or abundance of enzyme. Thus, for example, heart muscle is an excellent source of the enzymes of the tricarboxylic acid cycle because of the high number of mitochondria in this tissue. Large quantities of heart, brain or liver tissues are generally available from meat animals. On account of its economic importance in the baking and brewing industries, yeast (*Saccharomyces cerevisiae*) is available in large amounts and serves as an excellent source of many enzymes, especially those of the glycolytic pathway. In other cases it may be important to choose a tissue so that information is obtained which can be compared with that from a previously studied tissue in the same or another species, e.g. the lactate dehydrogenases from heart and skeletal muscle.

The situation has been transformed, however, by the advent of recombinant DNA technology, and an ever-increasing proportion of enzymes are being produced by this means either for detailed characterization or for commercial application. The basic requirements are the isolation of the gene encoding the enzyme of interest and the development of a suitable expression system (host cell) in which to express the gene. (It should be noted that in order to isolate the gene it is usually necessary to have purified at least a small quantity of the enzyme from its natural source, so that amino acid sequencing can be performed in order to design a suitable oligonucleotide probe to search a library). In general, expression is more likely to be successful when the host cell is closely related to the organism whose gene is being expressed.

Whatever the source, unless proteinases are themselves the object of interest, it may well be possible to avoid some potential problems (see Section 3.3) in enzyme extraction by a suitable choice of source. In the case of animal tissues, it would be wise to avoid liver, spleen, kidney and macrophages, which are rich in lysosomal proteinases, such as cathepsins. In the case of microorganisms, it may be possible to select or construct mutant strains that are deficient in certain proteinases. This approach has been successfully employed, for example, in yeast, *Escherichia coli* and *Bacillus subtilis* (6,7).

Prokaryotes offer potentially great advantages as host organisms because of their potential for rapid growth and their relatively simple nutritional requirements. The gene to be overexpressed is generally incorporated into a plasmid (extrachromosomal DNA) under the control of a strong promoter so that expression of the gene can be induced by addition of an inducer or by some other change in the culture medium. Disadvantages of bacteria when used to express eukaryotic proteins include the fact that they lack the correct machinery to carry out most post-translational modifications such as glycosylation. In addition, many overexpressed proteins, particularly large multidomain proteins, form insoluble 'inclusion bodies' (consisting of misfolded aggregated protein) in prokaryotes (8). Formation of the aggregated protein appears to result from the fact that in prokaryotes the folding of the polypeptide chain occurs post-translationally rather than co-translationally as in eukaryotes (9). Inclusion bodies are fairly well-defined with regard to their size and density and can generally be easily purified by low speed centrifugation (10, 11). For example, in the case of inclusion bodies formed in *E. coli*, the cells would be ruptured by enzymatic digestion (Section 2.3.4) and the inclusion bodies recovered by centrifugation at 8000 **g** for 20 min. (12). After washing the pellet, a strong denaturant such as guanidinium chloride or urea is usually added to solubilize the inclusion bodies. Refolding of the denatured protein can often be brought about by removal of the denaturant, for example by dialysis (12), although in some cases, it may be necessary to add other factors for satisfactory regain of biological activity (1, 11).

Particularly for large-scale production of proteins, formation of inclusion bodies can be a useful way of concentrating the protein and facilitating recovery and purification (11).

As an alternative to intracellular expression, it may be advantageous to modify the gene being expressed so that the product is secreted from the host, thus avoiding the possible formation of inclusion bodies. This approach is more feasible in Gram-positive (e.g. *B. subtilis*) than in Gram-negative (e.g. *E. coli*) bacteria (13).

For expression of eukaryotic proteins, a variety of host organisms can be employed. Lower eukaryotes such as yeasts have proved popular because of their good growth rates on simple media and because they are well understood at the genetic level. Incorporation of suitable signal sequences can allow the efficient secretion of proteins into the growth medium, though it should be noted that recovery and purification of such proteins is not always easy, particularly on a large scale. *S. cerevisiae* is not always an ideal host organism; it can be difficult to grow to high cell densities in continuous culture and it has a tendency to hyperglycosylate proteins. A number of other yeast species such as *Kluyveromyces lactis* and the methyltroph *Pichia pastoris* appear to offer a number of advantages in this respect and are finding increasing application. Another popular system is based on baculovirus-driven expression in insect cells, since this appears to use many of the modification, processing and transport systems of higher eukaryotic cells, thereby leading to high-level expression of functional proteins (3, 4).

An alternative approach is to express proteins as 'fusion proteins' in which the gene for the protein of interest is linked to that for another component such as glutathione-S-transferase. The main advantage of this approach is that the fusion protein can easily be purified by affinity chromatography, and then cleaved to generate the protein of interest (5).

2.2 Disruption of tissue and separation of cells

In some cases it may be desirable to disrupt tissue and prepare homogeneous populations of intact cells prior to disruption of these cells. Many types of cells from complex multicellular organisms can also be grown under defined conditions in culture. The preparation of isolated cells offers a number of advantages over intact tissues when, for example, investigating transport properties and responses to hormones. It may also be useful to separate the different types of cells from a tissue before performing extractions on these types, to allow comparative information to be obtained which may not be available if the whole tissue is studied, for example, the separation of parenchymal and non-parenchymal liver cells using either differential centrifugation or density gradient centrifugation (14).

Cell suspensions can be prepared from tissues by mechanical or enzymatic methods, or a combination of the two. Mechanical methods such as shaking or loose homogenization often damage the integrity of cells, so enzymatic methods are preferred (15). It is normal to include EDTA to chelate Ca^{2+} ions, which are often involved in cell adhesion; similarly, addition of bovine serum albumin is beneficial, possibly by complexing with free fatty acids, which might otherwise damage cell membranes. The enzyme that is most frequently employed is collagenase from *Clostridium histolyticum* at a concentration of 0.01–0.1% (w/v) for periods from 15 min to 1 h; other proteolytic enzymes such as trypsin, elastase and pronase have also found application. During the incubation, the tissue is seen to disintegrate and the isolated cells go into suspension.

The cells obtained by this type of procedure can be separated on the basis of a number of properties such as charge and antigenicity, but most commonly separation is performed on the basis of cell size and density by centrifugation (16). The media used for centrifugal density gradient separations must fulfil certain conditions, i.e. they must be non-toxic and non-permeable to cells, and form iso-osmotic gradients of the appropriate density; solutions of

Percoll, Ficoll or metrizamide have been widely used. Full details of the methods of isolation of homogeneous cell preparations from different tissues are given in a number of references (16–18).

2.3 Disruption of cells

A wide variety of methods is available to bring about disruption of cells (1); some of the principal procedures are listed in *Table 1*. Classification of the methods is broadly in terms of their harshness. Generally it is advisable to use a method which is as gentle as possible that extracts an acceptable yield of the enzyme of interest and also minimizes both damage to the enzyme and the release of degradative enzymes from subcellular organelles such as vacuoles or lysosomes. Some of the more important points are described below in connection with particular types of tissue.

2.3.1 Mammalian tissue

In general, the tissue is cut into small pieces and as much fat and connective tissue removed as possible. Soft tissue such as liver can be homogenized in a Potter–Elvehjem homogenizer in which a rotating pestle (Teflon piston attached to a metal shaft) fits into an outer glass vessel. The clearance varies from about 0.05 mm to about 0.6 mm in different types of homogenizers; too tight a fitting can lead to rupture of organelles. For tougher tissues such as skeletal or heart muscle, it is advisable to mince the chopped tissue in a Waring blender prior to homogenization; three or four bursts, each of 15 s, are normally sufficient to give a smooth extract. The extract is then stirred for about 30 min to allow further extraction of enzymes, before being centrifuged (10000 **g** for 20 min) to give a clear extract.

The solution used for homogenization will depend on the nature of the extract required. If it is important to isolate subcellular organelles, iso-osmotic sucrose or mannitol (0.25 M)

Table 1 Methods for disruption of cells

Method	Underlying principle
Gentle	
Cell lysis	Osmotic disruption of cell membrane
Enzyme digestion	Digestion of cell wall; contents released by osmotic disruption
Potter-Elvehjem homogeniser	Cells forced through narrow (0.05–0.6mm) gap between pestle and glass vessel. Cell membranes removed by shear forces
Moderately harsh	
Waring blender	Cells broken and sheared by rotating blades
Grinding with sand or alumina, or glass beads	Cell walls removed by abrasive action of sand or alumina particles
Vigorous	
French press	Cells forced through small orifice at very high pressure leading to cell disruption
Explosive decompression	Cells equilibrated with inert gas at high pressure. On release of the contents into atmospheric pressure disruption occurs and the contents are released
Bead mill	Rapid vibrations with glass beads lead to removal of the cell wall.
Ultrasonication	High-pressure sound waves cause cell breakage by cavitation and shear forces

For further details see reference (1).

lightly buffered with Tris, Hepes or Tes (5–20 mM) at pH 7.4 is generally used. When it is not important to isolate the intact organelles, the solution used should be chosen to give a good yield of the desired enzyme(s). Thus, most soluble enzymes, such as creatine kinase, are extracted from muscle using solutions of low ionic strength (0.01 M KCl). Myosin can be selectively extracted from muscle in solutions of high ionic strength (0.3 M KCl, 0.15 M potassium phosphate). Enzymes such as RNA polymerases, that are bound to chromatin in the nucleus, are extracted using 1.0 M NaCl (19).

2.3.2 Plant tissues

Plant tissues pose special problems when extracting enzymes, not only because of the presence of the tough cellulose cell wall, but because of the presence of vacuoles which occupy a large proportion of the total cell volume. Disruption of the vacuoles would lead to the release of proteinases and a lowering of the pH of the extract if it is not adequately buffered. An additional complication is caused by the presence of phenolic compounds in plant cells; in the presence of oxygen these are converted by phenol oxidases to polymeric pigments which can adsorb and inactivate enzymes in the extract. In order to minimize these effects it is usual to add a reducing agent such as 2-mercaptoethanol (Section 3.4) to inhibit the phenol oxidases, and a polymer such as polyvinylpolypyrollidone to adsorb the phenolic polymers.

The extraction of ribulose bisphosphate carboxylase/oxygenase from spinach leaves (20) involves homogenizing the leaves with 2 volumes of a buffer at pH 8.0 containing 50 mM bicine, 1 mM EDTA, 10 mM 2-mercaptoethanol and 2% (w/v) polyvinylpolypyrrolidone in a Waring blender for 40 s at low speed. The resulting extract is filtered through cheesecloth prior to centrifugation at 23000 **g** for 45 min.

2.3.3 Yeasts

Apart from problems caused by the presence of a tough cell wall, yeasts and other fungi contain large amounts of proteinases, which could damage enzymes during extraction. As mentioned in Section 2.1, it may be possible to select or construct mutants that are deficient in proteinase production, or to repress the synthesis or secretion of proteinases by growth on media that do not contain protein substrates. A number of methods have been used to extract enzymes from yeasts; two of the more important are outlined below:

(a) *Autolysis with toluene*. Incubation of yeast cake with toluene (6% (v/w)) and 2-mercaptoethanol (0.2% (v/w)) at 37°C for about 1 h leads to the formation of a smooth liquid due to extraction of the cell wall components. Subsequent addition of EDTA (15 mM) at pH 7.0, containing 5 mM 2-mercaptoethanol (10 times the original volume of toluene) and overnight incubation leads to the degradation of the cell wall by the action of endogenous enzymes. The extract can be clarified by centrifugation (15000 **g** for 30 min) (21).

(b) *Shaking with glass beads*. This technique involves shaking a suspension of yeast cells with small glass beads (1 mm diameter); subsequent centrifugation gives a clear extract (22). In the small-scale procedure, 1 ml of cell suspension is shaken at 2500 r.p.m. with 2.5 g glass beads; the temperature is maintained at 10 °C in this process. Maximum degrees of extraction (generally well over 90%) are obtained after 20 min shaking.

A freeze–thaw method of rupture of yeast cells has also been used successfully and the low temperatures involved have been claimed to be an important factor in minimizing damage caused by endogenous proteinases (23). If it is important to isolate intact organelles from yeasts, the preparation and lysis of sphaeroplasts is a useful method (24).

In most cases where the methylotrophic yeast *Pichia pastoris* is used as a host system (Section 2.1), the recombinant proteins are secreted into the culture medium from which they can readily be recovered. In cases where the recombinant protein has to be extracted from the *Pichia* cells, the latter are generally ruptured using the French press (25) or vortexing with glass beads (26).

2.3.4 Bacterial cells

Bacteria possess very tough cell walls and vigorous mechanical methods are usually necessary to break these down. Such methods include the French press, explosive decompression, ultrasonication, grinding with alumina, or bead milling. Apart from damage which might be done to the cellular contents, it is not always easy to scale-up these treatments to deal with large amounts of cells. A more gentle method of disruption involves the enzymatic breakdown of the cell wall. Gram-positive species (e.g. *Bacillus*, *Micrococcus*, *Streptococcus*) are readily susceptible to the action of lysozyme. Typical conditions involve incubation with hen egg-white lysozyme (0.2 mg/ml) at 37°C for 15 min (1).

Gram-negative bacteria (e.g. *E. coli*, *Klebsiella* spp., *Pseudomonas* spp.) are much less susceptible to the action of lysozyme in the absence of additional treatments. A detailed study (27) showed that the digestion by lysozyme could be made much more effective by incorporating, first, a preliminary washing of the cells in dilute detergent (0.1% (v/v) *N*-lauroyl-sarcosine), and secondly, a mild osmotic shock in which a cell suspension in sucrose (0.7 M), Tris (0.2 M), EDTA (0.04 M) is diluted with 4 volumes of distilled water. The first step may alter the permeability of the outer membrane and the second step involves a destabilization of the lipopolysaccharide-containing cytoplasmic membrane by the high concentration of Tris and EDTA. On dilution, lysozyme molecules are drawn osmotically into the murein layer of the cell wall, thus promoting digestion.

The release of DNA on cell lysis makes the resulting extract highly viscous and this can cause severe problems in subsequent purification steps. Deoxyribonuclease I (10 μg/ml) can be added to degrade the DNA; alternatively, nucleic acids can be precipitated by the action of protamine or the synthetic highly positively charged polymer, polyethyleneimine (1). Solutions of these compounds should be neutralized before addition to the extract.

2.3.5 The degree of cell breakage

Whatever method of disruption is used, it is important to have an estimate of the degree of cell breakage, so that the effectiveness of the procedure can be evaluated and the minimum degree of harshness required can be employed. For suspensions of single-celled organisms (e.g. yeasts, bacteria), an estimate of the degree of cell breakage can be easily obtained by analysis of the extract using a haemocytometer (22). A quantitative estimate of the release of cell contents can be obtained by measurement of the protein (see Appendix 2) that is not sedimented by centrifugation (at least 30000 **g**) relative to the total protein in the organism.

2.3.6 Membrane-bound enzymes

Many enzymes occur within cells physically associated with membranes. The methods used to extract different enzymes from membranes will depend on the mode and strength of the interactions involved.

Peripheral enzymes such as glyceraldehyde-3-phosphate dehydrogenase or aldolase can be extracted from erythrocyte membranes by treatment with EDTA (0.1 M) or KCl (0.7 M) plus NaCl (0.14 M) (28). The extraction of integral proteins requires more drastic treatments that disrupt the membrane structure, such as the use of detergents. (Organic solvents, chaotropic agents or hydrolytic enzymes are used to a lesser extent (20)). There is now a very wide range of detergents available which differ in terms of their charge characteristics (ionic, zwitterionic, non-ionic), critical micelle concentrations, effects on protein structure, and interference with subsequent steps in protein purification and characterization (1, 30, 31). In general, non-ionic detergents such as the Triton or Tween series have less effect on protein structure than do ionic detergents such as sodium dodecyl sulphate. Although some general guidelines are available, each extraction should be checked in preliminary experiments to optimize the procedure. As a rule, at least 2 mg of detergent is required for successful extraction of 1 mg membrane (1). It should be noted that extraction of a membrane-bound enzyme may make an assay of its activity very difficult, especially if the enzyme is involved in a vectorial transport process. Spectroscopic measures of structure such as circular dichroism have proved useful in assessing the degree to which native-type structure may be retained on extraction (32).

A classic example of the extraction of an integral membrane protein is provided by the NAD^+-dependent cytochrome b_5 reductase from calf liver microsomal membranes (33). Extraction of the enzyme with Triton X-100 leads to a form of the enzyme of molecular mass 43 kDa, whereas extraction by treatment of the microsomes with lysosomal proteinases (cathepsins) gives a smaller form of the enzyme (33 kDa). The latter form results from release of a hydrophobic portion of polypeptide chain which is responsible for anchoring the enzyme to the membrane. A number of enzymes are anchored to membranes either by hydrophobic portions of their polypeptide chains, or by covalent attachment to the N-terminal glycoinositol phospholipids or isoprenoid groups (34, 35).

3 Protection of enzyme activity

During the process of tissue and cell disruption or during subsequent treatment such as subcellular fractionation or chromatography, enzyme activity can be lost for a variety of reasons. It is therefore essential to consider strategies for protection of the activity; this section will consider some of the more important factors involved. It is important to ensure that the measures taken to protect activity do not interfere with the extraction of the enzyme or its subsequent assay. We have already noted (Section 2.3.6) that detergents used for extraction of membrane-bound enzymes must be chosen with a number of these factors in mind.

3.1 Control of pH

Many enzymes are only active within a fairly narrow range of pH and exposure to pH values outside this range can lead to irreversible loss of activity. It is thus advisable to ensure that a suitable buffer is used during the extraction process. The pH within certain subcellular organelles can differ markedly from neutral pH (e.g. the pH in the interior of vacuoles and lysosomes is estimated to be in the range 4.5–5.0) and the buffering capacity used must be sufficient to account for this if these organelles are ruptured. In addition, metabolic processes could continue within an extract affecting the pH. The continuing breakdown of glycogen in muscle extracts will lead to a decrease in pH due to the accumulation of lactate and pyruvate. A brief discussion of buffers is given in Appendix 1 to this chapter. Some of the more important factors involved in the choice of buffer are: (a) Over what range of pH is buffering required? (b) What ionic strength is required to provide adequate buffering and give optimum extraction of enzyme? (c) Does the pH of the buffer depend markedly on temperature or ionic strength? (d) Would the buffer affect the enzyme activity, e.g. by chelating important metal ions, or would it interfere with any subsequent purification procedures? (e) Would the buffer interfere with the assay method, e.g. by having a high absorbance at the wavelength used for assay?

3.2 Control of temperature

During cell disruption, especially using the harsher methods listed in *Table 1*, the temperature can rise considerably (by up to 30°C or more). In order to avoid such excessive rises in temperature it is advisable to use pre-cooled solutions and apparatus and, if necessary, take steps to dissipate heat generated during the extraction.

Although it is the usual practice to make sure that the temperature is kept low (near 4°C) during extraction in order to minimize the rate of denaturation of enzymes and reduce the activity of proteinases (23), it should be remembered that in some cases exposure of enzymes to low temperatures can lead to inactivation. This is generally due to dissociation of subunits, which are usually held together by predominantly hydrophobic forces (36). It is thus important to check the effect of temperature on the particular enzyme of interest.

3.3 Control of proteolysis

The control of the degradation of enzymes by endogenous proteinases during or after extraction represents one of the most difficult challenges in this type of work. Fuller discussions of the problems have been given (1, 6). Indications that proteolysis is a problem include:

(a) the isolation of a particular enzyme or protein in poor yield;

(b) instability of the enzyme on incubation;

(c) poor resolution of proteins on SDS-PAGE, reflecting heterogeneity in molecular mass (this would also be evident for example on mass spectrometric analysis of the purified protein);

(d) discrepancies between reported and observed values of proteins.

A number of strategies are available to minimize or suppress unwanted proteolysis. These include lowering the temperature to inhibit the action of proteinases and the use of proteinase-deficient strains or tissues if possible (Section 2.1). However, the most commonly employed method involves the addition of proteinase inhibitors during extraction and subsequent steps. The major types of proteinases in various tissues and the inhibitor 'cocktails' which can be used to inhibit them are listed in *Table 2*. (These are now available from Sigma

Table 2 Inhibitors used to control proteolysis

Type of tissue	Major types of proteinases	Inhibitors added	Stock solution (aqueous unless noted)
Animal tissues	Serine	PMSF (1 mM)*	0.2 M (MeOH)
	Metallo	EDTA (1 mM)	0.1 M
	Aspartic	Benzamidine (1 mM)	0.1 M
		Leupeptin (10 μg/ml)	1 mg/ml
		Pepstatin (10 μg/ml)	5 mg/ml (MeOH)
Plant tissues	Serine	PMSF (1 mM)*	0.2 M (MeOH)
	Cysteine	Chymostatin (20 μg/ml)	1 mg/ml (DMSO)
		EDTA (1 mM)	0.1 M
		E64 (10 μg/ml)	1 mg/ml
Fungi	Serine	PMSF (1 mM)*	0.2 M (MeOH)
	Aspartic	Pepstatin (15 μg/ml)	5 mg/ml (MeOH)
	Metallo (possibly)	Phenanthroline (5 mM)	1 M (EtOH)
Bacteria	Serine	PMSF (1 mM)*	0.2 M (MeOH)
	Metallo	EDTA (1 mM)	0.1 M

* If the use of PMSF is considered undesirable on safety grounds, 3,4-dichloroisocoumarin (DCI) can be used as an alternative. DCI is less toxic than PMSF and is more reactive towards a number of serine proteinases (45). DCI is, however, much more expensive than PMSF. The stock solution of DCI (10 mM) is prepared in DMSO; the final concentration in the extraction medium should be 0.1 mM. DCI is relatively unstable in aqueous solutions; the half-life at near neutral pH is 20–30 min (45). Abbreviations: DMSO, dimethylsulphoxide; PMSF, phenylmethanesulphonylfluoride; E64, L-trans-epoxysuccinyl leucylamido (4-guanidino) butane; MeOH, methanol; EtOH, ethanol; EDTA, ethylenediaminetetraacetic acid, disodium salt For further details see references (1) and (6).

and Boehringer as 'ready mixed' cocktails for particular types of tissues in either tablet or solution form).

Some particular points deserve comment:

(a) *Safety*. PMSF is reported to be highly toxic and should be handled with care. Gloves should be worn when handling both the solid and solutions. DCI could be used as an alternative to PMSF (see footnote to *Table 2*). DMSO should be handled with care as it is very easily absorbed through the skin.

(b) *Solubility*. Several of the inhibitors listed in *Table 2* are of only limited solubility in aqueous solvents, and are thus prepared as stock solutions in organic solvents. The volume of such stock solutions added during extraction should be kept to a minimum to avoid damage to the enzyme(s) of interest. The maximum solubility of PMSF in aqueous solutions is about 2 mM and it is important to note that this decreases markedly as the ionic strength of the solution increases.

(c) *Stability*. PMSF is unstable in aqueous solution with a half-life at 25°C, pH 7.0, of about 30 min (6). Successive additions of the stock solution of the inhibitor during the extraction procedure might be advantageous.

Using the data in *Table 2*, it should be possible to avoid many of the problems caused by proteolysis during extraction and purification of enzymes, although it must be emphasized that the 'cocktails' represent only general guidelines and should be tested by appropriate preliminary experiments in each case.

3.4 Protection of thiol groups

The thiol groups of cysteine side chains of proteins can be damaged during extraction. Within a cell, the prevailing reducing environment maintains cysteine side chains in the reduced (-SH) form; however, on cell rupture and exposure to oxygen, there is a tendency for the side chains to form either disulphide bonds or oxidized species such as sulphinic acids. Traces of heavy metal ions (e.g. Cu^{2+}) can catalyse this process by forming complexes with the otherwise rather unreactive oxygen molecule. Protection against such oxidative damage is normally provided by inclusion of a reagent containing a thiol group such as 2-mercaptoethanol or dithiothreitol. It would also be advisable to add EDTA at a low concentration (0.1 mM) to remove any heavy metal ions, and to use nitrogen-purged solutions for long-term storage of extracts.

2-mercaptoethanol is a liquid of density 1.12 g/ml with a most disagreeable odour and is toxic. It is usually necessary to add it to a final concentration of 10–20 mM to provide protection for thiol groups in proteins for up to 24 h (1). Dithiothreitol, however, because of its greater reducing power, can provide protection at lower concentrations (1 mM). (The standard redox potential of dithiothreitol at pH 7.0 is – 0.33 V, some 0.12 V more negative than that for cysteine (37)). Dithiothreitol is also much more convenient to handle; it is a white solid with little odour. The principal disadvantage of dithiothreitol is its cost; it is approximately 20 times more expensive to make a solution of 1 mM dithiothreitol than the equivalent volume of 20 mM 2-mercaptoethanol.

3.5 Protection against heavy metals

Heavy metal ions (such as those of Cu, Pb, Hg or Zn) can inhibit enzymes, usually by reacting with cysteine side chains. These metals can arise from the tissue used for extraction, the glassware or distilled water used, or can occur in the reagents employed. Inclusion of EDTA (< 1 mM) in the extraction medium will minimize any effects of these heavy metals, but it is still advisable to use high-quality reagents and highly purified water in the preparation of solutions used. It is important to check, however, that any EDTA added does not remove any essential metal ions that may be required for the activity of a given enzyme (e.g. zinc in alcohol hydrogenase or copper in a number of amine oxidases). If this is the case, it may be necessary either to add a lower concentration of EDTA, or to add a supplement to the extraction medium with the specific metal ion required (in as pure a form as possible).

3.6 Control of mechanical stress

During cell disruption by harsh techniques such as the French press or sonication, the cell contents are subjected to high pressure, which can lead to inactivation of a number of enzymes. Jaenicke (38) has shown that the effects of high pressures on enzymes can be complex. For many oligomeric enzymes such as lactate dehydrogenase or glyceraldehyde-3-phosphate dehydrogenase, the effect of moderate pressure (up to about 2 kbar) is to cause reversible dissociation to inactive monomers. At higher pressures the monomers can then aggregate to form a denatured polymer, a process that is normally irreversible.

In practice, therefore, it is important to control the period of time and pressure applied during vigorous disruption procedures in order to minimize potential damage to enzymes. The chosen conditions should, however, be consistent with the need to obtain adequate degrees of extraction of the enzyme(s) of interest.

3.7 Effects of dilution

When a tissue or cell is extracted, there can be a high degree of dilution of the enzymes and proteins within the cell. In some cellular compartments such as the mitochondrial matrix the protein concentration is estimated to be as high as 500 mg/ml. The concentration might be reduced to 5 mg/ml in a cell extract and to 5 μg/ml in a solution of pure enzyme used for assays of activity.

In practice, it has been found that many enzymes lose activity fairly rapidly on storage in dilute solution. This effect can often be overcome by inclusion of an 'inert' protein such as bovine serum albumin at a concentration of 1–10 mg/ml in the solution. It is possible that the added protein may help to prevent loss of enzyme by adsorption on the surface of the vessel, or it may act as a 'sacrificial' substrate for proteinases, thereby protecting the enzyme(s) of interest (1).

Alternative protective agents for enzymes include polyols such as glycerol, glucose or sucrose. These promote preferential hydration of the protein, thereby leading to stabilization (39). Glycerol at concentrations of 50% (w/v) or higher will lower the freezing point of aqueous solutions below the normal temperature of conventional laboratory freezers (−20°C). Such solutions are suitable for long-term storage of proteins, since freezing (which can be damaging to many enzymes) is avoided (1). However, it should be noted that concentrated solutions of glycerol are very viscous and are unsuitable for chromatographic procedures; concentrated solutions of sorbitol may be a practical alternative since they provide protection but are much less viscous.

An additional consequence of the dilution of cell contents on extraction can be the dissociation of cofactor (e.g. pyridoxal-5′-phosphate for aminotransferases). Apart from the requirement to add the cofactor during assays of enzyme activity, it is possible that the apoenzyme may be less stable than the holoenzyme. This appears to be the case for pyridoxal-5′-phosphate-dependent enzymes where the apoenzyme is more susceptible than the holoenzyme to the action of intracellular proteinases (40). If such an effect is suspected, it would be advisable to include the cofactor in the buffer used for extraction.

4 Assays of enzymes in unfractionated cell-extracts

The general principles involved in assaying enzymes have been described in earlier chapters in this volume. This section will highlight the particular considerations that should be borne in mind when performing assays on crude cell extracts. Such assays are useful in providing data on the fluxes through metabolic pathways and in helping to formulate and test theories of the regulation of these pathways (41). In addition, comparison of the kinetic properties of an enzyme in a cell extract with those of the corresponding purified enzyme can indicate any effects on the enzyme during the purification procedure. This may be important in order to assess the validity of conclusions about the enzyme in the cell which have been reached from studies of the purified enzyme. Some of the problems that arise in assays of crude extracts are listed below.

4.1 The presence of endogenous inhibitors

A crude extract may contain an inhibitor of the enzyme of interest, so that only a low rate is observed during the assay. As the purification proceeds the total activity will increase, corresponding to the removal of inhibitor. Inhibitors of low molecular mass (e.g. AMP, which acts as an inhibitor of fructose bisphosphatase in muscle extracts) can be removed by dialysis or gel filtration prior to assay. Inhibitors of high molecular mass are usually much more difficult to remove and some further purification of the extract may be required. For example, in the purification of the neutral proteinases from mycelial extracts of *Aspergillus nidulans*, an endogenous inhibitor is removed after a heat-treatment step or on prolonged storage of the extract at room temperature (42). The loss of inhibitor leads to a considerable (8-fold) increase in the total amount of activity compared with the crude extract.

4.2 Interference from other reactions

Other enzyme-catalysed reactions taking place in the extract could complicate the assays of the enzyme of interest. In such cases it may be possible to estimate the 'blank rate' in the absence of the specific substrate and then subtract this rate from the rate measured in the presence of this substrate. Assays of glycogen phosphorylase in muscle extracts are usually performed by measuring the phosphate released in the reaction shown below (in the presence of the activator AMP):

$$(\text{glycogen})_n + \text{glucose-1-phosphate} \rightarrow (\text{glycogen})_{n+1} + \text{P}_\text{i}$$

There could be interference, however, from the action of non-specific phosphatases present in the extract. In order to overcome this, 35 mM glycerol-2-phosphate was added to the buffer used for extraction. The contribution of the phosphatases was then assessed by measuring the rate of production of phosphate in the presence of glycogen and AMP, but in the absence of glucose-1-phosphate. By subtracting this rate from that observed in the presence of glucose-1-phosphate, the rate of the phosphorylase-catalysed reaction was estimated (43).

An alternative approach is to inhibit the interfering reaction. Assays of NAD^+-dependent dehydrogenases such as lactate dehydrogenase in muscle extracts can be interfered with by the activity of the electron transport chain which leads to oxidation of NADH. The latter activity can be eliminated by the addition of suitable electron transport chain inhibitors such as 1 mM potassium cyanide (*Care: poison!*) (44).

4.3 Removal of substrate

The presence of a competing reaction in an extract could reduce the concentration of the substrate available for the enzyme of interest so that the measured activity of that enzyme is reduced. If it is not feasible to inhibit the interfering activity, it may be possible to add a substrate-regenerating system so as to maintain the concentration. In assays of hexokinase in muscle extracts there is possible interference from the presence of ATPases, which would lower the concentration of ATP. Phosphocreatine and creatine kinase can be added to the assay system to replenish ATP (44). Assays of hexokinase are initiated by addition of glucose and control assays from which glucose is omitted are run in parallel.

4.4 Turbidity of extract

In any spectrophotometric assay, problems can arise if the extract to be assayed is turbid or contains high concentrations of an interfering absorbing species. It may be possible to overcome the first of these problems by clarifying the extract by centrifugation, assuming that the enzyme of interest is not sedimented by this procedure. Absorbing species of low molecular mass can be removed by gel filtration or dialysis, but addition of the extract could still give rise to large background absorbance or turbidity which requires the spectrophotometer to be 'backed off' so that changes in absorbance due to the enzyme-catalysed reaction are brought on to scale. A simple solution to the problem is to add less of the extract in the assay (to bring the absolute absorbance within the acceptable range), and then to increase the sensitivity of the spectrophotometer and/or recorder. It is important to check that under these conditions the instrument is still capable of giving accurate readings and that Beer's Law is still obeyed. For practical purposes, it is often more convenient to arrange to initiate the reaction by addition of a non-absorbing substrate to the solution already containing the extract, so that there will only be a minimal change in absorbance as the reaction is started. Problems with high background absorbance are more acute in assays performed in the far UV where more species are likely to interfere. Thus, for example, assays of enolase (or coupled assays of phosphoglycerate mutase involving enolase), which rely on the increase in absorbance at 240 nm on formation of phosphoenolpyruvate from 2-phosphoglycerate, are prone to interference from absorbance due to large quantities of protein in an extract.

5 Concluding remarks

In this chapter, some of the factors involved in extracting enzymes from tissues and cells have been discussed. Rather than presenting a series of 'recipes', the points to be borne in mind when devising experiments have been emphasized. As pointed out in Section 1, preparation of a cell extract is often only the first step towards the ultimate purification of an enzyme for structural or kinetic characterization. It is vital in this type of experiment to ensure that as much activity as possible is extracted in as intact and stable a state as possible (i.e. resembling the presumed native state). The considerations described in this chapter should help the investigator to achieve this goal.

References

1. Scopes, R. K. (1994). *Protein purification: principles and practice* (3rd edn). Springer, New York.
2. Harris, E. L.V. and Angal, S. (1990). *Protein purification: a practical approach*. IRL Press, Oxford.
3. Old, R. W. and Primrose, S. B. (1994). *Principles of gene manipulation* (5th edn). Blackwell, Oxford.
4. Walsh, G. and Headon, D. (ed.) (1994). *Protein biotechnology*. Wiley, Chichester.
5. Hurd, P. J. and Hornby,D. P. (1996). In *Proteins labfax* (ed. N. C. Price), pp.109–117. Bios Scientific Publishers, Oxford.
6. North, M. J. (1989). In: *Proteolytic enzymes: a practical approach* (ed. R.J. Beynon and J.S. Bond), pp. 105–124. IRL Press, Oxford.
7. Wu, X.-C., Lee, W., Tran, L. and Wong, S.-L. (1991). *J. Bacteriol.*, **173**, 4952.
8. Marston, F. A. O. (1986). *Biochem. J.*, **240**,1.

9. Netzer, W. J. and Hartl, F. U. (1997). *Nature*, **388**, 343.
10. Taylor, G., Hoare, M., Gray, D. R. and Marston, F. A. O. (1986). *Biotechnology*, **4**, 553.
11. Thatcher, D. R., Wilks, P. and Chaudhuri, J. (1996). In *Proteins labfax* (ed. N. C. Price), pp.119–130. Bios Scientific Publishers, Oxford.
12. Ishikawa,K., Matera, K. M., Zhou, H., Fujii, H., Sato, M., Yoshimura, T., Ikeda-Saito, M., and Yoshida, T. (1998). *J. Biol. Chem.*, **273**, 4317.
13. Harwood, C. R. (1992). *Trends Biotechnol.*, **10**, 247.
14. Ford, T. C. and Graham, J. M. (1991). *An introduction to centrifugation*, p.70ff. Bios Scientific Publishers, Oxford
15. Kula, M. R. and Schütte, H. (1987). *Biotechnol. Progress*, **3**, 31.
16. Brouwer,A., Barelds, R. J. and Knook, D. L. (1987). In *Centrifugation: a practical approach* (2nd edn), (ed. D.Rickwood), pp.183–218. IRL Press, Oxford.
17. Fleischer, S. and Packer, L. (ed.) (1974). *Methods in Enzymology*, **32**.
18. Hardman, J. G. and O'Malley, B. W. (ed.) (1975). *Methods in Enzymology*, **39**.
19. Siebert, G. and Humphrey, G. B. (1965). *Adv.Enzymol.*, **27**, 239.
20. Hall, N. P. and Tolbert, N. E. (1978). *FEBS Lett.*, **96**, 167.
21. Yun, S.-L., Aust, A. E., and Suelter, C. H. (1976). *J. Biol. Chem.*, **251**, 124.
22. Naganuma, T., Uzuka, Y.and Tanaka, K. (1984). *Anal. Biochem.*, **141**, 74.
23. Fell, D. A., Liddle, P. F., Peacocke, A. R. and Dwek, R. A. (1974). *Biochem. J.*, **139**, 665.
24. Schwenke, J., Canut, H. and Flores, A. (1983). *FEBS Lett.*, **156**, 274.
25. Rogl,H., Kosemund,K., Külbrandt, W. and Collinson, I. (1998). *FEBS Lett.*, **432**, 21.
26. Hult, M., Jörnall, H. and Oppermann, C. T. (1998). *FEBS Lett.*, **441**, 25.
27. Schwinghamer, E. A. (1980). *FEMS Microbiology Lett.*, **7**, 157.
28. Singer, S. J. (1974). *Annu.. Rev. Biochem.*, **43**, 805.
29. Penefsky, H. S. and Tzagaloff, A. (1971). *Methods in Enzymology*, **22**, 204.
30. Cogdell, R. J. and Lindsay, J. G. (1996). In *Proteins labfax* (ed. Price, N. C.), pp.101–107. Bios Scientific Publishers, Oxford.
31. Janson, J-.C. and Ryden, L. (1989). *Protein purification*, p. 8. VCH Publishers,New York.
32. Fasman, G. D. (1996). In *Circular dichroism and the conformational analysis of biomolecules* (ed. Fasman, G. D.), pp. 381–412. Plenum Press, New York.
33. Spatz, L. and Strittmatter, P. (1973). *J. Biol. Chem.*, **248**, 793.
34. Turner, A. J. (1994). *Essays Biochem.*, **28**, 113.
35. Hooper, N. M., Karren, E. H. and Turner, A. J. (1997). *Biochem. J.*, **321**, 265.
36. Creighton, T. E. (1992). *Protein folding*, p.106. Freeman, New York.
37. Cleland, W. W. (1964). *Biochemistry*, **3**, 480.
38. Jaenicke, R. (1981). *Annu. Rev. Biophys. Bioeng.*, **10**, 1.
39. Arakawa, T. and Timasheff, S. N. (1982). *Biochemistry*, **21**, 6536.
40. Katunuma, N., Kominami, E., Banno, Y., Kito, K., Aoki, Y. and Urata, G. (1976). *Adv. Enzyme Regulation*, **14**, 325.
41. Fell, D. (1997). *Understanding the control of metabolism*. Portland Press, London.
42. Ansari, H. and Stevens, L. (1983). *J. Gen. Microbiol.*, **129**, 1637.
43. Crabtree, B.and Newsholme, E. A. (1972). *Biochem. J.*, **126**, 49.
44. Zammit, V. A. and Newsholme, E. A. (1976). *Biochem. J.*, **160**, 447.
45. Harper, J. W., Hemmi, K. and Powers, J. C. (1985). *Biochemistry*, **24**, 1831.

Appendix 1 Buffers and the control of pH

Since the catalytic activity of most enzymes is sensitive to pH, the control of this variable during procedures of extraction, purification and assay is extremely important to ensure that reproducible results are obtained. This Appendix briefly examines some theoretical aspects of buffering of pH, and then outlines some of the practical considerations involved in the choice of a buffer in a particular situation. Some commonly used buffers are listed in *Table A1*. For more detailed accounts of buffer systems, references 1–4 should be consulted.

Table A1 Some commonly used buffers

Buffer	pK_a	dpK_a/dT	Saturated solution (M)	Other features to note
Acetic acid	4.64	0.0002	v. sol.	1
Citric acid (pK_3)	5.80	0	v. sol.	1, 4,10
Mes	6.02	−0.011	0.65	2, 3
Pyrophosphate (pK_3)	6.32	−0.01	0.1	1, 4, 5, 6
Phosphoric acid (pK_2)	6.84	−0.0028	0.2	1, 4, 5
Imidazole	6.97	−0.02	v .sol.	1, 4, 7
Hepes	7.39	−0.014	2.25	2, 3, 10
Triethanolamine	7.78	−0.02	v. sol.	1, 9
Tris	8.00	−0.031	2.4	1, 9, 10, 11
Borate	9.08	−0.008	0.05	1, 8
Ches	9.23	−0.029	1.14	2, 3, 10
Glycine	9.55	−0.025	4.0	1, 10,11
Carbonate (pK_2)	9.96	−0.009	0.8	1, 7

[1] Low price (< £20 ($30))/litre of 1 M buffer solution; [2] low UV absorbance, [3]low metal ion binding affinity; [4] high metal ion binding affinity; [5] tendency to precipitate Mg^{2+}, Ca^{2+},Fe^{3+} and other polyvalent cations; [6] inhibits kinases, some dehydrogenases and enzymes involving phosphate esters as substrates; [7] relatively low chemical stability; [8] complexes with *vic* diols, e.g. ribose; [9] forms imines with aldehydes and ketones; [10] interferes with Lowry protein estimation; [11] interferes with Coomassie blue protein estimation.

Buffers solutions comprise a weak acid (HA) and its conjugate base (A^-). If K_a represents the dissociation constant for the acid, the relative contributions of HA and A^- are given by the Henderson–Hasselbalch equation: (1):

$$pH = pK_a + \log ([A^-]/[HA]) \quad (1)$$

In most buffer solutions, HA and A^- are contributed to either by a weak acid and its salt with a strong base (e.g. acetic (ethanoic) acid and sodium acetate), or by the salt of a weak base with a strong acid and the weak base (e.g. triethanolamine-HCl and triethanolamine). The buffering capacity of any such solution is described as the 'buffer value', β, which is the amount of a strong base producing a resultant change in pH ($\beta = dB/d$ pH). β is related to C, the initial concentration of weak acid before any base has been added, and the degree of dissociation of the acid, α, by:

$$\beta = 2.303C\alpha(1 - \alpha) \quad (2)$$

A plot of β against pH shows that the maximum buffering capacity occurs when $\alpha = 0.5$, i.e. when pH = pK_a, and as a general rule, buffers should only be used over a pH range within 1 pH unit on either side of the pK_a. Equation 2 also shows that the buffering capacity is directly related to the concentration of buffer, and that the pH should be unaffected by dilution since the concentrations of HA and A^- are reduced in equivalent proportions. However, this is not always the case, since the ionic strength of the solution will affect the activity coefficients and thereby the activities of the ions in the solution. In practice the changes in pH arising from the dilution of a buffer consisting of a monovalent ion are small (e.g. dilution from 0.1 M to 0.05 M at pH = pK_a causes a change of 0.024 pH). By contrast, the effects can be appreciable if polyvalent buffering ions such as citrate or phosphate are involved, and it is sensible to avoid large degrees of dilution in such cases.

Temperature will have an effect on pH of a buffer; the sign and magnitude will depend on the enthalpy change on ionization of the acid (HA) involved. In general, carboxylic acid buffers are least sensitive to changes in temperature (e.g. acetate buffer adjusted to pH 4.5 at 25 °C will have a pH of 4.495 at 0 °C), whereas amine-containing buffers are most sensitive (e.g. Tris–HCl adjusted to pH 8.0 at 25 °C will have a pH of 8.78 at 0 °C). These considerations can be especially important when studying enzymes from thermophilic organisms, where assays, etc. can often be conducted at temperatures of 70 °C or above.

The practical considerations in the choice of buffer system can be listed as:.

(a) *What pH range is to be employed?* The buffer should be chosen to have a pK_a in this region. The possible effects of temperature and dilution of the pK_a should be taken into account.

(b) *What degree of buffering is required?* As a rule, the concentration of buffer used should be just sufficient to reduce any expected pH change to a negligible value (say less than 0.05 pH units). The concentration employed may well also depend on the cost of the buffer. The buffers introduced by Good and colleagues (5), based on organic amines and sulphonic acid derivatives, are considerably more expensive that those based on phosphate, citrate or borate, for example.

(c) *Will the buffer system interfere with the system under study?* For example, polyvalent anions such as citrate and phosphate can act as powerful chelators for metal ions, which may be required for enzyme stability or activity. The 'Good' buffers (5) cause fewer problems of this type. The absorbance of the buffer in the ultraviolet might preclude its use if that interfered with the determination of protein or the assay of enzyme activity, for example.

(d) *How convenient is the preparation of the buffer solution?* In general it is easier to make up solutions of crystalline solids (which should be stable and pure) than of liquids. Triethanolamine is a liquid which is prone to discoloration as a result of oxidation, but the hydrochloride is a crystalline solid. However, use of the hydrochloride will mean an increase in the ionic strength of the solution and the presence of chloride ions might be disadvantageous.

(e) *Are the buffers required for specialized applications?* In some processes, such as the concentration of proteins by freeze-drying or the analysis of peptides produced by proteolysis, it may be advantageous to use volatile buffers. Ammonium bicarbonate finds extensive use in this regard since it affords good buffering at pH values near neutrality. Details of other volatile buffers are given in reference (1).

Appendix 2 The determination of protein

Determination of the amount of protein present is important for a number of reasons. In unfractionated extracts it is important to know the total amount of protein present so that (a) the success of the extraction procedure can be assessed, (b) a balance sheet kept of the progress of a purification scheme, and (c) the capacity of chromatography columns can be chosen correctly. As the purification of particular protein proceeds, determination of the activity and protein present allows the specific activity of the protein in question to be calculated. Ultimately the characteristic properties of any purified protein (specific activity, stoichiometry of ligand binding, spectroscopic properties) all rely on an accurate knowledge of the amount of protein present in a sample.

Some of the more widely used methods for protein determination are listed in *Table A2*. For further details of these, references 6–8 should be consulted.

The choice of method to be used for the analysis of any particular sample will depend on factors indicated in *Table A2* (e.g. scale, need for absolute amount of protein, type of sample, equipment available, etc.), as well as other factors such as the possibility of interference from other components present in the sample. For instance, a large number of

Table A2 Some commonly used methods for the determination of protein

Method	Scale (μg)	Basis*	Preferred type of sample**
Biuret	500–5000	Absolute	All types
Lowry	5–100	Comparative	All types
BCA (bicinchoninic acid)	1–100	Comparative	All types
Near-UV absorbance	10–1000	Absolute	Purified protein
Far-UV absorbance	5–50	Absolute	Purified protein
Coomassie blue binding	1–100	Comparative	All types
Amino acid analysis	1–200	Absolute	Purified protein
Gravimetric	5000–20000	Absolute	Purified protein

*The term absolute indicates that the method is capable of giving the absolute (or very nearly absolute) mass of protein present. The term comparative means that the mass of protein in a sample is determined with reference to a standard protein such as bovine serum albumin. Although the biuret assay depends on the formation of a purple copper complex with adjacent peptide bonds and is relatively unaffected by the amino acid composition of the protein, a protein standard such as bovine serum albumin is normally used. **The term purified protein means that the technique is best suited to a purified protein, because of interference from other proteins or macromolecules in the mixture.

substances including aromatic amino acids, thiol compounds, 'Good' buffers (5), certain non-ionic detergents, sucrose and EDTA, interfere with the Lowry method, and special precautions are required to obtain reliable data. The dye-binding and BCA methods are less sensitive to interference and are now more widely used. Clearly, in crude extracts, the number of interfering substances is likely to be considerably greater than in more highly purified preparations.

References for Appendices

1. Perrin D. D. and Dempsey, B. (1974). *Buffers for pH and metal ion control.* Chapman & Hall, London.
2. Stoll, V. S. and Blanchard, J. S. (1990). *Methods Enzymol.*, **182**, 243.
3. Stevens, L. (1996). In *Enzymology labfax*, (ed. P. C. Engel), pp. 269–276. Bios Scientific Publishers, Oxford.
4. Dawson, R. M. C., Elliot, D. C., Elliot, W. M. and Jones, K. M. (1986). *Data for biochemical research* (3rd edn.), p.417. Oxford University Press, Oxford.
5. Good, N. E. and Izawa, S. (1972). *Methods Enzymol.*, **24**, 53.
6. Price, N. C. (1996). In *Enzymology labfax* (ed. P. C. Engel), pp.34–41. Bios Scientific Publishers, Oxford.
7. Harris, D. A. (1987). In *Spectrophotometry and spectrofluorimetry: a practical approach* (ed. Harris, D.A and Bashford, C. L.), Chapter 3. IRL Press, Oxford.
8. Scopes,R. K. (1994). *Protein purification: principles and practice* (3rd edn). Springer, New York.

Chapter 10
Determination of active site concentration

Mark T. Martin
Mirari Biosciences Inc., 6516 Old Farm Court, Rockville, Maryland 20852, USA

1 Introduction

Enzyme quantitation methods vary widely in complexity and reliability (*Table 1*). Methods that assume an enzyme is homogeneous and fully active (gravimetric or total protein assays) are far less reliable than methods that measure the concentration of *functional* enzyme (active site titrations[1] and substrate turnover rate assays). Of those methods that measure functional enzyme, active site titrations tend to be more reliable than activity assay methods. In contrast to activity assay methods, titrations do not require primary standards of 100% pure and active enzyme and are less likely to be affected by small variations in assay conditions (pH, ionic strength, temperature, etc.).

Despite the attractiveness of active site titration, there are a large number of well-known enzymes for which no titration protocols exist. One reason is that the design of titration methods can be quite challenging. Titrants must specifically target enzyme active sites in 1:1 stoichiometries. Moreover, reactions between the titrants and the active sites must result in measurable signals at the (often quite low) enzyme concentrations used. Nevertheless, a number of excellent titration protocols have been developed for a diverse collection of catalysts. The methods described in this chapter are included not only to be used as intended, but also to serve as guides in the design of new enzyme quantitation methods.

2 Areas of application

An active site titration is attractive in any work in which an accurate estimation of the operational enzyme concentration is required. As the number of known catalysts grows, so does the need for new titration methods. New enzymes are being discovered at a rapid pace by 'bioprospecting' (1, 2) and by *in vitro* protein evolution (3, 4). Active site titrations are also be valuable in studies of artificial enzymes such as designer ribozymes (5, 6), catalytic antibodies ('abzymes') (7, 8), functionalized polyethyleneimines ('synzymes') (9, 10), and molecularly imprinted polymers ('plastizymes') (11, 12). Finally, accurate methods are required to quantitate natural enzymes suspended in organic solvents (13, 14), and entrapped in porous plastic (15).

[1] Active site titration is the quantitation of *functional* enzyme by the addition of a compound that substantially perturbs the enzyme's activity in a 1:1 stoichiometry.

Table 1 Comparison of enzyme quantitation methods

Method	Advantages	Disadvantages
Weight	Simple, fast	Overestimates due to impurities and inactive enzyme, requires substantial quantities
Total protein (OD_{280}, Bradford, etc.)	Simple, fast	Overestimates due to protein impurities and inactive enzyme
Activity	Measures active enzyme	Requires 100% pure primary standard for comparison, often susceptible to variations in temperature, pH, etc.
Active site titration	Measures active enzyme, low susceptibility to variations in environmental conditions	Stoichiometric – need measurable signal at (often low) enzyme concentrations

3 Categories of titration methods

Design of an active site titration protocol requires some mechanistic or structural understanding of the enzyme to be quantitated. Universally, a titrant binds or reacts specifically and stoichiometrically in the intact active site. The titrant can be a substrate that undergoes a 1:1 stoichiometric 'burst' event during the catalytic pre-steady state. Alternatively, the titrant can be an inhibitor (e.g. tight-binding competitive) or an inactivator (affinity labels or mechanism-based inactivators) that blocks the enzyme active site in a titrable fashion. Finally, the titrant can be a simple protein-modifying reagent.

3.1 Activity bursts

Many enzymes have mechanisms with a rate-limiting step. When enzymes of this type are assayed, progress curves often show an initial curved pre-steady-state 'burst' followed by a linear steady- state phase (*Figure 1*). The burst represents the rapid ($>k_{cat}$) accumulation of a quasi-stable intermediate within during the first catalytic cycle. The linear steady state ($= k_{cat}$) often represents the slow rate of breakdown of this intermediate (the rate-limiting step). If the rate of formation of the intermediate greatly exceeds the rate of its breakdown, then the

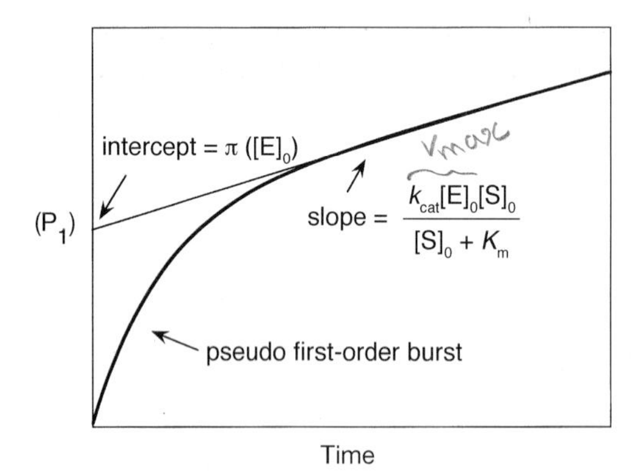

Figure 1 Typical progress curve for an enzyme reaction displaying a stoichiometric burst. The curved pre-steady-state phase is followed by a linear steady-state phase. In the case of serine proteases such as α-chymotrypsin, the burst typically results from the stoichiometric release of a chromophoric leaving group during the formation of the acyl-enzyme intermediate, k_2, in the first catalytic cycle. The burst phase is succeeded by the steady-state phase, which is controlled by the much slower deacylation rate, k_3 ($k_2 >> k_3$).

active site will be fully occupied by the intermediate during the burst, and the burst magnitude will be equal to the enzyme active-site concentration. The burst method of active-site titration was pioneered by Hartley and Kilby (16) on α-chymotrypsin (*vide infra*). Many other serine and cysteine hydrolases (e.g. chymotrypsin, trypsin, papain, elastase, subtilisin, acetylcholinesterase, thrombin, and factor Xa) (17, 18) have since been similarly be quantitated by their bursts. Other, mechanistically diverse, enzymes have been also quantitated by burst active site titration. They include: T7 DNA polymerase (20), HIV reverse transcriptase (21), aminoacyl-tRNA synthetases (22, 23), kinesin motor domain (24), and adenine glycosylase MutY (25).

The classic example of active site titration is that of serine proteases, which hydrolyse substrates via a covalent acyl-enzyme intermediate:

$$E + S \overset{K_s}{\rightleftharpoons} ES \xrightarrow{k_2} E\text{–}S + P_1 \xrightarrow{k_3} E + P_2$$

In the scheme, E is free enzyme, S is substrate, K_s is the dissociation constant of the Michaelis complex (ES), k_2 is the rate of acylation to form the covalent acyl intermediate (E–S), k_3 is the deacylation rate, and P_1 and P_2 are products. In many assays of enzymes of this type, the release of P_1 results in an observable signal change (often a chromophoric group such as *p*-nitrophenol). If the rate of deacylation (k_3) is negligible compared to the rate of acylation (k_2), progress curves show an initial pre-steady-state 'burst' in signal as a stoichiometric concentration of P_1 is rapidly produced in the first acylation. The burst is followed by a linear steady-state rate equal to k_3, which is the overall rate-limiting step.

The scheme shown above can be represented by the following equation under conditions of $[S]_0 >> [E]_0$ (17):

$$[P_1] = \frac{k_{cat}[E]_0[S]_0 t}{[S]_0 + K_m} + [E]_0 \left[\frac{\left(\frac{k_2}{k_2 + k_3}\right)}{1 + \frac{K_m}{[S]_0}} \right]^2 \times \left[1 - \exp \frac{-(k_2 + k_3)[S]_0 + k_3 K_s}{K_s + [S]_0} t \right] \quad (1)$$

where K_m is $(K_s + k_2/k_1)$, k_1 being the rate constant for the formation of the ES complex. When time, t, is large, the exponential term approaches zero and Equation 1 becomes a linear expression:

$$[P_1] = \pi + \frac{k_{cat}[E]_0[S]_0}{[S]_0 + K_m} t \quad (2)$$

where

$$\pi = \frac{[E]_0 \left(\frac{k_2}{k_2 + k_3}\right)^2}{\left(1 + \frac{K_m}{[S]_0}\right)^2} \quad (3)$$

Equations 3.1–3.3 show that if $k_2 >> k_3$ (deacylation is rate-limiting) and $[S_o] >> K_m$, then the linear portion of a progress curve ($[P_1]$ vs. t) will extrapolate to intercept the *y*-axis at $[P_1] = [E_o]$ ($\pi = [E]_o$) (*Figure 1*). If it is not technically feasible to assay the enzyme at $[S_o] >> K_m$, then the following transformation of Equation 2 can be used to determine enzyme concentration.

$$\frac{1}{\sqrt{\pi}} = \left(\frac{k_2 + k_3}{k_2}\right)\left(\frac{1}{\sqrt{[E]_0}}\right) + \frac{(k_2 + k_3)K_m}{k_2\sqrt{[E]_0}}\left(\frac{1}{[S]_0}\right) \quad (4)$$

Using Equation 4, a series of assays with $[E]_0$ held constant and $[S]_0$ varied, a plot of $1/[S]_0$ versus $1/(\pi)^{1/2}$ will give a straight line with an intercept of $1/([E]_0)^{1/2}$ (provided $k_2 >> k_3$).

Protocol 1

Quantitation of α-chymotrypsin by activity 'burst'

Equipment and reagents

- Thermostatted UV-vis spectrophotometer (335 nm) and 1.0 cm pathlength, 1.4 ml quartz cuvette
- Pipettes and tips
- α-chymotrypsin, bovine pancreas (Sigma)
- *N-trans*-cinnamoylimidazole (Sigma)
- Acetonitrile, HPLC grade (Aldrich)
- Sodium acetate buffer, 0.1 M, pH 5.0

Method

1. Prepare α-chymotrypsin stock solution; 50 mg protein in 1.0 ml buffer (2.0 mM)[a].
2. Prepare *N-trans*-cinnamoylimidazole stock solution; 2.3 mM in CH_3CN.
3. Mix 940 μl buffer and 30 μl α-chymotrypsin in cuvette. Record absorbance at 335 nm (A_0).
4. Add 30 μl *N-trans*-cinnamoylimidazole (a small excess over enzyme), mix rapidly, and immediately begin recording absorbance at 335 nm.
5. Continuously record absorbance decline ('burst') at 335 nm until steady state is reached ('*A*').
   This will take approximately 1–2 min.
6. Measure the total change (increase) in absorbance upon addition of a second aliquot (30 ml) of *N-trans*-cinnamoylimidazole stock solution. The absorbance change is designated ΔA_{max}.
7. Calculate the α-chymotrypsin concentration in the stock solution from the following equation;

$$[\alpha\text{-chymotrypsin}] = \frac{(A_0 + 0.97\Delta A_{max}) - (A_f - A_0)}{254}$$

[a] This method should be useful for α-chymotrypsin concentrations well into the micromolar range.

Adapted from refs. (17) and (25). See also (16), (18), and (26).

3.2 Inhibitor titration

One equivalent of a 'stoichiometric inhibitor' will inactivate one equivalent of enzyme. There are four general types of stoichiometric inhibitors:

- general reagents preferentially react with 'hyper-reactive' amino acids in certain perturbed active-site microenvironments
- competitive inhibitors form extremely tight-binding non-covalent complexes with the enzyme (e.g. transition state analogue inhibitors)
- affinity labels specifically bind to the enzyme active site, then react with some proximal functional group to form a covalent adduct.
- mechanism-based inactivators (suicide substrates) are transformed by the enzyme's normal catalytic mechanism into highly reactive species, which irreversibly inactivate the enzyme.

In all cases, inactivation can be measured either by direct detection of the inhibitor–enzyme complex, by stoichiometric release of a leaving group, or by measuring residual activity of unreacted enzyme. The inactivated enzyme can be detected, for example, by changes in absorbance or fluorescence, or by using a radiolabelled titrant.

3.2.1 Simple protein-modifying reagents

Simple-structured reactive compounds that have no steric specificity for the enzyme active site can sometimes be used to titrate enzyme active sites. Reaction rates between active site amino acids and protein chemical modification reagents can be greatly enhanced by perturbed active site microscopic environments (27, 28). Hyper-reactivity usually results from a shift in the p*K* of an amino acid side-chain (cysteine, glutamate, or other). In these cases, the active site residue will be substantially more reactive than the same type of residue elsewhere on the surface of the enzyme. Overall, the simple reaction between a general reagent and an amino acid side-chain can be described as follows, with no specific binding step:

$$\mathrm{E + R \longrightarrow E{-}R}$$

Cysteine proteases are a well-known class of enzymes that can be titrated using simple reagents. In these enzymes, the p*K* of an active-site essential cysteine is perturbed so that it reacts more rapidly with thiol reagents other cysteines on the surface of the enzyme. Titration can be carried out with 2,2′-dipyridyl disulphide (28–30). Other examples of enzymes that can be titrated using general reagents include amine oxidase (31) and tyrosine phosphatases (32). Titrations with simple reagents can be monitored by residual activity or by some associated physicochemical change.

In some instances, an enzyme may have a type of amino acid residue that exists only within the active site, not elsewhere on the surface of the molecule. Here, general reagents that react with that residue could be used to titrate the enzyme, regardless of microenvironmental considerations.

Finally, the active site concentration can be determined by 'subtractive labelling' of an amino acid known to be present in the active site. In this technique, the enzyme is reacted with an excess of an appropriate protein-modifying reagent in the presence and in the absence of an active-site-protecting inhibitor. The difference between the concentration of modifications (determined by some physical method) on the uninhibited (N) and inhibited enzyme ($N - 1$) is equal to the enzyme concentration. This method is workable provided the number of nonessential surface residues is not too large.

3.2.2 Tight-binding inhibitors and affinity labels

Affinity labels specifically bind to the active site, then covalently react with some nearby amino acid functional group to yield an inactive enzyme (27, 33, 34).

$$\mathrm{E + L} \underset{}{\overset{K_s}{\rightleftharpoons}} \mathrm{EL} \overset{k_2}{\longrightarrow} \mathrm{E{-}L}$$

In the above scheme, E is enzyme, L is affinity label, and E–L is a covalent complex between (inactive) enzyme and label. Affinity reagents typically have alkylating, acylating, or photoreactive groups. Affinity labels are especially useful for quantitation of dehydrogenases and kinases (33), enzymes that may not otherwise be amenable active site titration. Titrations can be carried out by adding sub-stoichiometric amounts of label to the enzyme solution, then measuring the residual enzymatic activity (35). Alternatively, protein-associated radioactivity (36), or some spectral change (37) can be measured.

Tight-binding inhibitors can also be used to titrate enzymes. Again, titration can be monitored by residual activity (38) or by some physiochemical property of the enzyme inhibitor complex (spectral change, radioactivity, etc.) (39)

3.2.3 Mechanism-based inactivators

Mechanism-based inactivators (suicide substrates) are relatively non-reactive substrate-resembling compounds that are chemically transformed by the enzyme's normal catalytic mechanism to form highly reactive species that inactive the enzyme (40, 41). As shown in the scheme below, the inactivator, I, specifically binds to the active site to form an EI complex, then undergoes an enzyme-accelerated transformation into a highly reactive inactivating species (E–I) which can either be released as product (P) or form a stable dead-end covalent adduct (E–I′).

$$E + I \underset{}{\overset{K_s}{\rightleftharpoons}} EI \xrightarrow{k_2} E\text{–}I \xrightarrow{k_3} E + P$$
$$E\text{–}I \xrightarrow{k_4} E\text{–}I'$$

Inhibition by mechanism-based inactivators is characterized by a 'partition constant', k_3/k_4 in the scheme above. The partition constant is the number of turnovers that the enzyme catalyses prior to becoming inactivated. Once the partition constant has been reliably determined (which requires enzyme quantitation by some other method!), mechanism-based inactivators can be used to titrate the enzyme in a similar fashion to other stoichiometric inhibitors listed above.

An alternative method for using mechanism-based inactivators is perhaps more reliable since it does not require knowing the partition constant. If the active-site-bound inactivator (E–I′ above) has some distinguishing spectral characteristics or is radioactive, then the inactivated enzyme active site could be directly observed.

3.3 Special techniques

Described above are methods for determining the concentration of enzymes under 'normal' assay conditions, focusing mainly on considerations of enzyme specificity and mechanism. Described in this section are primarily situational considerations that are independent of the type of enzyme under consideration. The questions addressed are:

- How can active sites of a poorly characterized catalyst be quantitated in the absence of a titrant?
- How can immunoassays be formatted to quantitate enzyme active sites?
- How does one determine active-site concentration in organic solvents?

3.3.1 Rate assays under single turnover conditions

Although the method described in this section is a rate assay, not an active-site titration, it is included in this chapter because it can be valuable when no titrant exists and the catalyst is poorly characterized. It can be especially useful in quantitating newly discovered enzymes, catalytic synthetic polymers, and polyclonal catalytic antibodies.

The method was described by Suh *et al.* (42) for the quantitation of catalytic sites on modified polyethyleneimine (see also 43). Two sets of experiments are run under two types of conditions; single turnover ($[E]_0 >> [S]_0$) and conventional ($[S]_0 >> [E]_0$). First, enzyme concentrations are used in great excess over substrate concentration, such that the catalytic reaction behaves 'opposite' to the usual Michaelis–Menten form:

$$k_{obsd} = \frac{k_{lim}[E]_0}{K_m + [E]_0} \quad (5)$$

Where $[S]_0$ is held constant, k_{obsd} is the experimentally observed pseudo-first order reaction rate constant, k_{lim} is the limiting first-order rate constant (at high $[E]_0$ where the substrate is saturated with enzyme), and K_m is the enzyme concentration at $k_{obsd} = k_{lim}/2$. In the general case of multiple catalytic sites per catalyst (such as in synthetic polymer catalysts):

$$k_{obsd} = \frac{k_{lim}n[P]_0}{K_m + n[P]_0} \quad (6)$$

where n is the estimated number of sites and $[P]_0$ is the estimated number of molecules. The linear transformation of Equation 6 is:

$$\frac{1}{k_{obsd}} = \left(\frac{K_m/n}{k_{lim}}\right)\frac{1}{[P]_0} + \frac{1}{k_{lim}} \quad (7)$$

Experiments are run in which $[P]_0$ is varied and k_{obsd} is measured. Values of K_m/n and k_{lim} are thus obtained.

Next, experiments are run under conventional Michaelis–Menten conditions of $[S]_0 >> [P]_0$;

$$k_0 = \frac{k_{lim}n[P]_0}{K_m + [S]_0} \quad (8)$$

where $k_0 = v/[S]_0$ (v is the observed rate) and k_{lim} is again maximal rate constant, this time at high $[S]_0$, at which enzyme is saturated with substrate). The linear transformation of Equation 8 is:

$$\frac{[P]_0}{k_0} = \frac{K_m}{nk_{lim}} + [S]_0 \frac{1}{nk_{lim}} \quad (9)$$

Experiments are run in which $[S]_0$ is varied and $k_0$is measured. Values of K_m and nk_{lim} can thus be obtained. Comparison of K_m/n and k_{lim} (Equation 7) with K_m and nk_{lim} (Equation 9) provides values of n (active site concentration), k_{lim}, and K_m. The method requires concentrations of P_0 and S_0 that can span K_m, otherwise K_m and k_{lim} cannot be accurately determined (9).

A potential limitation of this method is that it assumes that all sites that bind substrate are catalytic. An extension of the technique has been recently reported (44) for the situation when substrate-binding non-catalysts are present in solutions of catalysts. This can occur, for example, in polyclonal antibody preparations where total antibody consists of a heterogeneous mixture of catalysts, non-catalysts that bind substrate, and non-catalysts that do not bind substrate.

3.3.2 Active site directed immunoassays

Monoclonal antibodies can be used as specific reagents in active site titrations. One method involves using an antibody that specifically binds to an enzyme active site resulting in inactivation. In this way, the antibody can be thought of as a specific tight-binding competitive inhibitor and be used as described in Section 3.2.2. This is an attractive immunoassay since it does not require solid phase capture as do most immunoassay formats. However, the method has the drawback that it requires an antibody that binds only to the *functional* active site.

Titrations can be performed even with antibodies that bind to the enzyme at locations removed from the active site. For example, the enzymes of the fibrolytic and coagulation systems have been quantitated as follows (45). First, specific labelling of the enzyme active site is achieved using a biotinylated chloromethylketone peptide inhibitor. The biotin group protrudes from the inhibitor–enzyme complex, allowing the enzyme to be captured on an avidin-coated solid phase. A horseradish peroxidase-labeled anti-enzyme antibody is then used to generate a reporting signal. This elegant method has the advantage over many active site titration protocols in that it produces a strong non-stoichiometric signal (HRP catalytically produces a coloured signal). The potential drawbacks of the protocol include the need to develop a specific anti-enzyme antibody (often not a trivial feat), an appropriate capture inhibitor, and a reliable immunoassay protocol (46).

3.3.3 Titrations in organic solvents

The use of enzymes in unnatural media such as in organic solvents presents additional active site titration challenges. Unnatural environments can cause protein denaturation, rendering quantitation by weight or protein quantitation impractical (*Table 1*). In addition, activities of enzymes will be greatly perturbed in an unnatural environment, rendering activity-based quantitations useless. Thus, in research in which enzymes are used in organic solvents or other harsh environments, active site titrations are critically important methods. Three successful and unique concepts for quantitating enzymes in nonaqueous media are briefly described below.

In one method (47), α-chymotrypsin is incubated with the substrate *N-trans*-cinnamoylimidazole in buffer to form the stable acyl-enzyme. The acylated enzyme is lyophilized, then suspended in octane containing 1 M propanol. The deacylation of the acyl enzyme by propanol transesterification is followed by gas chromatography until the reaction is complete. The concentration of product is equal to the concentration of active enzyme.

A second method (48) involves two irreversible inhibitors in a two-step process (see Protocol 2). First, α-chymotrypsin is suspended in organic solvent with and without the covalent inhibitor phenylmethylsulfonyl fluoride (PMSF). Both suspensions are then lyophilized and redissolved in aqueous buffer. Active enzyme is then titrated with the chromogenic inhibitor 2-nitro-4-carboxyphenyl *N,N*-diphenylcarbonate (NCDC). Reaction of the enzyme with NCDC produces the coloured product hydroxynitrobenzoic acid, which is measured spectrophotometrically. The amount of enzyme active in organic solvents is determined by comparing the absorbance of the PMSF-treated sample with that of the PMSF-untreated sample.

In a third method, PMSF is used in a kinetic rather than an endpoint titration (49). This method obviates the long incubation time required for the PMSF-enzyme reaction to go to completion.

Protocol 2

Determination of the fraction of α-chymotrypsin active in organic solvent

Equipment and reagents

- Thermostatted UV-vis spectrophotometer (410 nm) and 1.0 cm pathlength, 1.4 ml quartz cuvette
- Lyophilizer
- Pipettes and tips
- α-Chymotrypsin, bovine pancreas (Sigma)
- Phenylmethylsulfonyl fluoride (PMSF, Sigma)
- 2-Nitro-4-carboxyphenyl *N,N*-diphenylcarbamate (NCDC, Sigma)
- Toluene (Aldrich)
- Tris buffer, 0.05 M, pH 7.6

Protocol 2 continued

Method

1. Lyophilize α-chymotrypsin from aqueous solution, then form an enzyme suspension in dry toluene.
2. Divide the suspension into two equal volumes. To one (sample 1), add an excess of PMSF, to the other (sample 2) add nothing. Allow PMSF and α-chymotrypsin to react to completion.
3. Remove enzyme from both suspensions (samples 1 and 2). Allow enzyme to dry to powder.
4. Dissolve enzyme in buffer containing an excess of NCDC, allow reaction to go to completion.
5. Spectrophotometrically measure hydroxynitrobenzoic acid (HNBA, molar extinction coefficient is 3910 $M^{-1}cm^{-1}$ at 410 nm (50)) produced in samples 1 and 2.
6. Calculate the fraction of α-chymotrypsin active sites that are active in toluene (PMSF-reactive) using the following equation;

$$\text{Fraction of a-chymotrypsin active sites active in toluene} = 1 - \frac{[\text{HNBA}]_{\text{sample1}}}{[\text{HNBA}]_{\text{sample2}}}$$

Adapted from refs. 48 and 50.

References

1. Short, J. M. (1997). *Nature Biotechnol.*, **15**, 1322.
2. Hough, D. W., and Danson, M. J. (1999). *Curr. Opin. Chem. Biol.*, **3**, 39.
3. Crameri, A., Raillard, S.-A., Bermudez, E., and Stemmer, W. P. C. (1998). *Nature*, **391**, 288.
4. Giver, L., Gershenson, A., Freskgard, P.-O., and Arnold, F. H. (1998). *Proc. Natl. Acad. Sci., USA*, **95**, 12809.
5. James, H. A., and Gibson, I. (1998). *Blood*, **91**, 371.
6. Amarzguioui, M., and Prydz, H. (1998). *Cell. Mol. Life Sci.*, **54**, 1175.
7. Martin, M. T. (1996). *Drug Discovery Today*, **1**, 239.
8. Schultz, P. G., and Lerner, R. A. (1995). *Science*, **269**, 1835.
9. Suh, J., and Hah, S. S. (1998). *J. Am. Chem. Soc.*, **120**, 10088.
10. Klotz, I. M. (1987). In *Enzyme mechanisms* (ed. M. I. Page and A. Williams), pp. 14–34. Royal Society of Chemistry, London.
11. Beach, J. V., and Shea, K. J. (1994). *J. Am. Chem. Soc.*, **116**, 379.
12. Leonhardt, A., and Mosbach, K. (1987). *Reactive Polymers*, **6**, 285.
13. Klibanov, A. M. (1997). *Trends Biochem. Sci.*, **15**, 97.
14. Aldercreutz, P. (1996). In *Enzymatic reactions in organic media*. (ed. A. M. P. Koskinen and A. M. Klibanov), pp. 9–42. Blackie, Glasgow.
15. Wang, P., Sergeeva, M. V., Lim, L., and Dordick, J. S. (1997). *Nature Biotechnol.*, **15**, 789.
16. Hartley, B. S., and Kilby, B. A. (1954). *Biochem. J.*, **63**, 288.
17. Bender, M. L., Begue-Canton, M. L., Blakeley, R. L., Brubacher, L. J., Feder, J., Gunter, C. R., *et al.* (1966). *J. Am. Chem. Soc.*, **88**, 5890.
18. Jameson, G. W., Roberts, D. V., Adams, R. W., Kyle, W. S. A., and Elmore, D. T. (1973). *Biochem. J.*, **131**, 107.
19. Patel, S. S., Wong, I., and Johnson, K. A. (1991). *Biochemistry*, **30**, 511.
20. Kati, W. M., Johnson, K. A., Jerva, L. F., Anderson, K. S. (1992). *J. Biol. Chem.*, **267**, 25988.
21. Fersht, A. R., Ashford, J. S., Bruton, C. J., Jakes, R., Koch, G. I. E., and Hartley, B. S. (1975). *Biochemistry*, **14**, 1.
22. Johnson, D. L. and Yang, D. C. H. (1981). *Proc. Natl. Acad. Sci,, USA*, **78**, 4059.
23. Gilbert, S. P. and Johnson, K. A. (1993). *Biochemistry*, **32**, 4677.
24. Porello, S. L., Leyes, A. E., and David, S. S. (1998). *Biochemistry*, **37**, 14756.

25. Schonbaum, G. R., Zerner, B., and Bender, M. L. (1961). *J. Biol. Chem.*, **236**, 2930.
26. Kezdy, F. J., and Bender, M. L. (1962). *Biochemistry*, **1**, 1097.
27. Plapp, B. V. (1982). *Methods Enzymol.*, **87**, 469.
28. Brocklehurst, K. (1979). *Int. J. Biochem.*, **10**, 259.
29. Brocklehurst, K., and Little, G. (1973). *Biochem. J.*, **133**, 67.
30. Baines, B. S., and Brocklehurst, K. (1982). *J. Protein Chem.*, **1**, 119.
31. Padiglia, A., Medda, R., Lorrai, A., Murgia, B., Pederson, J. Z., Agro, A. F., and Floris, G. (1998). *Plant Physiol.*, **117**, 1363.
32. Pregel, M. J. and Storer, A. C. (1997). *J. Biol. Chem.*, **272**, 23552.
33. Colman, R.F. (1990). In *The enzymes* (ed. D. S. Sigman, and P. D. Boyer), Vol. XIX, pp. 283–321, Academic Press, New York.
34. Singer, S.J. (1967). *Adv. Protein Chem.*, **22**, 1.
35. Potempa, J., Pike, R., and Travis, J. (1997). *Biol. Chem.*, **378**, 223.
36. Shirahata, A., Christman, K. L., and Pegg, A. E. (1985). *Biochemistry*, **24**, 4417.
37. Scoggins, R. M., Summerfield, A. E., Stein, R. A., Guyer, C. A., and Staros, J. V. (1996). *Biochemistry*, **35**, 9197.
38. Furfine, E. S., Harmon, M. F., Paith, J. E., Knowles, R. G., Salter, M., Kiff, R. J., *et al.*, (1994). *J. Biol. Chem.*, **269**, 26677.
39. Zhang, Y.-L., Keng, Y.-F., Zhao, Y., and Zhang, Z.-Y. (1998). *J. Biol. Chem.*, **273**, 12281.
40. Silverman, R. B. (1988). *Mechanism-based enzyme inactivation: chemistry and enzymology*, Vol. I, pp. 3–30. CRC Press, Boca Raton, FL.
41. Ator, M. A. and Ortiz De Montellano (1990). In *The enzymes* (ed. D. S. Sigman, and P. D. Boyer), Vol. XIX, pp. 213–282. Academic Press, New York.
42. Suh, J., Scarpa, I. S., and Klotz, I. M. (1976). *J. Am. Chem. Soc.*, **98**, 7060.
43. Hollfelder, F., Kirby, A. J., and Tawfik, D. S. (1997). *J. Am. Chem. Soc.*, **119**, 9578.
44. Resmini, M., Gul, S., Sonkaria, S., Gallacher, G., and Brocklehurst, K. (1998). *Biochem. Soc. Trans.*, **26**, S170.
45. Mann, K. E., Williams, E. B., Krishnaswamy, S., Church, W., Giles, A., Tracy, R. P. (1990). *Blood*, **76**, 755.
46. Davies, C. (1994). In *The immunoassay handbook.* (ed. D. Wild), pp. 3–47. Stockton, New York.
47. Zaks, A. and Klibanov, A. M. (1988). *J. Biol. Chem.*, **263**, 3194.
48. Paulaitis, M. E., Sowa, M. J., and McMinn, J. H. (1992). *Ann. NY Acad. Sci.*, **672**, 278.
49. Wangikar, P. P., Carmichael, D., Clark, D. S., and Dordick, J. S. (1996). *Biotechnol. Bioeng.*, **50**, 329.
50. Erlanger, B. F., and Edel, F. (1964). *Biochemistry*, **3**, 346.

Chapter 11

High throughput screening – considerations for enzyme assays

David Hayes and Geoff Mellor

GlaxoSmithKline Research and Development, Medicines Research Centre, Gunnels Wood Road, Stevenage, Hertfordshire SG1 2NY, UK.

1 Introduction

The rapidly increasing number of potential therapeutic targets is leading the pharmaceutical industry to radically alter its drug discovery strategy (for reviews, see 1–6). For these targets, there is often little or no structural information available for the rational design of putative drug candidates. Additionally, there is a desire to obtain 'lead' molecules of novel structure, to avoid patent infringement of known inhibitors. One approach to address these issues is to assay large numbers of compounds against the target, either in a random or sometimes targeted fashion, a process termed high throughput screening (HTS). The precise nature of these targets varies widely; for example they can be receptor–ligand, protein–protein, protein–carbohydrate, DNA and RNA interactions, ion-channels or enzymes. For this chapter we will consider only enzyme targets and will discuss the requirements for successful assay development and validation and subsequent screening. Details of two example assays are given, along with a snapshot of automation required to facilitate the process.

2 The drug discovery process

2.1 A historical perspective

Since ancient times man has made use of plant products for medicinal purposes. For many hundreds of years extracts such as morphine from the poppy, quinine from cinchona bark, digitalis from foxglove and ergot alkaloids have been used to treat a variety of maladies. Microorganisms as a potential source of pharmaceutical products became evident following the work of Fleming. The application of chemical dyes to treat malaria by Ehrlich started the science of medicinal chemistry. Early drug hunting was essentially non-selective, using whole organisms, cells or animals to test the effectiveness of putative drug candidates. In the main the molecular targets, for example the enzyme(s) that interacted with the compounds, were unknown. Although this approach led to the discovery of numerous drugs, many of the compounds had serious side-effects and dose-limiting toxicity. As the availability of compounds increased, advances in assay technology to accommodate this greater number of chemical entities lagged behind. Therefore, a more scientific approach to drug discovery evolved, termed rational drug design. The molecular target, for example an enzyme critical to the survival of a microorganism or one believed to be responsible for the underlying cause of disease, was identified. Detailed structural information was sometimes available and computer-aided drug design software allowed the docking of putative inhibitors in the active site. Often the design of the inhibitor was based on the natural substrate or known molecules that acted at that site. However, this imposed a lack of structural novelty and obtaining patents became increasingly difficult. The introduction of 96-well micro-titre plates and allied reading technologies allowed a vast increase in the number of assays which could be run in a given time for a target, and hence the birth of high throughput screening. The consistency and cost-effectiveness of the HTS process have been greatly improved by the introduction of automation and most major screening centres will at least have small, modular, automated screening systems and many, including GlaxoSmithKline, have large, integrated robotic systems in addition. HTS is a constantly evolving process and recent advances include plate miniaturization to 384 and 1536 wells and beyond, and the introduction of cooled charge-coupled dipole (CCD) imaging cameras capable of rapid and sensitive detection of assay read-outs.

2.2 A model of drug discovery

A typical drug discovery scheme is shown in *Figure 1*. A target is defined as an enzyme whose action is responsible for causing a disease or symptom, or supports a reaction vital to the survival of a microorganism. Targets are identified, for example, by literature precedent or at the site of action of a known compound, or by 'in-house' research. Increasingly the use of proteomics and genomics will lead to the discovery of enzymes associated with particular diseases. Confirmation that inhibition (or activation) of the target enzyme results in the desired effect, is in general required before resource is committed to a screen. This confidence may be gained by the use of compounds acting at the target site or by genetic techniques, e.g. expression of dominant negative mutant protein or knock out animals. Compound selection (20 000 to > 500 000) is dictated by the capacity of the assay used and by any automation (see below). At the end of the primary screen, compounds exhibiting good inhibition (or activation) are re-tested before being confirmed as hits (*Figure 2*). The IC_{50} value for each hit is then determined and the compounds may be ranked according to their potency. A selection of molecules is made based on a pre-determined strategy, for example, selectivity over closely related enzymes or lack of activity against the host enzyme in the case of anti-infective agents. Such molecules are termed leads and are the starting point for medicinal chemists who turn the leads into drug candidates. The desired properties of any drug depend on the exact target but are likely to include potency at the target enzyme, selectivity, activity in animal models of the disease, lack of toxicity, novelty (i.e. not previously described in the patent or other literature) and the requisite pharmacokinetic properties.

Before any compound can be used in man, rigorous safety and toxicity protocols are followed. Statutory bodies regulate this, e.g. the FDA (Food and Drug Administration). Phase I clinical trials (lasting around 1 year) are usually conducted using healthy volunteers to investigate pharmacokinetics in humans, such that the intended therapeutic dose is achieved and tolerated in the desired bodily fluid, i.e. plasma. Phase II clinical trials (lasting about 1 to 2 years) investigate the efficacy of the compound in the appropriate disease setting. These trials are typically located at one or two centres (hospitals or clinics) involving small numbers of patients. Much larger, often double blind, multi-centre trials are termed Phase III clinical trials (lasting 1 to 5 years). The purpose of this trial is to demonstrate clinical benefit of the compound and to lead to registration of the medicine. The drug may now be launched and the revenue raised starts to pay back the huge investment for the pharmaceutical company. The time taken from lead generation to product launch may take up to 10 years and cost in excess of £150 million. Most compounds fail to make it this far for a variety of reasons, e.g. unexpected toxicity, adverse side effects, poor pharmacokinetics in man. Therefore industry not only has to cover the cost of the successes but the many failures.

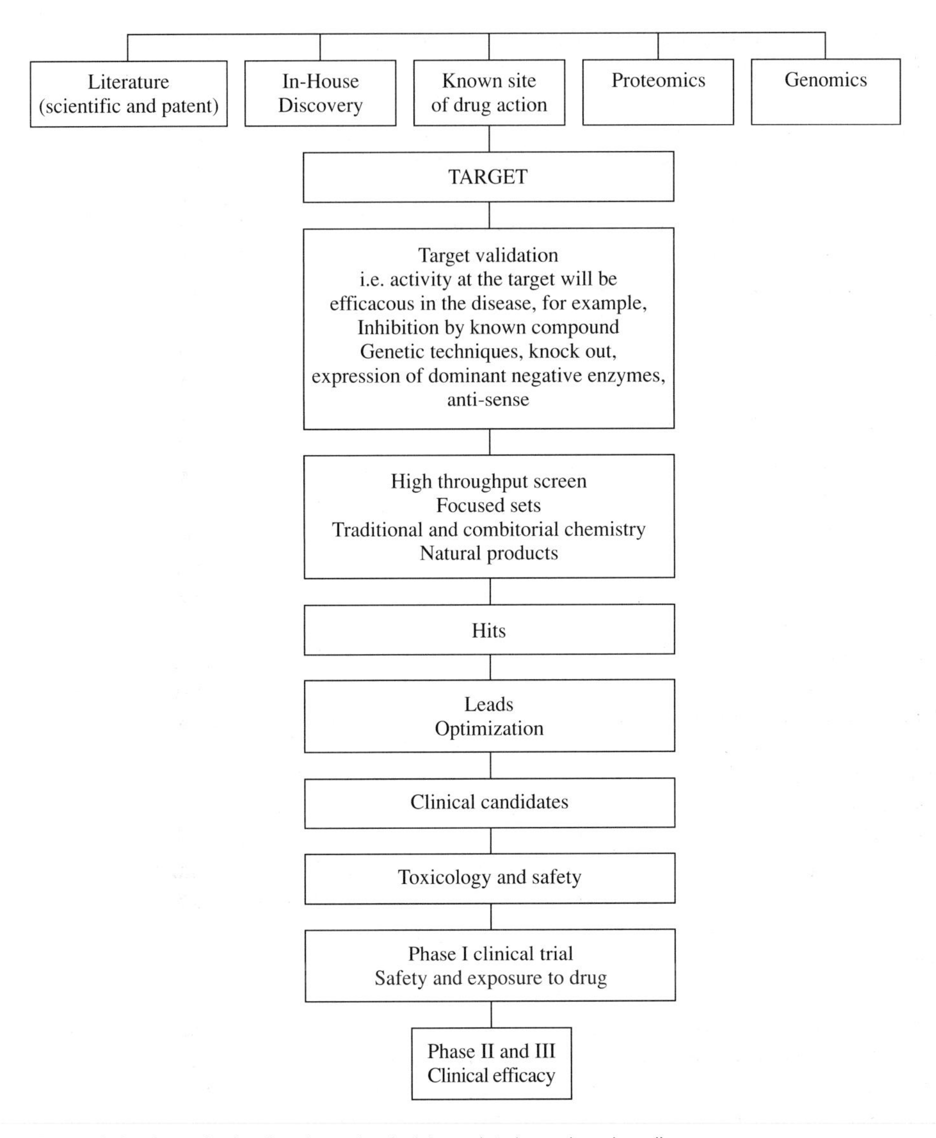

Figure 1 A schematic showing the major decision points in modern drug discovery.

3 High throughput screening

The modern approach to drug screening requires great efficiency to contain some of the costs (although the highest spend is during the clinical phases) and to speed the process from identification of a compound to the market. The selection of the target is rational and the selection of the molecules to be screened may be random, targeted or a combination of both. The premise is to expose the target to a wide variety of structural classes, hence maximizing the chance of achieving novelty. Most screens are designed around the micro-titre plate, and the traditional 96-well variety is now increasingly being replaced by 384-well plates or even higher densities (see below). The process is illustrated in *Figure 2*. The selected compounds

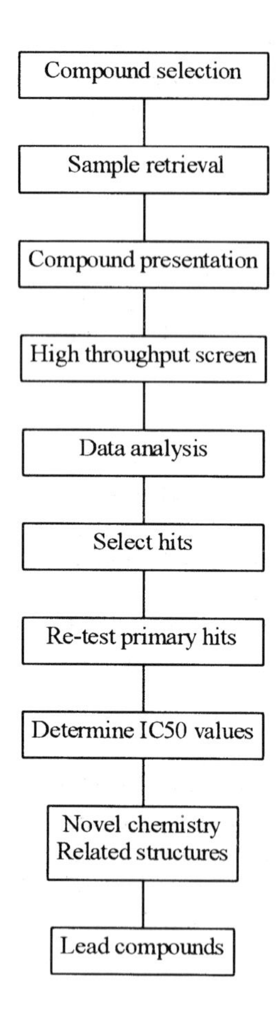

Figure 2 A schematic demonstrating the process for selecting lead compounds.

(next section) are held in a liquid store repository, dissolved in DMSO and stored at -20 °C in micro-titre plates. The store may be organized into sets of similar molecules or random sets. Natural products derived from plant, bacterial and fungal fermentation may be grouped according to the 'association' of the organism with specific diseases/symptoms. These stores are highly automated and computerized to enable fast and efficient dispensation of compounds and for the destination of each compound to be mapped to specific plates. Each plate will bear a bar-code that the screener uses to identify the compounds contained in that plate. The stock solutions (termed mother plates, at 2 to 5 mM) are thawed from − 20 °C and liquid handling robots dispense multiple copies (termed daughter plates; 1 to 20 μl per well) from each mother plate. At this stage a dilution step in DMSO may be effected to adjust the final concentration to that required in the screen. This is now the plate used to run the assay. Alternatively the stock concentration is delivered to the screen; a portion is transferred to a fresh plate for dilution and then to the screening plate to achieve the required final concentration.

Typically, the assay is performed in 96 or 384-well micro-titre plates with compounds presented in 80 or 320 wells. The remaining wells also contain DMSO (at the same final concentration) and are used as controls or blanks. Following the screen, data analysis software presents the result for each compound as a % inhibition value (based on the control and blank values used on their respective assay plate). The data are ranked and a hit rate is set, that is, greater than 70% inhibition is considered as a positive. These positives are re-tested to confirm their activity. Compounds that re-test positive are selected for the determination of IC_{50} values. Potent compounds against the target enzyme are termed hits. These hits are starting points for medicinal chemists to turn the hits into drug molecules.

3.1 Compounds for screening

The capacity for modern screen technology requires large numbers of diverse compounds. Large pharmaceutical companies may have millions of compounds available for screening. In general these are a mixture of natural products and discrete compounds obtained from specialist suppliers and libraries derived from various modern chemistries.

3.1.1 Natural products

As mentioned earlier, many medicines have their roots in natural products. It is often possible to link plant products to specific diseases based on their use by 'natural' practitioners. Broth from bacteria and fungi grown under a variety of conditions are excellent sources of diverse structures. In general, the biological material is extracted using different solvents and fractionated by HPLC. The extracts are dried down and reconstituted in DMSO, then entered into the compound store. The purpose of the fractionation is to reduce the number of individual chemical entities in the mixture. Once the mixture has been confirmed as active, the hard work of identifying the active component begins.

3.1.2 Medicinal chemistry

Historically, companies have identified a lead molecule (i.e. a compound with good activity against the target) and based chemical programs around the molecule. Hence companies will have stores of such molecules in their stores. In addition, many compounds can be purchased from commercial sources or from specialist supplier. However, there is a limit to the number of compounds that can be made using traditional chemistries in a timely manner, thus promoting the development of a new range of chemistries.

3.1.3 Combinatorial libraries

The new chemistries are highly automated and capable of producing > 40 000 molecules in a library (7–12). Typically, synthesis occurs on a solid support, for example, a TentaGel bead with an amine function on the surface. Such beads hold about 300 pmol of functionality and this can be expanded to about 600 pmol using a single lysine as a spacer. Monomers, a diverse range of compounds with a common functionality (e.g. a carboxylic acid) to react with the amine on the bead, are coupled to the bead as individual reactions. These are then pooled and split for the reaction with another monomer as individual reactions, hence generating a structurally diverse library. The compounds are released from the beads usually via a photoliable linker for transfer to the screen and a small volume is retained for analysis. The identification of any active compound depends on deconvolution, that is, several examples of each compound are dispensed throughout the library, and any positive well will contain a common molecule. However, encoded libraries are increasingly being used. In these libraries a unique tag is added to the bead and represents a specific monomer. Cleavage of the tag(s) and analysis by HPLC/mass spectrometry identify the specific monomer(s).

3.1.4 *In-silico* screening

Many screens are now using *in silico* methodologies as part of the overall strategy to identify novel inhibitors. If there are known inhibitors of an enzyme, their structures can be computed in three dimensions using energy-minimization software packages. These structures can be overlaid in 3D, to generate a space-filling model of the shape that interacts with the enzyme. This so-called pharmacophore is a composite fit of the individual inhibitors used. If any structural information is known, for example, X-ray crystallography data showing acid–base interaction between the protein backbone and an inhibitor or H-bonding sites, these can be added to the model. Thus the shape of an average inhibitor is generated. Numerous chemical structures may be able to achieve that shape. The chemistries outlined above can be used *in silico* to generate libraries comprising millions of molecules from the array of monomers chosen. The 3D space of each is calculated and their ability to fill the pharmacophore space computed. The chemists need only synthesize the actual molecules that fulfil the pharmacophore model. In addition, this analysis is often extended to the compounds held in the company's store or that are available from commercial suppliers. A theoretical disadvantage of this approach is that similar molecules to the known compounds used to obtain the pharmacophore are identified as hits in the screen. This may limit their novelty and hence the ability to file a patent. However, there is an excellent chance that novel structures will be found.

3.2 Considerations for high throughput assays

The process of screening large numbers of compounds described in the previous sections is dependent upon the design of a robust assay. Two specific examples of assays are given as protocols. Screens can be run on fully automated robotic systems, semi-automated systems or manually. The advantage of the robotic system is that it can work continuously, hence shortening the time of a screen. Semi-automated systems are dependent on humans for specific periods of time, and are thus less efficient. Clearly the least efficient and costly option is a screen run entirely by humans. The design of the assay often dictates the manner in which the screen is performed. However, stability of the enzyme or other biological reagents or the duration of the signal may override such niceties as running the screen on a robot 24 hours a day.

A review of the robotic systems and liquid handling robots available is beyond the scope of this chapter and would probably soon to be out of date. The first protocol describes a fully automated approach and the second protocol outlines a semi-automated screen. However, the considerations placed on assay design will remain largely similar and are described below.

3.2.1 Production of protein for HTS

Screening demands large amounts of biological material; for example, a 96-well assay using 500 000 compounds requires 610 000 wells worth of material for the campaign to be run. Most enzymes are isolated from various over-expression systems, e.g. transfected bacteria or baculovirus-infected insect cells. The purity of a protein required by the screen is discussed in the next section. If a highly purified sample of the enzyme is required then large amounts of the starting material may be needed, depending on the efficiency of the purification strategy. Often, the protein may be engineered to include a tag, e.g. GST, 6-histidines, as an aid to purification. The organisms used to express the enzyme are grown in bio-pilot plants using 10–500 litre fermentors.

If we take the Lck assay (Protocol 2) as an example using a 96-well format and a 500 000 compound campaign, then approximately 2 litres of a 100 000 g lysate is required to yield enough semi-purified kinase to run the screen. This equates to a 20 litre fermentor or $> 10^{10}$ insect cells.

Protocol 1

Rhinovirus 3C protease

Target

Rhinovirus (the common cold virus) 3C protease is a cysteine protease that cleaves the viral polyprotein at glutamine–glycine sites. The released proteins are assembled into infective virons ready for the next round of infection. The activity of 3C protease is essential for viral replication.

Expression and purification

The 3C protease was expressed in bacteria using an inducible promoter. A production run in a 10 litre fermentor yielded approximately 100 g of cell paste. Portions of the pellet were resuspended in 4 volumes of homogenization buffer per gram. The cells were broken by 5 rounds of sonication (30 s burst/2 min between rounds) on a salt–ice mixture. The 100 000 **g** supernatant was loaded on to a 45 ml Affi-Blue gel affinity column in buffer. The unbound fraction was eluted at 50 mM NaCl (in homogenization buffer) and 3C protease was eluted using 250 mM salt (in homogenization buffer). The active fractions were pooled, glycerol was added (10% final concentration) and aliquots stored frozen at −80 °C. The activity was stable for up to 6 months.

Reagents

- Homogenization and elution buffer: 50 mM Na acetate, pH 6.5, 1 mM EDTA, 5 mM DTT
- Assay buffer: 100 mM HEPES, pH 7.4, 1 mM EDTA, 0.1% BSA
- FRET substrate: 0.2 mM Dabcyl-Gly-Arg-Ala-Val-Phe-Gln-Gly-Pro-Val-(Edans)-Asp-NH_2 in DMSO (available from most custom peptide suppliers, e.g. SNPE/Neosystem). See below for description of the Fluorescence Resonance Energy Transfer (FRET) system.

Enzyme

Aliquots of the enzyme (~1 mg/ml) were thawed on ice, and diluted 1/300 in assay buffer. The activity was stable on ice for 24 hours.

Method

To 384-well plates containing compounds in columns 1–20, 1 μl DMSO was added to columns 21–24, assay buffer (50 μl) to columns 23–24, and diluted enzyme (50 μl) to columns 1–22. Thus columns 21–22 were controls (100% activity) and 23–24 were blanks (0% activity). The enzyme was incubated with the compound for 15 min at room temperature. The reaction was started by the addition of substrate (5 μl) and incubated for 30 min at room temperature. The fluorescence was read on a plate reader using 355 nm excitation and 535 nm emission. This screen was run on a fully automated system, the exact timing of additions determined by the scheduling software, such that each plate was pre-incubated for 15 min and read 30 min after the addition of the substrate. Alternatively the reaction can be performed as a stopped assay, by the addition of trifluroacetic acid (0.2% final concentration). The signal is stable for up to 12 hours, enabling reading off-line.

The robotic system

A complex robotic system was used to run the Rhinovirus 3C protease screen (*Figure 3*).

Explanation of Fluorescence Resonance Energy Transfer (FRET)

Fluorescence resonance energy transfer (FRET) is a distance-dependent interaction between the electronic excited states of two dye molecules, in which the excitation is transferred from a donor to an acceptor molecule without the emission of a photon. FRET is dependent on the inverse sixth power

Protocol 1 continued

of the intermolecular separation. Thus the donor and acceptor molecules must be in close proximity (10 to 100 Å), and the absorption of the acceptor molecule must overlap the fluorescence emission of the donor. In the case of the 3C protease assay, Edans is the donor (emission 495 nm) and Dabcyl is the acceptor (absorption 480 nm). Hence the action of the protease separates the donor and acceptor and a gain in fluorescence is observed.

Figure 3 Automated Rhinovirus 3C protease screening robotic system.This system, built by Scitec, is capable of running both cellular and non-cellular 96-well and 384-well assays in a fully automated fashion. Seen here is a CRS robotic arm which runs on a 5 metre linear track and is used for transferring microtitre plates from one piece of equipment to another, two Labsystems Multidrops (one of which was used to add enzyme) and a Matrix Platemate with 96-well head which was used to add substrate solution in neat DMSO. Also visible in this picture, but not used, are a Skatron plate washer and Tecan Genesis liquid handling system. Out of shot but present on the system is a Tecan Spectrafluor Plus used to read the fluorescence in the plates. Also the system has a Scitec plate hotel, a plate shaker and a CO_2 incubator, none of which were required for this assay. The whole system is enclosed in a cabinet for safety reasons. This system is designed to be modular, so it would be possible to exchange pieces of equipment with upgraded versions or to add more readers and/or liquid and cell handling devices as required.

Protocol 2

Lck tyrosine protein kinase

Target

Lck is a tyrosine protein kinase expressed in T and natural killer cells. It is essential for T cell receptor signalling. Following binding of the antigen, Lck is activated and phosphorylates a variety of proteins, which cascade down to cytokine production. Inhibition of this pathway may have utility in diseases involving inappropriate T cell activation, e.g. rheumatoid arthritis and psoriasis.

Protocol 2 continued

Expression and purification

Lck was expressed in insect cells using the baculovirus expression system. A 50 litre production fermenter run yielded approximately 400 g cell pellet. The cell pellet was resuspended in 4 volumes of homogenization buffer per g. The cells were broken by 3 rounds of sonication (15 sec burst/2 min between rounds) on a salt-ice mixture. The 100000 g supernatant was loaded onto a 45 ml ResourceQ ion exchange column in homogenization buffer. Unbound protein was washed off in homogenization buffer and, once a stable baseline was achieved, a salt gradient to 600 mM over 20 column lengths was applied. Active fractions were pooled and glycerol was added (10% final concentration) and aliquots stored frozen at −80 °C. The enzyme was stable for up to 12 months.

Stock solutions

- 400 mM HEPES, pH 7.4
- 25 mM ATP
- 10 mM peptide (biotin-Glu-Glu-Glu-Glu-Tyr-Phe-Glu-Leu-Val) in DMSO
- 500 mM $MgCl_2$
- 50 mM EDTA, pH 7.4
- 2 μM antiphosphotyrosine antibody labelled with Eu-cryptate (Wallac, CR32-100)
- 23 μM streptavidin-labelled allophycocyanin (Prozyme, PJ25S)
- *Homogenization buffer*: 50 mM Tris, pH 8.0, 25 mM NaCl, 1 mM DTT, plus protease inhibitors [leupeptin, aprotonin, pepstatin A (all at 2 μg/ml)]. iodoacetate (0.5 mM) and E64 (trans-epoxysuccinyl-leucylamido-(4-guanidino)butane, 5 μM))
- *Assay buffer*: 100 mM HEPES, pH 7.4, 25 mM $MgCl_2$, 0.5 μM peptide, 0.125 mM ATP
- *HTRF read buffer*: 40 mM HEPES, pH 7.4, 300 mM NaCl, 0.25% BSA, 100 nM streptavidin-allophycocyanin (APC), 1.36 nM antibody

Enzyme preparation

Aliquots of enzyme were thawed on ice and activated by autophosphorylation by the addition of 10 mM $MgCl_2$ and 0.1 mM ATP. Following a 30 min incubation on ice, the enzyme was diluted 1/60 in 100 mM HEPES. The enzyme was stable for about 1 hour on ice.

Method

To 96-well plates containing compound in columns 1–10, 2 μl DMSO was added to columns 11–12, and 25 μl EDTA to column 12 (the blank: 0% activity). Enzyme (30 μl) was added to all wells and incubated at room temperature for 15 min. The reaction was started by addition of the assay buffer (20 μl) and incubated for 30 min at room temperature. The reaction was stopped by the addition of EDTA (25 μl) to columns 1–11. The HTRF reagents were added (25 μl) and, following incubation, read on a HTRF plate reader (see below for description of Homogeneous Time Resolved Fluorescence (HTRF) system).

HTRF plate readers

Most fluorescent plate readers are capable of reading in the time-resolved mode and filter sets are widely available. In all but one reader, two reads must be obtained. In the current example, the first, from excitation at 340 nm and emission at 615 nm, is the europium cryptate signal and will be uniform across the plate. The second read is dependent on the concentration of phosphorylated peptide, and is taken from excitation at 340 nm and emission at 665 nm (APC). The Victor or Victor2 manufactured by Wallac can be fitted with red-sensitive photomultiplier tubes and filter sets optimized for HTRF. These machines are generally more sensitive than standard fluorescent plate readers. A reader capable of measuring both wavelengths at once is Packard's Discovery HTRF plate reader. Here a powerful laser excites the well at 337 nm; the emitted light is split and detected at 615 nm (europium) and 665 nm (APC). The software provides a ratio between the signal at 665 nm

Protocol 2 continued

and the signal at 615 nm. This serves as a quench correction function, since any compound effecting fluorescence will affect both channels equally. The Victor and Discovery readers are able to measure 96 and 384-well plates.

Robotic systems

An example of in-house benchtop automation system used for the Lck protease screen is shown in *Figure 4*. Using this system, it was easily possible for one person to process 100, 96-well microtitre plates per day, in 2 batches of 50. Enzyme, stop reagent and read buffer were added on this system and the finished assay plates were manually transported to and read on the Packard Discovery, an approach termed 'semi-automated screening'. This workstation can also be used for 384-well plate liquid handling.

Explanation of Homogeneous Time Resolved Fluorescence (HTRF)

Homogeneous time resolved fluorescence (HTRF) is a homogeneous method that uses FRET between two fluorophores in a time-resolved manner. The donor is europium cryptate that excites at 340 nm and emits at 615 nm. This emission decays over a long time period (~ 1000 μs) compared to prompt fluorescence (~ 100 ns). Therefore, if the acquisition of signal is delayed any background fluorescence will have decayed to zero. Typically reads are taken 400 μs after excitation and integrated for a further 400 μs. The acceptor is allophycocyanin (APC, covalently bound to streptavidin); this pigment absorbs over a broad range and emits at 665 nm. Typically one collects the signal after a 50 μs delay and integrates for a further 200 μs. In the kinase assay, a specific signal is generated when phosphorylated peptide binds to streptavidin-APC and the europium-labelled antiphosphtyrosine antibody binds to the phosphotyrosine moiety.

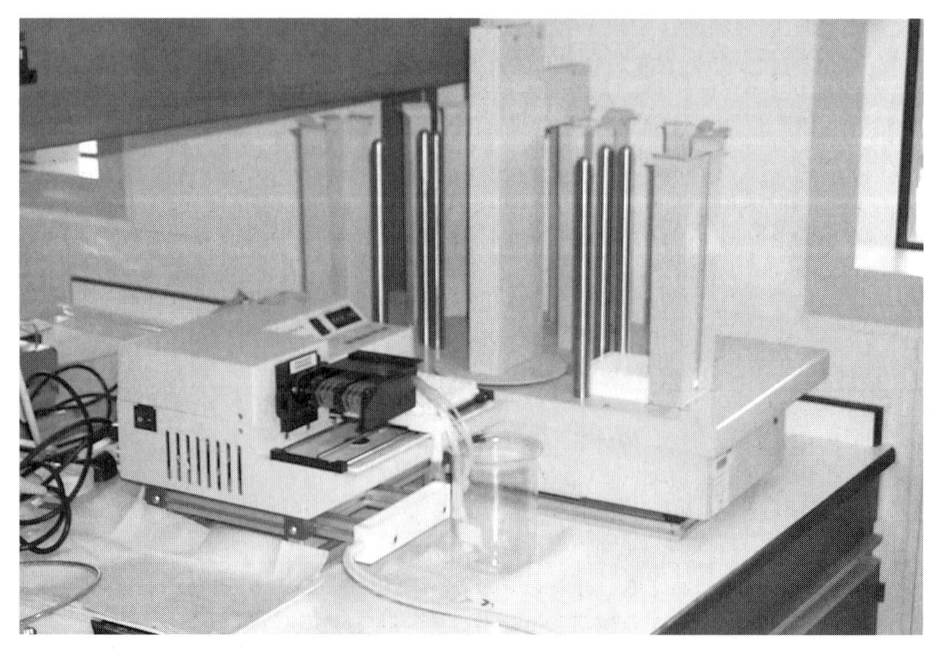

Figure 4 An in-house benchtop automation system used for the Lck protease screen. A Labsystems Multidrop (foreground) is integrated to a Zymark Twister (rear) for automated plate loading and liquid dispensing.

4 Enzymatic considerations

The degree of enzyme purity required to run a screen is influenced by several factors:

- Is the specific activity enough for detection?
- Are there any interfering reactions, i.e. that consume substrate or product or co-factor?
- Stability of the enzyme

The use of over-expression systems usually ensures that sufficient activity is present in a 100000 **g** supernatant. A partial purification may be required to increase the specific activity to achieve a good signal-to-noise ratio (see later). Sometimes there is/are reaction(s) that consume the substrate or destroy the product. Clearly, such competing activities must be absent from the final enzyme solution. In the case of 3C protease (Protocol 1) there was sufficient activity to run the assay in a 100000 **g** supernatant using a different peptide substrate to the FRET sequence. However, detection by HPLC was required, and hence is not an option for HTS. The FRET peptide was cleaved by a bacterial protease that caused a marked increase in fluorescence but not at the glutamine-glycine site. Therefore, the viral protease was purified yielding an enzyme preparation that cleaved at the desired glutamine-glycine site.

Obtaining a good signal-to-noise ratio is essential for screening. This ratio is defined as the control activity (i.e. 100%) divided by the blank (i.e. 0%). The blank is a series of wells that either lacks the addition of enzyme (Protocol 1) or contains a stopping reagent prior to the addition of enzyme (Protocol 2). In order to obtain a clear distinction between full, partial, and no inhibition, a ratio of greater than 8 is usually required.

In common with most assays, the HTS variant must be linear with time and the reaction must be read or stopped on the linear portion of the activity versus time plot. Likewise the rate must be linear with respect to added enzyme concentration. At fixed concentrations of reagent it is convenient to alter the concentration of the enzyme to achieve suitable linearity. The effect of buffers, salt, glycerol, added protein (e.g. BSA), pH, etc may be investigated to improve the stability of the enzyme. For example, by changing the buffer from acetate/EDTA to HEPES/EDTA (+ 0.1% BSA) improved the stability of the rhinovirus 3C protease from several to 24 hours on ice. Enzyme stability is important if the assay is running on a robotic system, thus one portion of enzyme is sufficient for a whole day of screening. The effect of solvents (e.g. DMSO in which the inhibitors are dissolved) used in the assay on enzyme function is assessed. A final concentration is selected that has no effect on enzyme activity; most enzymes tolerate up to 30% DMSO, although most screens use less than 10% final concentration of DMSO.

The concentration of substrate(s) is/are carefully selected, and listed below are some of the drivers that affect the selection of the final assay conditions.

- Cost of reagents
- Limits of detection
- Screening at K_m sensitizes the assay to competitive inhibitors
- Screening significantly below K_m enhances sensitivity
- Tolerance of the enzyme to solvent if any reagents require to be presented as non-aqueous solutions

The cost of any assay reagent and the concentration used in the screen are weighed up to give a balance between a good signal-to-noise ratio and the potential number of compounds that can be screened. The concentration of a substrate may be altered to achieve a reasonable signal-to-noise ratio. For example, the cost of SPA beads limits the amount that can be added to each well. In signal decrease protease assays, where a [^{3}H]biotinylated peptide is cleaved, the remaining, non-cleaved substrate is detected using streptavidin-coated SPA beads. Too little cleavage will not allow one to discriminated between blank wells (no cleavage) and controls (some cleavage). Therefore, a concentration of peptide substrate is selected that provides a sufficient window to allow one to distinguish between controls and wells containing inhibitor, at an affordable amount of beads.

In both Protocols the assay was run at substrate concentrations equal to K_m. These values were in the mid μM range, roughly around the concentration of compound selected for the screen. If a substrate for the target enzyme has a relatively high K_m, then the actual concentration of inhibitor present in the assay (typically 10 and 250 μM) may be an order of magnitude lower than the substrate. Assuming that many of the compounds screened act as competitive inhibitors, the likelihood of obtaining hits would be low if that screen were run at K_m. Therefore, the concentration of substrate must be reduced under these conditions.

Having optimized the assay using the criteria listed above, a final hurdle must be crossed before the HTS campaign commences: a robustness test. A representative selection of the compound database is screened against the target enzyme. This highlights any design faults, tests the assay under 'real' HTS conditions and gives a hit rate. If the hit rate is too high (> 1%, i.e. 5000 primary hits in a 500000 campaign) then either the screening concentration is lowered or a small redesign is carried out to decrease sensitivity. Large numbers of primary hits take a long time to re-test and determine IC_{50} values and may take longer than the screen!

5 Assay formats for enzymatic HTS

If an assay meets the criteria listed in above, it may still not be suitable as a format for HTS. The fewer additions to the microtitre plate the better. Therefore, plate-to-plate transfers and plate washing steps are generally avoided. Such stages add to the complexity of the screen and are inefficient in the use of time. However, if such steps cannot be avoided then a screen will still be run, but the daily throughput of compounds will be much reduced. ELISAs are example of this approach and several HTS assays have been published. Homogeneous formats are ideal for screening; the microtitre plates arrive at the screen pre-dispensed with compound, the solvent used to add the compound is diluted by addition of buffer containing enzyme, cofactors, etc., and the reaction is started by the addition of substrate(s). At the end of the incubation time the plate is either read or stopping/read reagents are added. Protocol 1 uses the former approach and Protocol 2 uses the latter. Some literature examples of assay formats run in HTS are shown in *Table 1*.

6 Automation

Automation is a rapidly-changing field and a detailed evaluation is beyond the scope of this chapter. However, the two examples chosen illustrate different ways of running a screen. The rhinovirus 3C protease activity was stable on ice for 24 h. Therefore, the assay was run on a bespoke fully-automated robotic system. The pre-incubation time with compound, the assay time and read were dealt with by the scheduling element within the software. The software 'works out' the optimal use of time ensuring that each plate is read exactly 45 min after the addition of the enzyme.

The Lck kinase protein was far less stable, and therefore unsuitable for the bespoke system. The semi-automated system dealt with all the liquid handling steps but humans were required to transfer the plates to the reader.

Whilst the two examples shown gave a flavour for the technology, the principles outlined in this chapter will still apply.

7 Developments

7.1 Higher density plates

Significant cost reductions have been achieved by moving to 384-well plates (assay volume 30 to 50 μl) from 96-well plates (assay volume 100 to 250μl). Cost savings of about 2- to 3-fold are typical. Thus for a given budget more screens can be performed, or the same number of screens with an increased range of compounds. Therefore, the drive to reduce the assay volume further is an attractive idea. The advent of the 1536-well plate (assay volume 1 to 10μl) along with the necessary liquid handling equipment means cheaper and faster screening. A variety of plate readers capable of reading 1536 wells are now available. It is likely that cooled CCD cameras and image analysis software will deal with SPA and luminescent read-outs.

References

1. Burbaum, J. J. and Sigal, N. H. (1997). *Curr. Opin. Chem. Biol.*, **1,** 72.
2. Houston, J. G. (1997). *Methods Find. Exp. Clin. Pharmacol.*, **19**(Suppl. A), 43.
3. Houston, J. G. and Banks, M (1997). *Curr. Opin. Biotechnol.*, **8,** 734.
4. Kubinyi, H. (1995). *Pharmazie*, **50,** 647.
5. Persidis, A. (1998). *Nature Biotechnol.*, **16,** 488.
6. Sittampalam, G. S., Kahl, S. D., and Janzen, W. P. (1997). *Curr. Opin. Chem. Biol.*, **1,** 384.
7. Baldwin, J. J. (1996). *Mol. Divers.*, **2,** 81.
8. Schullek, J. R., Butler, J. H., Ni, Z., Chen, D., and Yuan, Z. (1997). *Anal. Biochem.*, **246,** 20.
9. Hardin, J. H. and Smietana, F. R. (1995). *Mol. Divers.*, **1,** 270.
10. Hassan, M., Bielawski, J. P., Hempel, J. C., and Waldman, M. (1996). *Mol. Divers.*, **2,** 64.
11. Dolle, R. E. (1996). *Mol. Divers.*, **2,** 223.
12. Takeshita, N., Kakiuchi, N., Kanazawa, T., Komoda, Y., Nishizawa, M., Tani, T., and Shimotohno, K. (1997). *Anal. Biochem.*, **247,** 242.
13. Lehel, C., Daniel-Issakani, S., Brasseur, M., and Strulovici, B. (1997). *Anal. Biochem.*, **244,** 340.
14. Braunwalder, A. F., Yarwood, D. R., Sills, M. A., and Lipson, K. E. (1996). *Anal. Biochem.*, **238,** 159.
15. Seethala, R. and Menzel, R. (1998). *Anal. Biochem.*, **255,** 257.
16. Levine, L. M., Michener, M. M., Toth, M. V., and Holwerda, B. (1997). *Anal. Biochem.*, **247,** 83.
17. Kolb, A. J., Kaplita, P. V., Hayes, D. J., Park, Y-W., Pernell, C., Major, J. S., and Mathis, G. (1998). *Drug Discovery Today*, **3,** 333.
18. Taliana, M., Bianchi, E., Narjes, F., Fossatelli, M., Urbani, A., Steinkuhler, C., *et al.*(1996). *Anal. Biochem.*, **240,** 60.
19. Lerner, C. G. and Saiki, A. Y. C. (1996). *Anal. Biochem.*, **240,** 185.
20. Baum, E. Z., Johnston, S. H., Bebernitz, G. A., and Gluzman, Y. (1996). *Anal. Biochem.*, **237,** 129.
21. Kyono, K., Miyashiro, M., and Taguchi, I. (1998). *Anal. Biochem.*, **257,** 120.
22. Hayes, D. J. and Waslidge, N. B. (1995). *Anal. Biochem.*, **231,** 354.

Chapter 12
Statistical analysis of enzyme kinetic data

Athel Cornish-Bowden
Institut Fédératif 'Biologie Structurale et Microbiologie,' Bioénergétique et Ingénierie des Protéines, Centre National de la Recherche Scientifique,
31 chemin Joseph-Aiguier, B.P. 71, 13402 Marseille Cedex 20, France.

1 Introduction

All serious kinetic investigations of enzymes include some data analysis, and the methods used can be classified into three categories: (i) graphical analysis; (ii) best-fit analysis with general-purpose commercial packages; (iii) best-fit analysis with specialized programs designed with enzyme kinetic experiments in mind. The near-universal availability of small computers —a major change from the context in which the chapter corresponding to this one in the first edition of this book (1) was written—has greatly increased the use of the second of these approaches, but the others continue to be widely used.

The gradual disappearance of graphical methods has been confidently predicted for many years, especially since computer programs were first made available to enzymologists, and some suppose that this has already happened. For example, Gutfreund (2) asks rhetorically 'Is there anyone still doing enzyme kinetics who is not using a PC which can run Grafit, Micromath Scientist or some other similar program for the analysis of kinetic data?', going on to admit that 'the reviewer has been assured by a number of practising biochemists that plots of linearized data are still widely used'. Actually it is hardly necessary to rely on the opinion of other biochemists, as examination of any current issue of a journal of biochemistry, such as the *European Journal of Biochemistry* vol. 259, No. 3 (1999), provides examples of the use of graphical methods (3–5), and of papers where the authors have not made it clear what methods they used (6), in addition to the increasing numbers of papers where commercial software (7–11) or non-linear regression (12–13) was used.

One may wish that graphical methods would disappear completely, but until they do even experimenters who fit all their own data by computer still need some knowledge of graphical methods, if only to assess the results published by others. In any case, for some purposes graphs remain an essential part of the analyst's armoury, because they are much better for some tasks than any computer program can be.

In reality, each of the three approaches mentioned has some dangers. Graphical analysis is subjective, and whenever lines are drawn on a graph there is an implicit weighting that the user may not be conscious of. Commercial packages may circumvent the subjectivity, but they render users very dependent on the competence of the programmer, and some essential details of the fitting procedure may be hidden. All fitting calculations incorporate some assumptions, but if these are not explicitly stated it is difficult to judge whether they are appropriate for the data at hand. Finally, programs written specifically for the analysis of enzyme kinetic data may eliminate some of the dangers of the first two approaches, but with a greater risk of programming errors: few computer programs that do anything more sophisticated than display 'Hello World' on the screen are completely free from errors, but errors are more likely to be noticed and eliminated when they occur in mass-circulation commercial programs than when they occur in purpose-built programs that have small numbers of users. This estimate of the frequency of errors in computer programs may seem unduly pessimistic, but if one couples the observation in computer science that expert programmers cannot achieve an average of more than 10 lines of error-free code per day (14) (in whatever language they may be writing) with the fact that modern programs typically contain many tens of thousands of lines of code it will be evident that it is just realistic.

Unfortunately 'blind' use of commercial software is becoming increasingly common, that is to say use of such software without considering whether the default options about weighting are appropriate for the data, or even whether the model fitted actually fits the data. This has always occurred, of course, but in the past when data analyses were always accompanied by one or more graphs it was easy for the reader to make a judgement about whether the analysis had been done correctly. The current literature contains many examples of papers that state that a particular program was used, but give no information about how the weighting was done, no mention of alternative models, and no plots to illustrate the results. In such cases it is impossible for the reader to know whether the kinetic parameters reported later in the paper have any validity or not.

All of this means that it remains desirable for anyone who analyses enzyme kinetic experiments to have some knowledge of the principles that underlie the analysis. It is also important to realize that there are two primary objectives, parameter estimation and model discrimination, that are by no means equivalent. The experimental designs appropriate for one are not the best for the other, and the assumptions implicit in the analysis are not the same in the two cases.

2 Derivation of relationships

In this chapter I shall not in general derive results from first principles, as this would be rather lengthy and it is more convenient to refer to other sources. Much of the background is to be found in my book (15), and more specific references will be given where appropriate.

3 Defining objectives

The objective of data analysis is normally either parameter estimation or model discrimination, i.e. either determination of the best values of the parameters of the equation considered to explain the data, or identification of the physical model that best explains the data. A parameter value may be of interest for its own sake in a study of the effect of a mutation (natural or engineered) on the activity of an enzyme, or for understanding how the enzyme fulfils its physiological role, but in an investigation of its mechanism parameter estimation is mainly just a step on the route to model discrimination. In a two-substrate enzyme reaction, for example, we are more likely to be interested in whether the enzyme follows a ternary-complex mechanism or a substituted-enzyme mechanism than in whether the K_m value for one of the substrates is 3.85 mM or 4.33 mM. In other words, we are more interested in how K_m or another kinetic parameter changes with changing conditions than in its actual value under any particular conditions. In practice, therefore, we estimate parameter values as a way of summarizing the results of a large number of measurements in a few numbers, with the hope that study of the values will eventually allow us to determine the real mechanism of the enzyme.

However, even if the ultimate objective is usually model discrimination, it will often be too remote an objective to address immediately, and in consequence many experiments are designed and analysed as if determination of the numerical parameters were the aim in itself. From the point of view of the statistical analysis it is important to make the distinction, because neither the underlying assumptions nor the range of methods available are the same. When estimating parameters we must normally assume that we are fitting the right model. This is a very strong assumption, because in effect it means that we are assuming at the outset the information that in the long term we want to obtain. In other respects, however, the assumptions needed for parameter estimation are weaker than those needed for model discrimination, because classical tests for goodness of fit are heavily dependent on assumptions about error distribution whereas much weaker assumptions suffice for parameter estimation. These will be considered in the next section.

4 Basic assumptions of least squares

Most methods for estimating parameters and testing hypotheses involve minimizing a function that measures the difference between the observations and the model that is supposed to fit them. This function is often, though not necessarily, calculated as the sum of the squared differences between observed and calculated values of a variable, and thus the process of fitting is called the *method of least squares*. The justifications usually offered for using this approach are threefold: it is computationally easier than most alternatives; it is said to provide *minimum-variance* estimates of the parameter values; and it is said to be equivalent to *maximum likelihood*. The first of these is the least often invoked, but it is also the least open to argument. The other two are valid in some circumstances, but they are much less universally valid than is sometimes claimed.

Considering first the question of minimum variance, a least-squares estimate of a parameter satisfies this only if the following conditions are met:

1. The observations must be correctly weighted when calculating the function to be minimized; this implies knowledge of how the reliability of the measurements varies from observation to observation.
2. The errors in the observations must follow a normal (Gaussian) distribution. This assumption is not necessary if the model is linear and we confine attention to estimates that can be expressed as linear functions of the observed values. However, as enzyme kineticists rarely have occasion to deal with linear models (the Michaelis–Menten equation is not a linear model, for example, and nor are any of the other equations commonly used in enzyme kinetics) and virtually never use linear estimates of parameters, the normal distribution is an absolutely necessary assumption for applying least squares in enzyme kinetics. Even with linear models it is perfectly possible for a non-linear estimate to have smaller variance than the minimum-variance linear estimate if the error distribution is not normal.

For the least-squares estimate of a parameter to satisfy the principle of maximum likelihood we need to start with a likelihood function, i.e. a function that tells us, on the basis of some assumptions about the distribution of errors, how likely it is that we should obtain any particular value when we try to observe some property. For example, if we believe that we have an unbiassed dice and we toss it 600 times an obvious probability calculation will lead us to think that the most likely result is that we shall toss 100 sixes. Although this is the most likely result it is only very slightly more likely than tossing 99 sixes, so if we obtained 99 sixes, or even 85 or 115, we would not be particularly surprised, and probably would not feel that the assumption of an unbiassed dice was seriously brought into question. If, however, we observed 350 sixes in 600 tosses we would certainly think this was sufficiently unlikely for an unbiassed dice to suggest that we were dealing in reality with a biassed dice, and if we wanted a maximum-likelihood estimate of the probability of throwing a six with that particular dice we should calculate this as 350/600, or 0.583.

Maximum-likelihood estimation in a kinetic experiment follows exactly the same idea: we try to identify the probability function that will maximize the likelihood that the rates we ought to observe in particular conditions turn out to be the rates that were in fact observed. Notice, however, that this probability function embodies not only the information that interests us—the values of the kinetic parameters, K_m, V, or whatever—but also statistical assumptions that probably do not much interest us and, more important, are probably unknown. Gauss (16) did not derive the normal distribution from first principles and then deduce that the arithmetic mean would be the best average to use (or, more generally, that the method of least squares would be the best method of estimating parameters); he did exactly the opposite, taking the conclusion that he wished to reach as axiomatic and then deducing what assumptions about the underlying probabilities would allow him to reach it. More discussion of this point may be found elsewhere (15); for the present purpose the essential is to realize that even though assumptions exist that make least-squares estimation equivalent to maximum-likelihood estimation there is no certainty in the real world that these assumptions apply.

In general, it is safest to say that the widespread use of the method of least squares derives principally from the fact that it is computationally the easiest to apply and that efforts to justify it in more theoretical terms require more assumptions about the unknown than it is wise to make.

5 Fitting the Michaelis–Menten equation

The Michaelis–Menten equation provides a convenient starting point for considering the ideas that will be needed for more complicated models. It may be written as follows:

$$v = \frac{Va}{K_m + a} + e \tag{1}$$

in which v is the rate at a concentration a of the substrate, V and K_m, the limiting rate and Michaelis constant respectively, are the two parameters, and e is the difference between the value of v observed and the value calculated from the parameter values and substrate concentration.

Although this equation is familiar to all biochemists it is worthwhile making two points about it. First of all, the error term e is essential: without it there is no statistical analysis to discuss, and it becomes impossible to understand why different graphical methods give different results. Second, the equation is linear with respect to one parameter, V, but it is non-linear with respect to K_m, the other. This means that given the value of K_m the best-fit value of V can be calculated directly by simple linear regression, but not vice versa. In practice, therefore, non-linear methods have to be used for determining the two parameters.

As an aside, the fact that the equation is linear in V means that in principle the fitting can be done with a one-dimensional search, calculating V exactly at each of a series of K_m values until some criterion of best fit is satisfied. This is perfectly feasible as a method of fitting, but is rarely if ever used, because it is not in practice any less laborious than methods that ignore the linearity with respect to V. Moreover, when the Michaelis–Menten equation is generalized to more complex equations this nearly always involves introducing additional non-linear parameters, so it is only in the simplest possible case that a one-dimensional search is feasible.

The commonly used graphical methods use the transformation of the Michaelis–Menten equation into a straight-line form. In the case of the double-reciprocal plot this is as follows:

$$\frac{1}{v} = \frac{1}{V} + \frac{K_m}{Va} + e' \tag{2}$$

It is important to realize that despite the impression given by most elementary textbooks this equation cannot be derived from Equation 1, because the 'derivation' involves ignoring the error term e in Equation 1 and then introducing a *different* error term e' into Equation 2. This sleight of hand might not matter very much if e' was approximately the same as e, or approximately proportional to e, but in reality they are very different. Simple algebra shows that

$$\frac{e}{e'} = -\frac{Vav}{K_m + a} = -\hat{v}v \tag{3}$$

where $\hat{v}$ represents the value of v calculated from a, V and K_m. As v and $\hat{v}$ are approximately the same if the errors are moderate and V and K_m are good estimates, it follows that the ratio of the error in v divided by the error in $1/v$ is approximately proportional to v^2. Moreover, as fitting methods are usually based on the squared errors it is more appropriate to note that the square of e/e' is proportional to v^4. This is an extremely steep function, so that even if the v values span a modest three-fold range the values of $(e/e')^2$ are spread over nearly two orders of magnitude. This means that the locations of the points on a double-reciprocal plot in relation to the line drawn provide almost no intelligible information about how well the line fits the points.

As a result it is almost impossible to judge how to draw the line on a double-reciprocal plot so that it represents a good fit to the original data. It is sometimes argued that this is not important if the plot is just used for illustration, the actual parameter estimation being done independently. The result of this will be a 'best-fit' line that looks to the observer as if it fits badly, requiring a special explanation. Moreover, a convincing illustration needs to look right as well as to be right; otherwise it will not convince anyone. Even with an understanding of the mechanism of estimating the best-fit parameters it is very difficult to inspect the resulting line to make a judgement about which observations fit it well and which ones may be anomalous in some way. In a general way it is clear that the points near the ordinate axis (high v or low $1/v$) can be almost on the line but fit the model badly, whereas points far from the axis may lie noticeably off the line but fit the model well, but translating this general statement into a precise interpretation of the characteristics of a particular observation is entirely another matter.

These points are illustrated by the example in *Figure 1*, which shows double-reciprocal and direct plots of the same data analysed in two ways. When the direct plot looks right (*Figure 1(b)*), the corresponding double-reciprocal plot (*Figure 1(a)*) appears to ignore the point at $1/a = 10$, whereas when the double-reciprocal plot (*Figure 1(c)*) looks right the corresponding direct plot (*Figure 1(d)*) fails obviously to fit the observations at the three highest a values.

Similar difficulties arise to a lesser degree with the other two commonly used linear plots of Michaelis–Menten data. The plot of a/v against a is based on the following transformation of Equation 1:

$$\frac{a}{v} = \frac{K_m}{V} + \frac{a}{V} + e'' \tag{4}$$

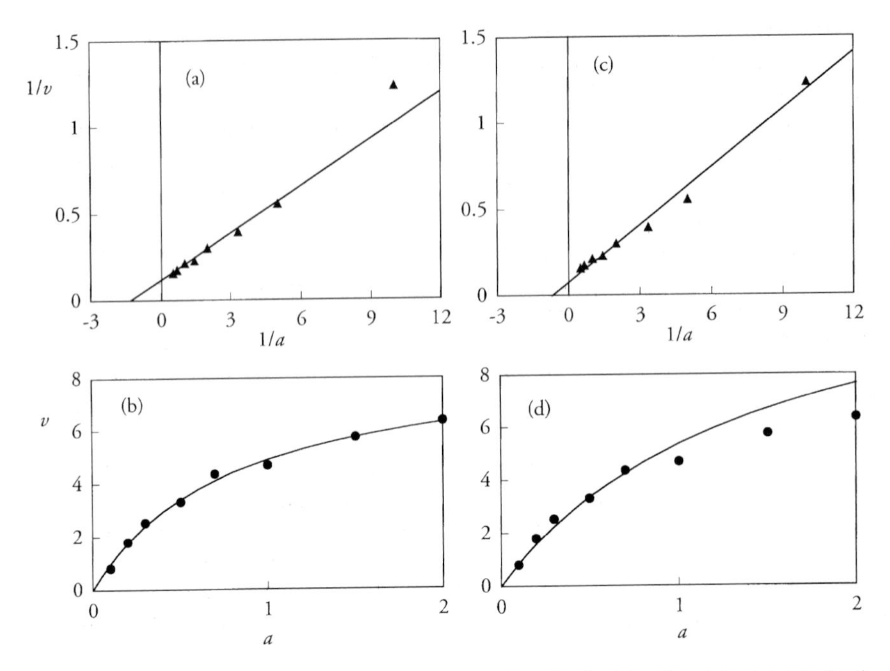

Figure 1 The Figure shows double-reciprocal (a, c) and direct (b, d) plots of the following data (a, v) = [(0.1, 0.81), (0.2, 1.80), (0.3, 2.53), (0.5, 3.31), (0.7, 4.36), (1.0, 4.69), (1.5, 5.74), (2.0, 6.35)]. In the left-hand panels (a, b) the lines are drawn for $(\hat{K}_m, \hat{V}) = (0.794, 8.802)$, the values obtained by non-linear regression of the rate equation assuming a uniform variance for errors in v (i.e. Wilkinson's method (17)); in the right-hand panels (c, d) the lines are drawn for $(\hat{K}_m, \hat{V}) = (1.468, 13.225)$, the values obtained by unweighted linear regression of the double-reciprocal plot.

In this case a derivation similar to that above shows that again the error e'' in this expression is not the same as the error e in v, but now the ratio is

$$\frac{e}{e''} = -\frac{Vv}{K_m + a} = -\frac{v\hat{v}}{a} \qquad (5)$$

Although this is not a constant it varies very much less over ordinary ranges of a and v than the ratio defined by Equation 3. For example, in the range $a = 0.2K_m$ to $a = 5K_m$ it is 0.14 at both extremes and does not rise above 0.25 at intermediate values: compare the expression for e/e' from Equation 3, which varies from 0.028 to 0.69 in the same conditions. It follows that even though the deviations in the plot of a/v against a do not give a perfectly correct idea of the corresponding deviations in v they come far closer to doing this than do the deviations in the double-reciprocal plot.

The third of the straight-line plots of the Michaelis–Menten equation is the plot of v against v/a. In this case the effects of experimental error are more difficult to analyse, because errors in v affect both coordinates, with the result that deviations do not occur parallel with the ordinate axis but towards or away from the origin. This complication aside, the general view is that the statistical problems associated with this plot are of the same order of gravity as those for the plot of a/v against a and far less serious than those for the double-reciprocal plot. In addition, the plot of v against v/a has two major merits that could be considered to outweigh any statistical considerations.

The first of these is that the entire observable range of v values from 0 to V is mapped onto a finite range of paper, with the result that it is virtually impossible to hide the experimental design from the observer. With the other two plots it is possible to choose axes and scales so that a badly designed experiment, such as one in which all of the substrate concentrations are between K_m and $2K_m$, looks quite acceptable unless one examines the plot carefully. With the plot of v against v/a it is immediately obvious how much of the potential range of v has been explored in the experiment.

The second merit is that systematic deviations from Michaelis–Menten kinetics typically produce more obvious deviations from linearity in the plot of v against v/a than they do with either of the other straight-line plots. It is thus much more obvious if one is attempting to explain data in terms of a model that is not in fact correct. In a double-reciprocal plot, for example, it is quite easy for a mild degree of substrate inhibition to be hidden or confused with experimental error, as it produces deviations close to the $1/v$ axis in a region where even quite large errors can appear negligible.

This hiding of imperfections in the experimental design, the data, or the model used for fitting them, can have serious consequences, especially if the experimenter is not expert in enzyme kinetics and is quite unaware that results that look very satisfactory when plotted in double-reciprocal form may actually not satisfy the proposed explanation at all well. If faults in the experimental design are not obvious there will be no incentive to improve it; if badly fitting points close to the $1/v$ axis pass unnoticed there will be no incentive to check for the existence of substrate inhibition, and so on.

Once the problems with the straight-line transformations of the Michaelis–Menten equation are recognized it becomes clear that one should try to fit the equation directly. This is not as straightforward as it may sound, however, because of the need to take account of the distribution of errors in the rates (or to use a 'distribution-free' method: see below). There are two basic assumptions that can be made as starting points (though the reality may be more complicated):

1. If each error in v comes from the same distribution, so that it has the same standard deviation when measured in rate units (e.g. in μmol l^{-1} s^{-1}), then the variance (squared standard deviation) $\sigma^2(v)$ is a constant σ_0^2:

$$\sigma^2(v) = \sigma_0^2 \tag{6}$$

In this case it is appropriate to give equal weight to each v value and to minimize a sum of squares SS defined as

$$SS = \Sigma\,(v - \hat{v})^2 \tag{7}$$

in which v and $\hat{v}$ are the observed and calculated values respectively of the rate, and the summation is over all observations. (In statistical calculations summations are normally made over all observations, and it is usually unnecessary to show the limits explicitly in the summation sign). Minimizing this function cannot be solved in a single step, and requires an iterative calculation, but this is not a problem with modern computer facilities and many programs are capable of finding the solution. Details of the method were given originally by Wilkinson (17). An approximate solution, useful not only as a starting point for the full iterative process but also for checking the results given by a program, is as follows:

$$\hat{K}_m = \frac{\Sigma v^4 \Sigma (v^3/a) - \Sigma (v^4/a) \Sigma v^3}{\Sigma (v^4/a^2) \Sigma v^3 - \Sigma (v^4/a) \Sigma (v^3/a)} \tag{8}$$

$$\hat{V} = \frac{\Sigma(v^4/a^2)\Sigma v^3 - [\Sigma(v^4/a)]^2}{\Sigma(v^4/a^2)\Sigma v^3 - \Sigma(v^4/a)\Sigma(v^3/a)} \tag{9}$$

where the symbols $\hat{K}_m$ and $\hat{V}$ (with circumflexes) represent the best-fit values of K_m and V respectively. If desired one can use the same equations again to obtain better parameter estimates, replacing v^4 wherever it occurs by $\hat{v}^3v$, and replacing v^3 by $\hat{v}^3$, where $\hat{v}$ is the value of v calculated with the current parameter estimates, continuing until the parameter estimates do not change from one iteration to the next. This is not Wilkinson's method (17), but it is easier to understand and it leads to exactly the same result with approximately the same amount of computation. (The reason for replacing v^4 by $\hat{v}^3v$ rather than more obvious choices such as $\hat{v}^4$ or $\hat{v}^2v^2$, and similarly for replacing v^3 by $\hat{v}^3$, is primarily that these substitutions give the right answer, but the explanation of why they give the right answer (15, 18) is rather subtle and does not need to be discussed here).

2. If each error comes from the same relative distribution, i.e. if each has the same coefficient of variation (or the same standard deviation when expressed as a percentage of the true value), then the true standard deviation is not a constant but is proportional to the true value of v, and the variance is proportional to its square:

$$\sigma^2(v) = \hat{v}^2\sigma_2^2 \tag{10}$$

where σ_2^2 is a constant of proportionality. As the true value of v is always unknown we cannot use it for calculation, and so we replace it with the best available estimate, the value $\hat{v}$ calculated from the best available model. In this case it is appropriate to give a weight of $1/\hat{v}^2$ to each v value and to minimize a sum of squares SS defined as

$$SS = \Sigma\,(1 - v/\hat{v})^2 \tag{11}$$

This minimization does not require iteration as the solution can be calculated exactly in a single step:

$$\hat{K}_m = \frac{\Sigma v^2\Sigma(v/a) - \Sigma(v^2/a)\Sigma v}{\Sigma(v^2/a^2)\Sigma v - \Sigma(v^2/a)\Sigma(v/a)} \tag{12}$$

$$\hat{V} = \frac{\Sigma(v^2/a^2)\Sigma v^2 - [\Sigma(v^2/a)]^2}{\Sigma(v^2/a^2)\Sigma v - \Sigma(v^2/a)\Sigma(v/a)} \tag{13}$$

Equations 7 and 11 are special cases of a general expression for the weighted sum of squares:

$$SS = \Sigma\, w(v - \hat{v})^2 \tag{14}$$

where the *weight* w for each observation is the reciprocal of the variance $\sigma^2(v)$ of v. The equation is accordingly sometimes written (e.g. by Gutfreund (19)) in a form resembling the following:

$$SS = \Sigma \frac{1}{\sigma^2(v)}(v - \hat{v})^2 \tag{15}$$

and coupled with a statement implying that when replicate measurements of v are available at each concentration a these can be used to calculate the sample variance to be used as an estimate of $\sigma^2(v)$; indeed, this has sometimes been proposed explicitly, e.g. by Ottaway (20). This sort of statement can be highly misleading if it is not clearly understood that the value of $\sigma^2(v)$ that we require is the *true* population value and that the value that can be estimated from the sample variance is not an adequate estimate of it unless the sample size is very large. Thus writing Equation 15 even more explicitly as follows:

$$SS = \frac{\Sigma (r - 1) (v - \hat{v})^2}{(v - \bar{v})^2} \tag{16}$$

where $\bar{v}$ is the mean of r replicate determinations of v, is most emphatically *not* equivalent to Equation 14 and gives poor results. Studies with simulated data (21) showed that if $r = 2$ (i.e. if each v value is measured in duplicate) the results from this method are even worse in general than those obtained by unweighted linear regression of the double-reciprocal equation, Equation 2. Although the method improves rapidly as r increases, it requires at least five measurements of each v before it approaches the quality of methods based on reasonable assumptions about the proper weighting function. To summarize this paragraph, what is needed in Equation 14 or 15 is a function that defines how the variance varies with the conditions, not a series of disconnected variance estimates at each set of conditions.

Setting the weight w to 1 in Equation 14 for every observation produces Equation 7, and setting it to $1/\hat{v}^2$ produces Equation 11. The truth, of course, may lie between these two extremes, or even, though less plausibly, beyond one or other of them. The simplest way to conceive of intermediate behaviour is to assume an equation similar to Equation 10

$$\sigma^2(v) = \hat{v}^\alpha \sigma_\alpha^2 \tag{17}$$

in which the exponent α is not required to be 0, as in Equation 7, or 2, as in Equation 11. Putting $\alpha = 1$ implies that the standard deviation of v increases steeply from a value of zero at $v = 0$ but flattens out as v increases, but this is not necessarily the most likely kind of intermediate behaviour. One might find it more plausible to suppose that a zero rate cannot be measured with perfect accuracy, but that at large rates the coefficient of variation tends towards a constant. This behaviour is not consistent with Equation 17, but can be obtained by combining Equations 7 and 11 additively:

$$\sigma^2(v) = \sigma_0^2 + \hat{v}^2 \sigma_2^2 \tag{18}$$

Although Equations 17 and 18 may appear to be quite different (and to make qualitatively opposite predictions about what happens at very high or very low rates), the results they give in practice are much more similar than one might expect. Each contains two unknown constants, but only one of these is needed for weighting purposes: α in the case of Equation 17 or the ratio σ_0^2/σ_2^2 in the case of Equation 18.

Returning now to the extreme cases, several studies (21–24) suggest that Equation 11 applies, at least approximately, more often than Equation 7, with the implication that it is safer to use Equations 12–13 than Equations 8–9. None of these investigations is very recent, however, so there is only a weak basis for assuming that their conclusions apply to current experimental techniques. The ideal is to determine the proper weighting system by experiment, but this requires a great deal of effort, which many are likely to feel could be better spent in other ways. One might think that a typical experiment with a small number of observations would not contain enough information to allow a weighting choice to be made on the basis of internal evidence, but studies of simulated data (25) suggest that this is too pessimistic and that even with as few as ten observations one can deduce an adequate estimate of α or σ_0^2/σ_2^2, and hence an appropriate weighting scheme, with a fair degree of success. The method proposed there has now been incorporated into the computer program Leonora that allows robust fitting not only of the Michaelis–Menten equation but also of most other equations used in steady-state enzyme kinetics (15).

6 Equations with more than two parameters

The Michaelis–Menten equation is useful for introducing the problem of fitting non-linear equations to experimental data, but we also need to consider equations with more than two parameters. Many of these are generalizations of the Michaelis–Menten equation. For example, the equation for a two-substrate reaction following a ternary-complex mechanism is as follows:

$$v = \frac{Vab}{K_{iA}K_{mB} + K_{mB}a + K_{mA}b + ab} + e = \frac{V^{app}a}{K_m^{app} + a} + e \tag{19}$$

It is just the Michaelis–Menten equation (Equation 1) written in terms of apparent parameters V^{app} and K_m^{app} that are themselves Michaelis–Menten-like functions of the concentration of the other substrate, B:

$$V^{app} = \frac{Vb}{K_{mB} + b} \tag{20}$$

$$\frac{V^{app}}{K_m^{app}} = \frac{(V/K_{mA})b}{(K_{iA}K_{mB}/K_{mA}) + b} \tag{21}$$

Similar correspondences arise with other equations for multi-substrate reactions, and with equations for inhibition, pH dependence, etc. The question therefore arises of whether it is appropriate to compute with an adaptation of the commonly used graphical technique (e.g. 26) of using a primary plot to estimate the apparent Michaelis–Menten parameters and secondary plots to extract the real parameters. It is certainly possible to adapt this approach to computation, but although it allows the use of simple programs designed just for fitting the Michaelis–Menten equation it appears to have no other advantages; in general, it is much better to analyse all the observations simultaneously, as originally recommended by Cleland (27) and subsequently confirmed with simulated data to yield more precise estimates of kinetic parameters (29). Programs capable of handling equations of several parameters (e.g. V, K_{mA}, K_{iA}, and K_{mB} in Equation 19) and data with two or more independent variables (e.g. the two concentrations a and b in Equation 19) have been available for many years (27), and these requirements ought to be well within the capability of any computer program for current use.

The technique described above for fitting the Michaelis–Menten equation applies with virtually no changes to these cases, and thus requires little discussion (though more detail can be found if required in ref. 15). The definitions of the weights to be used in different circumstances and the different definitions of the sum of squares that these imply are exactly the same as above, and the method of minimizing it is in principle exactly the same, though more laborious as each iteration requires the solution of three or more simultaneous equations, rather than the pairs of simultaneous equations whose solutions are given in Equations 8–9 and 12–13.

7 Detecting lack of fit

If the ultimate objective of data fitting is to determine which model accounts best for a set of experimental results it is evident that we need a way of judging whether one model fits better than another or even, in the absence of an alternative model, whether the model of choice fits well enough for it to be unnecessary to seek an alternative. Essentially two kinds of statistical test are available, and these should be supplemented with the use of residual plots, which are discussed later. The statistical tests have the advantage that they can be automated, so they can be done automatically by the same program that is used for fitting the data, but they have the disadvantage of being dependent on assumptions about the distribution of error which may not be true, and in any case are rarely if ever known to be true. Residual plots cannot be automated, as they require the active participation of the analyst as observer, but they readily allow detection of behaviour that may be completely unforeseen and hence not taken into account by an automated approach.

The first kind of test does not require replicates, but it does require a choice of models, because it is essentially a test of whether one model fits a set of data better than another. It is illustrated in *Table 1*. Despite the extensive footnote in the table there are several points that may be obscure without further explanation. The first is that as the parameters V and K_s have the same meanings in the two models, it is appropriate to give them the same symbols, but K_s cannot be written as K_m in the second model because K_m is specifically a parameter of the Michaelis- Menten equation and the second model is not the Michaelis–Menten equation. So we write it as K_s in both cases.

Looking at the standard errors we might be inclined to prefer the Michaelis–Menten equation without further study, as the standard errors of V and K_s increase substantially when the new parameter is added, and the new parameter itself has so large a standard error that we might doubt whether it is significantly different from infinity. However, it is quite normal for the standard errors of the existing parameters to increase substantially when an additional parameter is introduced, so this is not a sufficient reason for rejecting the more complex model. However, the fact that the sum of squares decreases on introducing the third parameter is not a sufficient reason for saying that it is better, because introducing a new parameter to a model *always* causes a decrease in the sum of squares.

What we must compare, therefore, is not the sum of squares in each case but the *mean square*, which is effectively a variance estimate as it is the sum of squares divided by the number of *degrees of freedom*. This does not necessarily decrease when a parameter is added, but even if it does decrease we still need a statistical test to assess whether the decrease is more than we could expect from chance alone. The number of degrees of freedom that appears in the denominator is the number of observations minus the number that have been 'used' for extracting particular pieces of information. In the example, the total number of observations is 6, but we 'use' two of these to calculate the two first parameters V and K_s. If we wanted we could begin by first testing the hypothesis that the rates are different from zero (i.e. that the enzyme really does catalyse a reaction), and that they are not all the same, but the truth of these two hypotheses is rarely in doubt in enzyme kinetic experiments and so we do not have to waste time testing them formally but start with the hypothesis that the true model is at least as complicated as the Michaelis–Menten equation.

Table 1 Statistical tests for lack of fit when comparing two models

Source of variation	SS	df	MS	*F*
Total (corrected for $\hat{V}$, $\hat{K}_s$)	0.02773227	4		
$\hat{K}_{si} \mid \hat{V}, \hat{K}_s$	0.01460553	1	0.01460553	3.338
Residual	0.01312674	3	0.00437558	

The calculations were done after fitting the set of data (a, v) = [(1, 1.34), (2, 2.36), (4, 3.52), (6, 4.12), (8, 4.65), (10, 4.77)] to the Michaelis–Menten equation, $v = Va/(K_s + a)$ and to the equation for substrate inhibition, $v = Va/[K_s + a(1 + a/K_{si})]$, in both cases assuming a uniform variance, i.e. minimizing the sum of squares defined by Equation 7. In the first case the parameter values were $\hat{V} = 6.65 \pm 0.20$, $\hat{K}_s = 3.66 \pm 0.28$ and the sum of squares was 0.02773, and in the second case the parameter values were $\hat{V} = 8.08 \pm 1.03$, $\hat{K}_s = 4.88 \pm 0.91$, $C_{si} = 50.9 \pm 35.3$ and the sum of squares was 0.01313. The analysis of variance allows one to conclude that although the mean square is smaller in the latter case it is not *significantly* smaller, i.e. one cannot reject the hypothesis that the Michaelis–Menten equation is the true model. The symbols used to head the columns are standard for this type of table and are as follows: SS, sum of squares; df, number of degrees of freedom; MS, mean square (i.e. SS/df); *F*, variance ratio (Fisher's *F*).

This means that the 'total' variation in the first line of the table is not the real total Σv^2, which measures the variation of the data from zero, and is not normally of any interest, but this total $\Sigma (v - \hat{v})^2$ after it has been corrected for the two parameters that we assume at the outset to be needed. Thus we have a sum of squares of 0.02773227 for variation that is unexplained by the first two parameters. In principle it may be due just to experimental error, but the objective is to see whether it can be explained better by supposing that a better model exists. Introducing K_{si} decreases the sum of squares to 0.01312674, and the difference 0.01460553 is the improvement brought about by introducing a third parameter. This improvement has one degree of freedom (as it refers to one parameter), so the residual sum of squares has three degrees of freedom (six for the six observations minus three for the three parameters). So we end by comparing the mean square of 0.01460553 for the improvement with the mean square of 0.00437558 for residual error (variation not explained by the better model). In the past one would have compared their ratio of 3.34 with the value of Fisher's $F = 10.13$ for 1 degree of freedom in the numerator and 3 in the denominator at 95% confidence, which one may find in standard tables (28). This tabulated value means that a value of F calculated in an analysis of variance could be as large as 10.13 just by chance, and as 3.34 is smaller than 10.13 the improvement is not significant. Modern computer programs may dispense with the need for standard tables by calculating the probability of observing by chance an F value at least as large as the one found; for example, Leonora (15) yields a value of 0.1652 for this probability, indicating that more than 16% of the time F will be as big as 3.34 from chance alone.

8 Estimating pure error

To test the adequacy of a model in the absence of an alternative model the data must contain some replicate observations, i.e. measurements repeated under exactly the same conditions as the original ones. This is much more difficult to achieve than it seems, because 'exactly the same' implies more than just the same concentrations of reagents, the same nominal pH and the same nominal temperature. It means the same degree of denaturation of the enzyme, the same ambient temperature, the same degree of tiredness and attentiveness of the experimenter, the same amount of background noise from the conversation of the other people in the laboratory, and so on.

Moreover, and very important, it is not just the replicate observations that must be made under the same conditions: *all* of the observations must be made under the same conditions (apart of course from the specific condition whose effects are being tested). As this is in reality impossible to achieve it means that even in the most carefully executed experiments there will be some degree of inaccuracy in the assumptions that underlie any statistical calculations, and hence some degree of uncertainty about the correctness of the conclusions that they yield. The best one can do in practice is to randomize the order in which the different conditions are tested.

To take a simple (and possibly over-simplified) example, suppose the plan is to study the rate of a reaction at all possible combinations of four concentrations of a substrate and four concentrations of an inhibitor, with replicate measurements of four of the 16 combinations, as indicated schematically in *Figure 2(a)*. However, the particular choice of replicates shown here is unsatisfactory, because only the lowest substrate concentration is represented, and consequently the replicate observations provide no information about the error behaviour at higher substrate concentrations. *Figure 2(b)* is better, as the whole ranges of both substrate and inhibitor concentrations are spanned, but it is still one-dimensional, and thus provides information only about error behaviour when substrate and inhibitor concentrations are varied

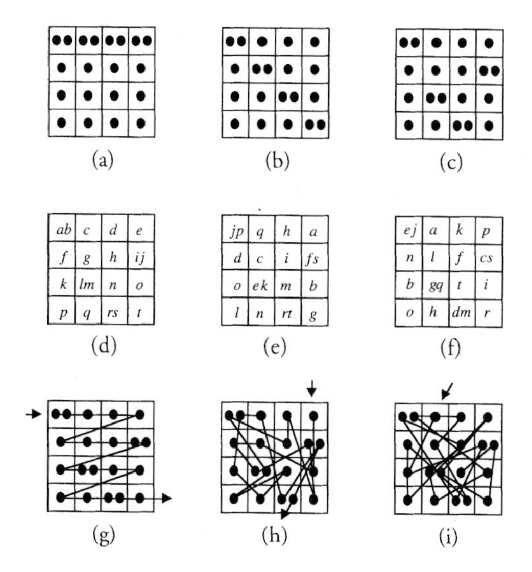

Figure 2 Illustration of the possibilities for experimental design for 16 combinations of four values each for two concentrations, one of a substrate and the other of an inhibitor, with four measurements in duplicate. In each square the four rows represent four values of the substrate concentration, and the four columns represent four values of the inhibitor concentration. In panels (a–c) the symbols represent different design points without regard to the order of exploring them. In panel (a) all of the replicates are made at the same substrate concentration; in panel (b) the replicates span all values of each variable, but in an excessively systematic way; in panel (c) they are distributed more haphazardly. All of the other panels (d–i) refer to the design of panel (c), but in addition define specific sequences for exploring the design points. In panel (d) this order, represented by the alphabetic sequence of letters from *a* to *t*, is completely systematic; in panel (e) the order was determined with the aid of random number tables; in panel (f) it was chosen informally to 'look' random. Panels (g–i) show the same designs as those in panels (d–f) respectively in a more schematic way. Notice that it is obvious which order is systematic, but not at all obvious which is genuinely random and which was designed to look random.

according to a specific plan. What is needed, therefore, is a design as in *Figure 2(c)* in which there is no obvious arrangement.

Once the choice of design points is decided there still needs to be a decision about the order in which they are explored. Again, a systematic order, as in *Figure 2(d)*, is unsatisfactory because the replicate observations are not separated from one another in time, and the design does not allow effects due to gradual deterioration of the enzyme (or substrate) to be distinguished from effects due to changes in the substrate concentration. Ideally the experiments should be done in an order determined with tables of random numbers, as in *Figure 2(e)*, but in practice the result is not likely to be grossly different if one just selects an order that 'looks' random, as in *Figure 2(f)*.

Provided that the replication is properly done, replicate observations now allow a distinction to be made between *pure error* and *lack of fit*, and the type of calculation needed is illustrated in *Table 2*, using the same data as *Table 1* apart from the introduction of three replicate observations. The principles of the calculation are similar to those in *Table 1*, with the additional point that now we can use the mean $\bar{v}$ for a group of replicate observations as an estimate of the true value that does not require any assumption about the correctness of any kinetic model. All we assume is that if we make the same measurement twice we should get the same result twice, any difference being due to experimental error and not to any wrong

Table 2 Statistical tests for lack of fit when replicates are available

Source of variation	SS	df	MS	F
Residual	0.05601148	7		
Lack of fit	0.05250036	4	0.01312509	11.214
Pure error	0.00351112	3	0.00117037	

The calculations were done as in *Table 1*, but using the set of data $(a, v) = [(1, 1.34), (1, 1.31), (2, 2.36), (4, 3.52), (4, 3.58), (6, 4.12), (8, 4.65), (10, 4.77), (10, 4.82)]$. This is the same set of data as in *Table 1* with three additional observations $[(1, 1.31), (4, 3.58), (10, 4.82)]$. As each of these replicates an existing observation it is now possible to make an estimate of pure error and compare it with the lack of fit. The line labelled 'Residual' is the one that would be called 'Total (corrected for $\hat{V}$, $\hat{K}_S$)' if no calculation of pure error were being made, i.e. it is the sum of squares $\Sigma\ (v - \hat{v})^2$. The sum of squares for pure error is calculated as $SS = \Sigma\ (v - \bar{v})^2$, using only the six v values that occur in groups of replicates, and taking $\bar{v}$ as the mean rate for the group of replicates. The sum of squares for lack of fit is calculated by subtraction, i.e. 0.05601148 − 0.00351112 = 0.05250036. Symbols are as in *Table 1*.

assumption about the kinetics. Each calculation of a mean consumes one degree of freedom, so the estimate of pure error has three degrees of freedom, leaving four for lack of fit. This calculation now gives an F value of 11.214: we can conclude that this is significant, either by comparing it with the tabulated value of 9.12 for 4 and 3 degrees of freedom (28), or by noting that it has a probability of 0.0378 of occurring by chance alone (15).

9 Distribution-free methods

All of the statistical methods considered to this point have been classical or 'parametric' methods, that is to say methods that depend on an assumed type of distribution of the experimental errors. (Note that the parameters referred to here are the parameters that define the shape of the distribution curve, not the parameters such as K_m and V that the biochemist may wish to estimate.) These are the best known and probably the most widely used statistical methods, being the only ones considered in most elementary textbooks of statistics, but they are not the only ones available. They are open to the objection that in practice experimenters rarely have any real knowledge of the type of distribution curves that define their experimental errors, and so methods that rely on an assumed adherence to a normal or Gaussian distribution curve rely to some degree on wishful thinking. The alternative is to replace the assumption that a particular type of distribution curve is followed with the much weaker assumption that any given error is as likely to be negative as to be positive. Methods that start from this assumption are called *distribution-free methods* (because they do not assume a particular type of distribution) or *non-parametric methods* (because they do not make assumptions about the parameters that define the distribution curve—not quite the same thing, but similar).

Distribution-free methods have both advantages and disadvantages with respect to parametric methods. We may consider the latter first, as these are the points that will be emphasized by people who disapprove of them. In the event that the classical assumptions are true, the best distribution-free method will always be less efficient than the corresponding parametric method: it is based on weaker assumptions, so it leads to weaker conclusions, which translates into the experimental consequence that it requires more experimental effort to arrive at any given degree of precision in the answer—typically more experimental observations need to be made. This objection is valid but its importance is often exaggerated. Even in the worst case a good distribution-free method will usually achieve about 70% of the efficiency of an ideal (best possible) parametric method, i.e. in the worst case it may require about 40% more experimental work to reach the same degree of precision in the answer.

To set against this one disadvantage there are numerous advantages of the distribution-free approach. First of all, a distribution-free method will usually be much *more* efficient than a poorly chosen parametric method, which means that the much-vaunted efficiency of parametric methods is an illusion unless much more effort is devoted to choosing proper weights, checking the form of the error distribution curve, etc., than is usual in biochemistry or most experimental sciences. If outliers (abnormally poor observations) are present or the weighting is badly chosen a parametric method may perform very badly indeed, whereas the corresponding distribution-free method may be affected little or not at all by such considerations (30). Another point, trivial for people who are content to follow recipes blindly, but important for those who like to understand the methods they use, is that the theory of distribution-free statistics is in general quite simple—similar in many respects to analysis of coin-tossing experiments, whereas the theory of classical statistics is sufficiently difficult that it is virtually never taught to science students, and only in advanced courses to students of mathematics.

In the context of enzyme kinetics the archetypal distribution-free method is the computer analogue of the direct linear plot (31). Both the plot (26) and its theoretical background (15) are extensively discussed in current books, as well as the chapter corresponding to this one in the first edition of the present book (1). Only a brief account will therefore be given here. For any observation $v = v_i$ at $a = a_i$ that follows the Michaelis–Menten equation, Equation 1, without experimental error (i.e. with $e = 0$), the equation can be rearranged to show the value of V that corresponds to any assumed value of K_m:

$$V = v_i + \frac{v_i}{a_i} \cdot K_m \tag{22}$$

and the set of possible parameter values can be expressed graphically by drawing axes for K_m (abscissa) and V (ordinate) and plotting a straight line with slope v_i/a_i and intercept v_i on the ordinate (or more simply, in the sense that no calculation is needed, with intercepts $-a_i$ and v_i on the K_m and V axes respectively), as illustrated in *Figure 3*. Exactly the same may be done with a second observation with $v = v_j$ at $a = a_j$, and the point where the two lines cross has (K_m, V) coordinates that correspond to the only parameter values that satisfy both observations:

$$K_{m,\,ij} = \frac{v_j - v_i}{\dfrac{v_i}{a_i} - \dfrac{v_j}{a_j}} \tag{23}$$

$$V_{ij} = \frac{a_i - a_j}{\dfrac{a_i}{v_i} - \dfrac{a_j}{v_j}} \tag{24}$$

In the absence of experimental error this would be all there was to it: Equations 23–24 would give consistent results for any combination of i and j for any number n of observations. In practice, however, experimental error causes these equations to give inconsistent results, and for n observations at different values of a Equation 23 will provide $n(n - 1)/2$ different estimates of K_m and Equation 24 will provide $n(n - 1)/2$ different estimates of V, as illustrated for a very simple case in the inset to *Figure 3*, with $n = 4$. (When replicate observations are present, i.e. some a values are repeated, the number of separate estimates is decreased, but we shall not consider this complication here). It is important to realize that some of the estimates obtained in this way may be very bad estimates—typically those that result from lines

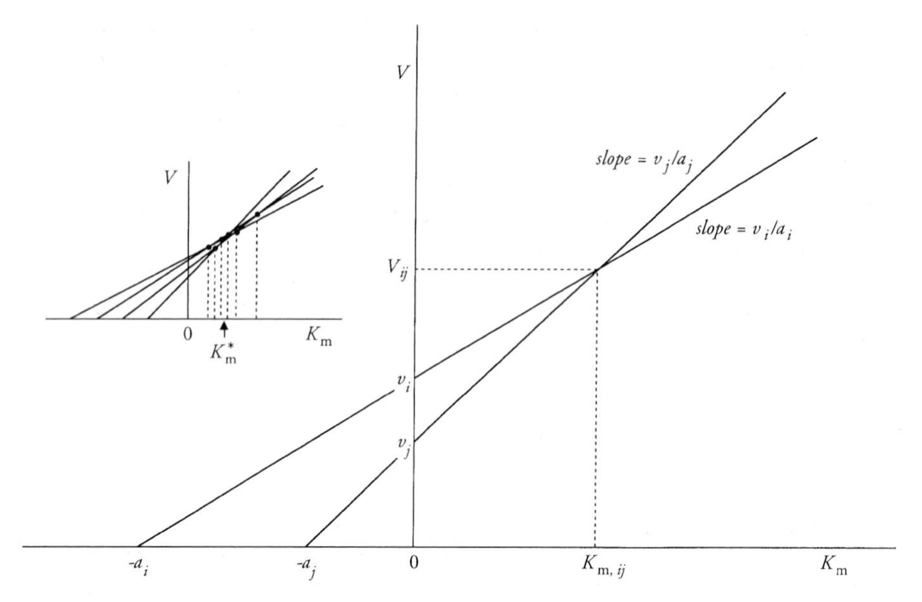

Figure 3 Direct linear plot. The plot is drawn with axes of K_m (abscissa) and V (ordinate), and for any observation (a_i, v_i) that satisfies Equation 1 exactly a straight line making intercepts $-a_i$ on the K_m axis and v_i on the V axis relates all (K_m, V) pairs that satisfy the observation. A second line drawn in the same way for another observation (a_j, v_j) has the same meaning for that observation. The point of intersection of the two lines is the only point that lies on both lines and thus defines the only (K_m, V) that satisfies both observations. When there are more than two observations experimental error causes the points of intersection to disagree with one another, as illustrated in the inset, and so there is a family of intersection points instead of a unique point. In this case the best estimate of K_m is taken as the median of all the estimates, i.e. as the middle one if they are arranged in order and there are an odd number of them, or, as in the case illustrated, the mean of the middle two if there are an even number of intersection points. The best value of V is estimated similarly.

on the plot that are almost parallel—and so what one must *not* do is take any kind of mean as a best estimate. Instead it is appropriate to arrange the values in order (which is done automatically by the direct linear plot) and take the median (middle) value of the set of $K_{m,ij}$ values as the best estimate of K_m and the median of the set of V_{ij} values as the best estimate of V.

10 Residual plots

The statistical tests for detecting lack of fit discussed earlier provide ways of looking in detail at the differences between observations and the theoretical model that is intended to explain them. These differences are called *residuals*, which is short for residual errors or residual deviations, because they constitute the residue of unexplained variation in the measurements after everything that can be explained by the model is taken into account. Although the tests can be very powerful, they are limited to testing the kinds of deviation that have been envisaged as likely. By contrast, the graphical equivalent of these tests is capable of allowing immediate recognition of anomalous behaviour that is entirely unexpected, such as the problems that arise from premature and over-enthusiastic rounding (32).

This graphical equivalent is to plot the residuals themselves as a function of some suitable variable, which may be the calculated rate, or one of the independent variables, or the time at which the particular measurement was made, etc., depending on what sort of behaviour one is hoping to explain. All of these are useful in particular circumstances, but here we shall consider only the case where the abscissa is the calculated rate $\hat{v}$; this is the most generally useful choice, to the point that any serious modern computer program for data analysis ought to display a residual plot with $\hat{v}$ as abscissa whether the user requests it or not.

In the simplest case of the simple difference $v - \hat{v}$ between observed and calculated rates is appropriate as ordinate variable, but if, as will usually be the case, the residuals are plotted after a least-squares calculation has been done then strictly the residual ought to be weighted by the square root of the weight used already. In other words if the sum of squares is defined as in Equation 14 as a sum of $w(v - \hat{v})^2$ values the plotted residual ought to be $w^{1/2}(v - \hat{v})$. This is especially important if the objective of the plot is to inspect the quality of the weighting that has been used, rather than, say, to detect lack of fit to the model. A more subtle complication arises from the fact that any *outliers*, or observations with abnormally large residuals, in the data may require a choice of ordinate scale that compresses most of the other points so close to the abscissa axis that it becomes difficult to see any trend in their behaviour. The obvious solution to this problem is to omit the outliers from the plot, but this is not a good solution because outliers may contain important information about the system (for example, they may hint at a serious failure of the model or of the measuring apparatus in a certain range of values). It is better to use a stabilizing transformation that has little effect on small residuals but decreases the values of large ones. In practice good results are obtained by plotting $\arctan[w^{1/2}(v - \hat{v})/2\bar{r}]$, where $\bar{r}$ is the mean of all the $w^{1/2}|v - \hat{v}|$ values, instead of the raw residual $w^{1/2}(v - \hat{v})$ itself. This is tedious to do by hand, but easy to incorporate into a computer program.

Figure 4 illustrates some of the results one may observe in such plots. All of the examples shown are idealized, in the sense that the conclusions emerge more cleanly than they often

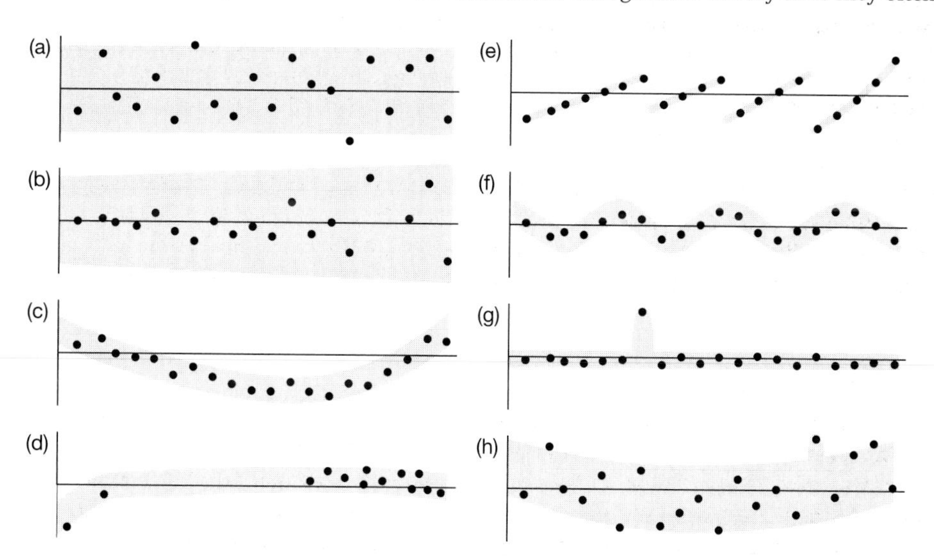

Figure 4 Residual plots. In each case the ordinate represents the weighted difference between the observed and calculated rates and the abscissa represents the calculated rate. No labelling is used in order to avoid distracting the eye with information that is irrelevant to the purpose of the plot. If numerical information is needed for a published residual plot, such as the units and the numerical ranges of values, this should be stated in the legend, not labelled on the plot. The eight cases illustrated are discussed in the text; only brief characterization is noted here: (a) ideal case; (b) inappropriate weighting; (c) incorrect model; (d) poor experimental design; (e) over-aggressive and premature rounding; (f) local correlation; (g) anomalous point; (h) weak suggestions.

do in reality, when two or more effects may be present simultaneously. Even in the ideal case, expected after fitting the correct model with correct weights and with no anomalies of any sort, the results may be less clear in reality than those illustrated in *Figure 4(a)*, because a genuinely random distribution often looks less random to the eye than an artificially constructed illustration. In this ideal case the points should be scattered in a parallel band symmetrically placed about the abscissa axis, as shown by the grey shading, and all the points should be inside this band or just outside it.

Figure 4(b) illustrates the result of fitting the correct model within correct weights. The points are still scattered symmetrically about the axis, but the band is no longer parallel. In the particular case illustrated the implication is that greater weight should be given to the smaller rates (for example, because Equation 7 was used when Equation 11 ought to have been used), but others are possible.

In *Figure 4(c)* the behaviour is quite different: there is no suggestion of incorrect weighting, but instead the points follow an obvious trend. This is a clear case of fitting the wrong model. The trend is too obvious to be missed in a residual plot with many points, but is still easily detectable in many experiments with fewer points, especially if qualitatively the same suggestion of a trend is visible in several experiments of the same kind. Real residual plots are very rare in the biochemical literature, but one can usually form an idea of what a residual plot would look like by looking along a conventionally plotted line with the eye close to the paper. Such 'virtual residual plots' are abundant in the literature, and a recent example may be found in Figure 4 of a biochemical study of nitric oxide (33): although for any one of the three lines plotted the trend of residuals is no more than suggestive, the same suggestion in three different lines indicates the presence of a real effect that was overlooked by the authors.

Figure 4(d) may show another example of lack of fit, but it is much less convincing than the previous example. It is the sort of plot that may result from an experiment that has been so poorly designed that one cannot deduce whether the behaviour at low $\hat{v}$ is evidence of a real trend or just is a consequence of inappropriate weighting or the sampling variation expected in a very small sample. Although the diagnosis of poor experimental design ought to be clear in a conventional plot, it is often even clearer in a residual plot, and in this case the remedy is obvious: one needs a better distribution of points, with more at low $\hat{v}$ where a major trend may exist, and at least some in the large gap between the two groups.

Figure 4(e) illustrates behaviour that is probably (and ought to be) extremely rare, but is included to illustrate the fact that a residual plot makes it easy to recognize an anomaly that may be totally unexpected and that may pass unnoticed in a conventional plot. In this case the points appear to follow not a single trend but several different trend with breaks between them, yet along each trend line the fit is almost exact. The explanation of this is discussed elsewhere (32); here it will suffice to say that it shows the effect of excessive rounding of the primary observations at an early stage in the analysis.

Figure 4(f) could be taken as evidence of a very complicated trend (for example, the need for a model *much* more complicated than the one fitted), but it is more likely to arise from anomalous correlations between neighbouring points. In the absence of systematic error the magnitude of any residual ought to be completely unpredictable from knowledge of the magnitude of the residual for an observation at a similar $\hat{v}$, but that is clearly not the case in *Figure 4(f)*. There are many possible explanations of this behaviour (for example, progressive deterioration of stock solutions coupled with an insufficiently random order of doing the experiments), and the point here is not to describe all of these, but simply to emphasize that a residual plot resembling *Figure 4(f)* is evidence of a serious problem with the experimental design or with the analysis that needs to be investigated before proceeding further.

As a last example, nearly all the residuals may be distributed in a narrow band, but with a small minority, one in the case illustrated in *Figure 4(g)*, lying well outside it. In most respects this is a very satisfactory result, similar to *Figure 4(a)* and indicating that the correct model has been fitted with correct weights, but the point that lies outside the main band is evidence of a serious anomaly that requires careful investigation. If left uncorrected it may cause severe bias in the estimated parameters (as seen in the illustration by the fact that the main band is not arranged symmetrically about the axis for zero residuals) and may suggest that the precision of the measurements was lower than it was in fact. There are at least three possible causes for such an anomaly: it may be the result of a typing error when entering the data (for example, 105 might be typed instead of 10.5); it may be a genuine outlier, i.e. a measurement that for some unknown reason was badly measured; it may be a symptom of more complicated behaviour of the enzyme than was expected. A good computer program can recognize the presence of such an anomaly and can draw the experimenter's attention to it; it can even correct for its harmful influence on the parameter estimation; however, it cannot determine whether the cause is trivial and easily corrected (a typing error), unknown and possibly unknowable (a genuine outlier), or diagnostic of potentially important information about the enzyme (unexpectedly complicated behaviour). As the difference between these three is potentially crucial for a correct analysis it follows that an analysis must never be left entirely to a computer program but that human participation is essential.

In all the examples discussed, the illustrations have been drawn so that the characteristics of each plot are more or less obvious. In reality the results are usually less clear, for example because different effects are superimposed or because there are not enough data points to convey a definite message. In *Figure 4(h)*, for example, there is a suggestion of lack of fit, but it is far from overwhelming and just indicates that more experiments need to be done; there is a suggestion of an outlier, but not nearly so obvious as in *Figure 4(g)*, so one cannot be confident that it is anything more than an error at the high end of the expected distribution.

Two other points need to be emphasized about residual plots. The first is that all of the plots in *Figure 4* are unlabelled, which was not an oversight but deliberate. Most residual plots are, or ought to be, drawn for private study in the laboratory rather than for publication, and should be accompanied by the minimum of extraneous information that may distract the eye from the visual impact made by the distribution of points. If any labelling is needed (for example, in a publication) it should be stated in the legend rather than drawn directly on the plot (if the journal permits it, of course). The second point is that although careful draughtsmanship is usually considered important for plotting data it has little value in a residual plot. A residual plot drawn by hand and eye on plain paper without use of a ruler or calculator will look much the same as one drawn by a laser printer with points placed exactly to within 0.1 mm, and will convey much the same message.

11 A note about rounding

It is widely and rightly emphasized to students that one of the easiest ways to make one's work look silly is to express the numerical results with much more precision than they can reasonably be supposed to have, for example to quote a K_m value as 4.8134668 mM. An unfortunate consequence of the demise of the slide rule and its replacement by the pocket calculator has been to increase this tendency, and although specific guidelines need to be adjusted to specific circumstances it is usually reasonable to write about three significant figures for most measured values (or four significant figures if the first one is 1, for example, 10.37 is acceptable when 31.87 would be regarded as unreasonably precise). If standard errors are available these should be written with two significant figures (or with three if the first one is

1), and the number they refer to should have the same number of decimal digits after the decimal point as the standard error.

These guidelines may seem to have been violated rather wildly in some of the values given in this chapter, and as it is not my aim to look silly it may be helpful to note that in statistical calculations it is usual to retain many more digits in the intermediate stages than one expects to retain at the end. The reason for this is that once a digit is discarded it is lost for ever, and if one decides later on that maybe it was significant after all it cannot be recovered without going back and repeating the calculation from the point at which it was discarded. Once the end of the calculation is reached one can make a judgement of how much precision to retain and act accordingly. *Figure 4(e)* illustrates one possible consequence of rounding too much and too soon.

References

1. Henderson, P. J. F. (1992). *Enzyme assays: a practical approach*, 1st edn. (ed. R. Eisenthal and M. J. Danson), pp. 277–316. IRL Press, Oxford.
2. Gutfreund, H. (1996). *J. Memb. Biol.*, **13,** 61–62.
3. Benen, J. A. E., Kester, H. C. M., and Visser, J. (1999). *Eur. J. Biochem.*, **259,** 577–585.
4. Sakaki, T., Sawada, N., Takeyama, K., Kato, S., and Inouye, K. (1999). *Eur. J. Biochem.*, **259,** 731–738.
5. Kunugi, S., Yanagi, Y., and Oda, K. (1999). *Eur. J. Biochem.*, **259,** 815–820.
6. Wang, Z.-X. (1999). *Eur. J. Biochem*, **259,** 609–617.
7. Page, J. P., Munagala, N. R., and Wang, C. C. (1999). *Eur. J. Biochem.*, **259,** 565–571.
8. Puri, V., Arora, A., and Gupta, C. M. (1999). *Eur. J. Biochem.*, **259,** 586–591.
9. Kato, R. and Kuramitsu, S. (1999). *Eur. J. Biochem.*, **259,** 592–601.
10. Seto, N. O., Compston, C. A., Evans, S. V., Bundle, D. R., Narang, S. A., and Placic, M. M. (1999). *Eur. J. Biochem.*, **259,** 770–775.
11. Keng, Y.-F., Wu, L., and Zhang, Z.-Y. (1999). *Eur. J. Biochem.*, **259,** 809–814.
12. Spychala, J., Chen, V., Oka, J., and Mitchell, B. S. (1999). *Eur. J. Biochem.*, **259,** 851–858.
13. Turk, B., Dolenc, I., Lenarcic, B., Krizaj, I., Turk, V., Bieth, J. G., and Björk, I. (1999). *Eur. J. Biochem.*, **259,** 926–932.
14. Upstill, C. (1988). *Nature*, **333,** 613–614.
15. Cornish-Bowden, A. (1995). *Analysis of enzyme kinetic data*. Oxford University Press, Oxford.
16. Gauss, K. F. (1809). *Theoria motus corporus coelestium in sectionibus conicis solem ambientium* (trans. Davis, C. H.), section 177, pp. 257–259. Dover, New York, 1963.
17. Wilkinson, G. N. (1961). *Biochem. J.*, **80,** 324–328.
18. Cornish-Bowden, A. (1982). *J. Mol. Sci. (Wuhan)*, **2,** 107–112.
19. Gutfreund, H. (1995). *Kinetics for the life sciences*. Cambridge University Press, Cambridge.
20. Ottaway, J. H. (1973). *Biochem. J.*, **134,** 729–736.
21. Storer, A. C., Darlison, M. G., and Cornish-Bowden, A. (1975). *Biochem. J.*, **151,** 361–367.
22. Siano, D. B., Zyskind, J. W., and Fromm, H. J. (1975). *Arch. Biochem. Biophys.*, **170,** 587–600.
23. Askelöf, P. Korsfeldt, M., and Mannervik, B. (1976). *Eur. J. Biochem.*, **69,** 61–67.
24. Nimmo, I. A. and Mabood, S. F. (1979). *Anal. Biochem.*, **94,** 265–269.
25. Cornish-Bowden, A. and Endrenyi, L. (1981). *Biochem. J.*, **193,** 1005–1008.
26. Cornish-Bowden, A. (1995). *Fundamentals of enzyme kinetics*, pp. 144–147. Portland Press, London.
27. Cleland, W. W. (1963). *Nature (Lond.)*, **198,** 463–465.
28. Snedecor, G. W. (1956). *Statistical methods*, 5th edn, pp. 246–249. Iowa State University, Ames (reproduced in many statistics textbooks and reference works).
29. Nimmo, I. A. and Atkins, G. L. (1976). *Biochem. J.*, **157,** 489–492.
30. Cornish-Bowden, A. and Eisenthal, R. (1974). *Biochem. J.*, **139,** 721–730.
31. Eisenthal, R. and Cornish-Bowden, A. (1974). *Biochem. J.*, **139,** 715–720.
32. Cárdenas, M. L. and Cornish-Bowden, A. (1993). *Biochem. J.*, **292,** 37–40.
33. Yoneyama, H., Kosaka, H., Ohnishi, T., Kawazoe, T., Mizoguchi, K., and Ichikawa, Y. (1999). *Eur. J. Biochem.*, **266,** 771–777.

List of suppliers

Amersham Pharmacia Biotech UK Ltd, Amersham Place, Little Chalfont, Buckinghamshire HP7 9NA, UK
(see also Nycomed Amersham Imaging UK; Pharmacia).
Tel: 0800 515313
Fax: 0800 616927
URL: http//www.apbiotech.com/

Anachem Ltd. 20 Charles Street, Luton, Bedfordshire, LU2 0EB, UK.

Anderman and Co. Ltd, 145 London Road, Kingston-upon-Thames, Surrey KT2 6NH, UK.
Tel: 0181 5410035
Fax: 0181 5410623

Beckman Coulter (UK) Ltd, Oakley Court, Kingsmead Business Park, London Road, High Wycombe, Buckinghamshire HP11 1JU, UK.
Tel: 01494 441181
Fax: 01494 447558
URL: http://www.beckman.com/

Beckman Coulter Inc., 4300 N. Harbor Boulevard, PO Box 3100, Fullerton, CA 92834-3100, USA.
Tel: 001 714 8714848
Fax: 001 714 7738283
URL: http://www.beckman.com/

Becton Dickinson and Co., 21 Between Towns Road, Cowley, Oxford OX4 3LY, UK.
Tel: 01865 748844
Fax: 01865 781627
URL: http://www.bd.com/

Becton Dickinson and Co., 1 Becton Drive, Franklin Lakes, NJ 07417-1883, USA.
Tel: 001 201 8476800
URL: http://www.bd.com/

Bio 101 Inc., c/o Anachem Ltd, Anachem House, 20 Charles Street, Luton, Bedfordshire LU2 0EB, UK.
Tel: 01582 456666
Fax: 01582 391768
URL: http://www.anachem.co.uk/

Bio 101 Inc., PO Box 2284, La Jolla, CA 92038-2284, USA.
Tel: 001 760 5987299
Fax: 001 760 5980116
URL: http://www.bio101.com/

Bio-Rad Laboratories Ltd, Bio-Rad House, Maylands Avenue, Hemel Hempstead, Hertfordshire HP2 7TD, UK.
Tel: 0181 3282000
Fax: 0181 3282550
URL: http://www.bio-rad.com/

Bio-Rad Laboratories Ltd, Division Headquarters, 1000 Alfred Noble Drive, Hercules, CA 94547, USA.
Tel: 001 510 7247000
Fax: 001 510 7415817
URL: http://www.bio-rad.com/

Boehringer-Mannheim (see Roche)

Clandon Scientific Ltd. Aldershot, Hampshire, GU12 5QR, UK.

CP Instrument Co. Ltd, PO Box 22, Bishop Stortford, Hertfordshire CM23 3DX, UK.
Tel: 01279 757711
Fax: 01279 755785
URL: http//:www.cpinstrument.co.uk/

Dupont (UK) Ltd, Industrial Products Division, Wedgwood Way, Stevenage, Hertfordshire SG1 4QN, UK.
Tel: 01438 734000
Fax: 01438 734382
URL: http://www.dupont.com/

Dupont Co. (Biotechnology Systems Division), PO Box 80024, Wilmington, DE 19880-002, USA
Tel: 001 302 7741000
Fax: 001 302 7747321
URL: http://www.dupont.com/

Eastman Chemical Co., 100 North Eastman Road, PO Box 511, Kingsport, TN 37662-5075, USA.
Tel: 001 423 2292000
URL: http//:www.eastman.com/

Fisher Scientific UK Ltd, Bishop Meadow Road, Loughborough, Leicestershire LE11 5RG, UK
Tel: 01509 231166
Fax: 01509 231893
URL: http://www.fisher.co.uk/

Fisher Scientific, Fisher Research, 2761 Walnut Avenue, Tustin, CA 92780, USA.
Tel: 001 714 6694600
Fax: 001 714 6691613
URL: http://www.fishersci.com/

Fluka, PO Box 2060, Milwaukee, WI 53201, USA.
Tel: 001 414 2735013
Fax: 001 414 2734979
URL: http://www.sigma-aldrich.com/

Fluka Chemical Co. Ltd, PO Box 260, CH-9471, Buchs, Switzerland
Tel: 0041 81 7452828
Fax: 0041 81 7565449
URL: http://www.sigma-aldrich.com/

Gibco-BRL (see Life Technologies)

Grant Instruments (Cambridge) Ltd. Barrington, Cambridge, CB2 5QZ, UK.

Hybaid Ltd, Action Court, Ashford Road, Ashford, Middlesex TW15 1XB, UK.
Tel: 01784 425000
Fax: 01784 248085
URL: http://www.hybaid.com/

Hybaid US, 8 East Forge Parkway, Franklin, MA 02038, USA
Tel: 001 508 5416918
Fax: 001 508 5413041
URL: http://www.hybaid.com/

HyClone Laboratories, 1725 South HyClone Road, Logan, UT 84321, USA
Tel: 001 435 7534584
Fax: 001 435 7534589
URL: http//:www.hyclone.com/

Invitrogen Corp., 1600 Faraday Avenue, Carlsbad, CA 92008, USA
Tel: 001 760 6037200
Fax: 001 760 6037201
URL: http://www.invitrogen.com/

Invitrogen BV, PO Box 2312, 9704 CH Groningen, The Netherlands
Tel: 00800 53455345
Fax: 00800 78907890
URL: http://www.invitrogen.com/

Life Technologies Ltd, PO Box 35, 3 Free Fountain Drive, Inchinnan Business Park, Paisley PA4 9RF, UK.
Tel: 0800 269210
Fax: 0800 243485
URL: http://www.lifetech.com/

Life Technologies Inc., 9800 Medical Center Drive, Rockville, MD 20850, USA.
Tel: 001 301 6108000
URL: http://www.lifetech.com/

Merck Sharp & Dohme Research Laboratories, Neuroscience Research Centre, Terlings Park, Harlow, Essex CM20 2QR, UK.
URL: http://www.msd-nrc.co.uk/

Metrohm (UK) Ltd. Unit 2, Top Angel, Buckinghamshire Industrial Park, Buckingham, MK18 1TH, UK.

MSD Sharp and Dohme GmbH, Lindenplatz 1, D-85540, Haar, Germany.
URL: http://www.msd-deutschland.com/

Millipore (UK) Ltd, The Boulevard, Blackmoor Lane, Watford, Hertfordshire WD1 8YW, UK.
Tel: 01923 816375
Fax: 01923 818297
URL: http://www.millipore.com/local/UKhtm/

Millipore Corp., 80 Ashby Road, Bedford, MA 01730, USA.
Tel: 001 800 6455476
Fax: 001 800 6455439
URL: http://www.millipore.com/

New England Biolabs, 32 Tozer Road, Beverley, MA 01915-5510, USA.
Tel: 001 978 9275054

Nikon Inc., 1300 Walt Whitman Road, Melville, NY 11747-3064, USA.
Tel: 001 516 5474200
Fax: 001 516 5470299
URL: http://www.nikonusa.com/

Nikon Corp., Fuji Building, 2-3, 3-chome, Marunouchi, Chiyoda-ku, Tokyo 100, Japan.
Tel: 00813 32145311
Fax: 00813 32015856
URL: http://www.nikon.co.jp/main/index_e.htm/

Nycomed Amersham Imaging, Amersham Labs, White Lion Rd, Amersham, Buckinghamshire HP7 9LL, UK.
Tel: 0800 558822 (or 01494 544000)
Fax: 0800 669933 (or 01494 542266)
URL: http//:www.amersham.co.uk/

Nycomed Amersham, 101 Carnegie Center, Princeton, NJ 08540, USA.
Tel: 001 609 5146000
URL: http://www.amersham.co.uk/

Perkin Elmer Ltd, Post Office Lane, Beaconsfield, Buckinghamshire HP9 1QA, UK.
Tel: 01494 676161
URL: http//:www.perkin-elmer.com/

Pharmacia, Davy Avenue, Knowlhill, Milton Keynes, Buckinghamshire MK5 8PH, UK (also see Amersham Pharmacia Biotech).
Tel: 01908 661101
Fax: 01908 690091
URL: http//www.eu.pnu.com/

Promega UK Ltd, Delta House, Chilworth Research Centre, Southampton SO16 7NS, UK.
Tel: 0800 378994
Fax: 0800 181037
URL: http://www.promega.com/

Promega Corp., 2800 Woods Hollow Road, Madison, WI 53711-5399, USA.
Tel: 001 608 2744330
Fax: 001 608 2772516
URL: http://www.promega.com/

Qiagen UK Ltd, Boundary Court, Gatwick Road, Crawley, West Sussex RH10 2AX, UK
Tel: 01293 422911
Fax: 01293 422922
URL: http://www.qiagen.com/

Qiagen Inc., 28159 Avenue Stanford, Valencia, CA 91355, USA.
Tel: 001 800 4268157
Fax: 001 800 7182056
URL: http://www.qiagen.com/

Radiometer Ltd. The Manor, Manor Court, Crawley, West Sussex, RH10 2PY, UK.

Roche Diagnostics Ltd, Bell Lane, Lewes, East Sussex BN7 1LG, UK.
Tel: 0808 1009998 (or 01273 480044)
Fax: 0808 1001920 (01273 480266)
URL: http://www.roche.com/

Roche Diagnostics Corp., 9115 Hague Road, PO Box 50457, Indianapolis, IN 46256, USA.
Tel: 001 317 8452358
Fax: 001 317 5762126
URL: http://www.roche.com/

Roche Diagnostics GmbH, Sandhoferstrasse 116, 68305 Mannheim, Germany.
Tel: 0049 621 7594747
Fax: 0049 621 7594002
URL: http://www.roche.com/

Schleicher and Schuell Inc., Keene, NH 03431A, USA.
Tel: 001 603 3572398

Shandon Scientific Ltd, 93-96 Chadwick Road, Astmoor, Runcorn, Cheshire WA7 1PR, UK.
Tel: 01928 566611
URL: http//www.shandon.com/

Sigma-Aldrich Co. Ltd, The Old Brickyard, New Road, Gillingham, Dorset SP8 4XT, UK.
Tel: 0800 717181 (or 01747 822211)
Fax: 0800 378538 (or 01747 823779)
URL: http://www.sigma-aldrich.com/

Sigma Chemical Co., PO Box 14508, St Louis, MO 63178, USA.
Tel: 001 314 7715765
Fax: 001 314 7715757
URL: http://www.sigma-aldrich.com/

Stratagene Inc., 11011 North Torrey Pines Road, La Jolla, CA 92037, USA.
Tel: 001 858 5355400
URL: http://www.stratagene.com/

Stratagene Europe, Gebouw California, Hogehilweg 15, 1101 CB Amsterdam Zuidoost, The Netherlands.
Tel: 00800 91009100
URL: http://www.stratagene.com/

United States Biochemical (USB), PO Box 22400, Cleveland, OH 44122, USA.
Tel: 001 216 4649277

Enzyme Index

As many assays, or references to them, are given in the text, the editors felt that inclusion of an Enzyme Index would be useful. Wherever possible, EC numbers are shown for individual enzymes. Many of the protocols described in the text are generally applicable to a group of enzymes, and these groups have been given separate listings. A subject index follows this listing.

General Index